ELECTRICAL DISTRIBUTION ENGINEERING

ELECTRICAL DISTRIBUTION ENGINEERING

ANTHONY J. PANSINI, EE, PE

Life Fellow IEEE; Member ASTM
Engineering and Management Consultant
Waco, Texas

McGRAW-HILL BOOK COMPANY

New York   St. Louis   San Francisco   Auckland   Bogotá
Hamburg   Johannesburg   London   Madrid   Mexico
Montreal   New Delhi   Panama   Paris   São Paulo
Singapore   Sydney   Tokyo   Toronto

Library of Congress Cataloging in Publication Data

Pansini, Anthony J.
    Electrical distribution engineering.

    Includes index.
    1. Electric power distribution.   I. Title.
TK3001.P28   1983   621.319'2   82-20814
ISBN 0-07-048454-6

1 2 3 4 5 6 7 8 9 0   KGP/KGP   8 9 8 7 6 5 4 3

ISBN 0-07-048454-6

The editors for this book were Diane Heiberg and
Richard Mickey, the designer was Elliot Epstein, and the
production supervisor was Sally Fliess. It was set in
Baskerville by J.M. Post Graphics, Corp.

Printed and bound by The Kingsport Press.

CONTENTS

PREFACE

This book is meant principally as a source of information for electrical distribution engineers in their approach to the daily and long-range problems that confront them. It should also find much application by engineers involved in electrical matters, including industrial engineers, many of whose problems are quite similar. Nor does its scope limit its usefulness only to those actually engaged in industry; it should prove an excellent college course text on both undergraduate and postgraduate levels.

The material presented is meant to update general information regarding electric utility distribution systems. The author makes no warranty, expressed or implied, nor assumes any legal liability for the methods and techniques described herein. References to specific products and services should not be considered endorsements.

This work reflects the author's experiences in this field covering over half a century. These included long-term operating assignments with Consolidated Edison Company and its predecessor companies in New York City and with Long Island Lighting Company. More than a decade of consulting experience included work with public service companies in Sonora and Sinaloa in Mexico, the last to be nationalized by the government of that country; some rural electrification systems in Texas; and a half dozen other utilities, all presenting specialized problems to which solutions were provided. Presentation of papers at sessions of technical groups and in industry technical journals, participation in training and teaching programs, both in house and in associated colleges, and the publication of several technical and semitechnical books all supplemented the long and varied experiences amassed in over fifty years.

Acknowledgment is made to the many people within and outside the industry with whom the author was privileged to be associated, and it is to them he dedicates this book. As an early professor once remarked, no one is born with any knowledge. We build on the work of each other, a process that fortunately is never-ending—a debt we acknowledge to our predecessors. Efforts such as this impose on the lives of my friends and family; for their kindness, patience, and inspiration, my deepest gratitude and appreciation.

I would be remiss, however, if I did not single out for a special expression of gratitude Harold M. "Jolly" Jalonack for his unstinting help and encouragement; the many former associates in the Long Island Lighting Company—but especially Arthur C. Seale—for their patience in replying to my questions and requests, and for their comments and criticisms; my editors at McGraw-Hill, particularly Richard Mickey, for his very constructive criticism; and Dan Howard, friend and neighbor, for his critical review of the manuscript.

Anthony J. Pansini

HISTORY AND
DEVELOPMENT

While much attention is focused on electric power generating plants, their necessary adjuncts, electrical distribution systems, receive relatively scant attention from the public and investors—a phenomenon reflected in many engineering schools and among managements of many utility companies. This may perhaps be because electric lines on poles in streets and alleys and along rear property lines often go unnoticed; indeed, they are sometimes installed out of sight beneath the ground.

In comparison with power plants, expenditures for distribution systems are usually made in relatively small increments—another reason for the rather meager treatment sometimes accorded them. Until a decade ago, of every dollar spent by utility companies for electric facilities, 50 cents was spent for the distribution systems. Escalating costs for generation and reduced costs of distribution equipment have lowered this proportion to 30 cents, still a substantial amount.

With society, in all walks of life, becoming more dependent for its successful functioning on a good supply of electric energy, the link between the source and the consumer, the distribution system, assumes an ever more critical role. It is not only called upon to deliver ever greater quantities of electric energy, but the demand for ever higher standards of quality imposes on it requirements that become ever more stringent.

Higher quality is not limited to better regulation of voltage, to narrower bands of almost flickerless voltage variations. Though not closely associated with electrical distribution, a very high degree of maintaining alternating current frequencies has been sought. The awareness of faults and other contingencies, their identification and location, and the means of service restoration are important factors involved. These may be accomplished by the installation of additional devices operating automatically or manually.

These objectives may also be affected by such "nonrelated" items as better-trained personnel; improved transportation and communication facilities, including tools and equipment; quicker access to records, including use of computers; adequate stocks of materials; liaison with other sources of assistance; preventive maintenance programs; and various continually updated procedures for handling a variety of contingencies. All of these are reflected in carrying charges and operating expenses, and ultimately in the consumers' bills. These "nonrelated" items will not be explored further in this presentation; they are mentioned generally to illustrate other important subjects that should be given consideration in arriving at overall solutions to problems affecting electric distribution systems.

* * *

In the early days of the electric power industry, the distribution systems were often mere appendages to the power generating plants. Their designs, if such they may be called, sometimes were predicated almost entirely on expediency and practicality. With little study, their installation and operation were considered more of an art than a science. The areas served and the number of consumers were relatively small; individual usages were not very large, generally limited to few applications. Quality, in terms of voltage regulation and service reliability, was almost nonexistent. Other means of taking care of people's lighting and power needs were readily available.

With the expansion in the use of electricity, the demands on the distribution systems became greater and more complex. They not only had to serve greater numbers of consumers, but had to supply their greater individual loads that now required closer supervision of voltage variations at the consumers' terminals. Further, consumers demanded a reliability in their service that could tolerate only fewer interruptions of shorter duration.

At this point, the design, construction, maintenance, and operation of distribution systems became a science involving technical and economic disciplines not only in the field of electrical engineering, but in mechanical, civil, chemical, and almost all other fields of engineering as well.

From the early, simple, "radial" circuit, i.e., a feeder supplied from one source, other more sophisticated designs evolved. Radial circuits were provided with sectionalizing points which enabled a faulted section of the circuit to be disconnected. This enabled the remainder of the circuit beyond the faulted section to be reenergized by connecting it to other sources, usually adjacent circuits. These "emergency" tie points, specifically provided for this purpose, also enabled loads to be transferred conveniently from one circuit to another.

Other designs provided for duplicate feeds, with manual or automatic throwover from one circuit to another. Circuits were formed into loops, operating open at some point or as a closed loop. In areas of more important and greater load densities, circuits were interconnected into a mesh or network.

Original distribution systems supplied direct current at the low distribution voltages. The advent of the transformer and the economics of serving larger and larger loads more and more distant from the sources of supply soon had alternating current systems supplant the direct current distribution systems almost universally, although some declining ones still survive. Larger loads could now be supplied over longer distances at higher voltages and then lowered to utilization voltages to supply a consumer or group of consumers.

Requirements for electric service became geared to the different types of consumers served: residential, including urban, suburban, and rural; commercial, including individual stores, shopping centers, and office buildings; and industrial, including manufacturing and service plants of varying sizes. Further, other considerations sometimes made the underground installation of distribution systems desirable; such systems present problems very different from the simpler overhead systems.

*   *   *

Parallel with the development of the electric distribution circuits was the development of more suitable materials, electric apparatus, tools, and equipment, which permitted new and more efficient methods of construction, maintenance, and operation, a process that continues to this day.

Rough-hewn raw-wood poles have given way to well-turned, well-shaped, well-preserved poles of selected woods, including hard, strong wallaba for special applications. These, in turn, may give way to reinforced concrete, steel, and aluminum alloys. Experimentation continues with poles made of special plastic compounds.

Conductors, originally always made of copper, now also include those made of aluminum and copper-clad steel; during World War II, steel and silver were also used to replace scarce materials needed for the war effort. More recently, experimental conductors made of sodium and other materials have been installed for test purposes.

Porcelain insulators, originally made in one piece and almost exclusively used, are now also made as modular suspension-type units capable of being added together to accommodate almost any voltage. Glass and Pyrex have also been used extensively, while work now progresses with insulators made of plastic compounds.

Similarly, rubber insulation for cables, the initial material almost solely used, with limited ability to withstand higher voltages as well as age, has given way for the higher voltage ratings to varnished cambric, oil-impregnated paper, and plastic compounds. Research, which has extended the use of plastic compounds to voltages in the 138-kV category, continues.

Transformers have become smaller and more efficient, as well as less costly. New forms and kinds of steel cores have materially reduced magnetizing losses, while new types of insulation have not only affected their life spans, but noticeably increased their capacity size for size. Further, associated protective devices are now included within the same enclosure, making for improved appearance, easier handling, and better coordination of such devices. For some smaller sizes, epoxy-encapsulated units to replace oil-filled tanked transformers are in experimental use. Research continues for better cores and insulation.

Secondary mains have been streamlined into cabled conductors, or completely eliminated; and fewer cross arms are being installed in many areas. Capacitors have been applied to improve voltage and reduce losses, complementing or supplanting voltage regulators. Mechanical connectors have almost completely replaced manually constructed splices; better electrical contacts result as well as more uniform, safer, and more easily made installations. Street lighting now employs photoelectric cell-actuated relays for control.

Underground cables, formerly using lead almost exclusively for waterproof sheathing, now employ plastic compound coverings for that purpose as well as for insulation. Fiber, tile, wood, concrete, steel, and asbestos-based and plastic ducts are, in many cases, dispensed with and cables buried directly in the ground.

Sufficient examples have been cited to indicate changes and progress in the development of materials, methods, and equipment. The greatest development, however, has been in the realm of standardization, notably in transformer ratings, voltages, types, etc., but extending also to poles, conductors, fuses, and almost every element of electric distribution systems.

$$*\quad *\quad *$$

Concurrent with progress in the development of the several elements making up the electrical distribution system has been the improvement in means of transportation, communication, and tools and equipment.

The horse-drawn truck has been replaced by specially designed and constructed vehicles powered by internal combustion engines capable of speeds limited only by safety considerations and local speed laws. Messenger, mail, and telegraph services have been replaced by telephones, to which later were added shortwave two-way mobile radio units, making for very rapid communication with personnel and crews in the field. More recently, such radio and telephone communications have included the installation of cathode ray tubes (CRTs) in both field vehicles and operating offices, made possible by developments in electronics and miniaturization. These enable data recorded in the computer to be made almost instantly available to those people.

Bucket-type line trucks are making the lineman's work safer and easier. Vibrating plows and horizontal boring machines make possible the relatively deep burial of cable; in many instances, this is accomplished by one unit in one operation. These developments represent significant factors in preventing or holding down the duration of interruptions and other contingencies, resulting in overall greater reliability of electric service.

$$*\quad *\quad *$$

Despite some prevailing views, distribution engineers have always been conscious of appearance and other environmental factors. It is true that a pole line can really look beautiful only to distribution engineers, though it must never be forgotten that the use of such construction made possible the rather inexpensive supply of electric energy to almost everyone, not only in this country, but in most other countries as well.

It is equally true, however, that the distribution engineer has given recognition to those environmental factors even earlier than recent local ordinances would suggest.

Designs were adopted in many cases that attempted to make the appearance of such lines less obtrusive. From locations in the street, many were placed out of sight along rear property lines. The shapes, sizes, and color of poles were designed to be more pleasing to the eye, and their numbers, as well as the number of prominent cross arms, were reduced as much as practical. Often such lines were built through trees, even though continual tree trimming and the use of covered and insulated conductors resulted in additional expense. Agreements were reached to place power, communication, and other facilities on a common pole line to avoid cluttering the landscape with too many pole lines. In many cases, facilities were placed underground at much greater cost to allay objections in certain particular areas. All of these were done in the interest of better public relations, all without benefit of a host of rules and regulations.

Changes in labor practices have also greatly influenced the design of distribution systems (as well as other utility operations). Where in earlier days (the 1920s) the labor component of an installation accounted for only some 20 (or less) percent compared to 80 percent for material, today that ratio has been reversed with labor constituting some 80 percent of the cost and material only 20 percent. Thus in designing an economical distribution system, the engineer could now make more ample use of material, e.g., by calling for larger-size conductors, insulators, transformers, and other components. The net result is a more reliable system requiring fewer emergency operations because of overloads, installations generally providing for larger (and longer) future demands, and a reduction in losses on such systems.

*   *   *

The problem of losses in the distribution system assumes greater importance with the price of fuels no longer a relatively minor factor in the supply of electric energy. It is difficult to measure the actual energy losses in such a system, as many other factors are included in the difference between the energy consumed by each of the consumers connected to it and the energy sent out by the power plant. Educated estimates, however, place these losses at from 10 to 20 percent of the energy sent out.

Since the losses, in general, vary with the square of the current flowing through the conductors, whether in a line or in electrical equipment, holding down the value of such current will reduce losses. Many means have been employed to achieve this, the principal one being that of raising the voltages of circuits, thereby reducing current values for given loads. Increasing conductor sizes and shortening circuit lengths, by reducing resistance values, have also been employed. In alternating current systems, the installation of capacitors at strategic locations, by improving the power factor and thereby reducing the current flow for given loads, has also been used.

Since current flow is a measurement of the demand for electric energy by a consumer, efforts have been directed toward holding down the *demand* for electricity by attempting to even out more uniformly the consumption of energy. This has been termed *energy management.* Devices, mostly electronically controlled, cut off and on electric supply to the various loads and appliances connected to separate circuits, so that while the same end results are accomplished, peaks and valleys are reduced and load curves tend toward a continuously uniform flow. Special metering arrangements and rate schedules are provided to encourage and police such arrangements. Designs of distribution systems can also contribute to the realization of this goal.

In addition, the reduction of demands and currents can also result in the same facilities' carrying greater amounts of energy, delaying, if not making unnecessary, additional installations of power generating and transmission facilities (including substations), as well as of distribution facilities. This can have important impacts on the financial requirements of a utility.

* * *

Many of the features described for improving the quality of electric service as well as for reducing losses lend themselves to automatic operation of the distribution systems. Advances in electronics and miniaturization (much of it fallout from the space program) now make such controls feasible, both technically and economically. A simple example is the control of street lighting through relays actuated by photoelectric cells. Instead of being turned on and off on some time schedule, streetlights are permitted by such relays to operate when they are needed because of darkness. Not only are circuits simplified and a smaller investment made, but losses can be minimized and better public relations achieved.

Through other types of electronically controlled relays, switches can be remotely operated automatically (opened and closed) as desired, capacitors switched on and off, loads divided more equitably between circuits as demands vary, and, during contingencies, emergency switching-off of faulted portions and reenergization of unfaulted portions from other sources accomplished quickly and automatically without manual intervention.

Remote reading and billing of consumer meter readings has been in the experimental stage for some time. Moreover, more rapid and positive operation of relays that can accommodate more sensitive settings can result in substantial savings in the installation of protective and control equipment.

* * *

There are many other factors that influence the design, construction, and operation of distribution systems, many not of a technical nature. Economics plays a most important part, but associated with such considerations are also those of financing, interest rates, rates of inflation, future worth of present expenditures as well as present worth of future expenditures, taxes, patterns of future growth, government regulations at all levels, consumer relations, public images, employee relations, availability of skilled personnel and training programs, and a host of other considerations, not excluding the more important and universal ones of weather and climate.

This work will attempt to limit itself to technical considerations, though at times it may be necessary to refer to some nontechnical factors where these may bear on the subject. In discussing the distribution system, no details on the operation of electric circuits or of such equipment as transformers and capacitors will be included (except where they may be pertinent), as such are generally contained in standard basic electrical engineering texts. In general, it will be assumed the reader is familiar with such theory and the mathematics covered in college-level courses.

Moreover, it is to be noted that the basic fundamentals of distribution engineering are well established, while its practice has been changing and continues to develop rapidly, employing more and more the results of research and development in other disciplines.

Further, distribution system designs are often affected by extraneous factors. For example, sometimes improvement or modernization of a circuit cannot be justified technically or economically. Often, however, advantage is taken of other considerations, such as road widening or other construction, to rebuild, revamp, or replace lines, the opportunity being afforded to make desirable changes that otherwise would not be considered for some time.

The normal sequence in the installation or expansion of distribution systems begins with the planning and design of facilities, then proceeds to their construction, and finally includes their maintenance and operation. The interrelationship of these factors, their effect one upon the other, is of the utmost importance in achieving an eminently satisfactory, if not optimum or maximum, operation.

Because the distribution engineer has to deal with existing systems whose vintages may vary, it has been thought desirable to describe present installations and practices, and also some past changes and variations that have taken place. Also, as overhead systems still predominate—and it appears they will continue to do so for some time despite the proliferation of underground construction—discussion of such overhead systems will appear to predominate, though discussion of underground systems will be more than adequate. It is to be hoped that some thoughts for future developments may also be found in these discussions. At any rate, distribution systems will continue to develop as demands and requirements change, and as technologies develop to meet them.

*   *   *

No discussion of any segment of the electric power system is complete without at least some peering into the future. The economics of energy supply will indeed have a marked effect on almost any and all endeavors, not only in this country, but throughout the industrialized world. Its effect on power systems, and particularly the distribution portion, may be profound. At one end of the spectrum, there may be a trend toward the complete electrification of consumers' energy requirements and their supply from a central source. The other end may well call for the dismantling of distribution systems as we know them today. Either extreme signals almost revolutionary changes that will present enormous problems to the distribution engineer.

Inexpensive oil and natural gas supplies appear to have seen their heyday and to have given place to other sources of energy. For the near future, coal (in some form or other) and nuclear fuels would seem to have a priority of sorts. For the longer term, other forms of energy—perhaps some new chemical storage cells, alcohol or other fuels from agricultural products, solar energy directly from the sun, wind power, or a combination of these—appear to hold some promise. The "ultimate" may be nuclear power packs, with life spans of several decades or longer, installed at each consumer's premises, and with the demise of central power supply, including the distribution systems. What compromises may occur over what period of time is open to wide speculation.

For the shorter period, however, the period of most concern to present distribution engineers, signs point to maximizing the use of coal and nuclear fuels. Both of these, from environmental and conservation viewpoints, would promote the almost complete electrification of consumers' requirements. Thus the distribution systems would need to be reinforced very considerably. At the same time, however, there would be an almost equal need to hold losses to a minimum.

These two factors, fortunately, are not incompatible. With labor costs escalating continually, greater use of materials is indicated, which is in the direction of reducing losses.

Efforts at holding down maximum demands of consumers and maximum coincidental demands will be expanded. Load management (mentioned earlier) will be a requirement, with punitive rate schedules, and perhaps tax schedules, used to enforce it; the limiting of demand would also have the effect of holding down the size and cost of new facilities required.

On the other hand, the almost total dependence on electricity implied would almost certainly have consumers seeking a better degree of service reliability. The achievement of this improvement, while at the same time holding down costs, will no doubt tax the skills and ingenuity of distribution engineers—it would not be the first of such challenges successfully met by them in previous decades.

*   *   *

Engineering, as has been observed, is a combination of science and art. The scientist, the researcher, establishes facts and laws, discovers or creates new materials, all of which are subject to rigid interpretations and descriptions. On the other end is the "pure" artist who creates and imagines things and situations, often with no conscious regard to the realms of practicability and possibility.

More and more, engineers find it necessary to add art to their métier. While the scientist and artist operate with almost no consideration of cost, the engineer is almost always firmly wedded to economics. Indeed, it has often been observed that it is the engineer's job to do for one dollar what others can do for ten—or even two!

The electrical distribution engineer faces problems that are seldom exactly, or even approximately, the same. And the solutions proposed are often not "perfect" but the "best available" solutions. Often improvisation and compromise must be used, so that any work on this subject cannot be exact, nor provide all the answers to all the questions that may arise. All that this work can purport to do is to lend some direction and to point the way to where we have been, where we are, and perhaps where we are going. It has succeeded if the student, the practicing distribution engineer, and others having an interest in this subject find it useful in their daily endeavors.

ELECTRICAL
DISTRIBUTION
ENGINEERING

PART ONE

THE DISTRIBUTION SYSTEM

CHAPTER 1

THE DISTRIBUTION SYSTEM: DESCRIPTION

Although there is no "typical" electric power system, a diagram including the several components that are usually to be found in the makeup of such a system is shown in Fig. 1-1; particular attention should be paid to those elements which will make up the component under discussion, the distribution system.

While the energy flow is obviously from the power generating plant to the consumer, it may be more informative for our purposes to reverse the direction of observation and consider events from the consumer back to the generating source.

Energy is consumed by users at a nominal utilization voltage that may range generally (in the United States) from 110 to 125 V, and from 220 to 250 V (for some large commercial and industrial users, the nominal figures are 277 and 480 V). It flows through a metering device that determines the billing for the consumer, but which may also serve to obtain data useful later for planning, design, and operating purposes. The metering equipment usually includes a means of disconnecting the consumer from the incoming supply should this become necessary for any reason.

The energy flows through conductors to the meter from the secondary mains (if any); these conductors are referred to as the consumer's *service*, or sometimes also as the *service drop*.

Several services are connected to the secondary mains; the secondary mains now serve as a path to the several services from the distribution transformers which supply them.

At the transformer, the voltage of the energy being delivered is reduced to the utilization voltage values mentioned earlier from higher *primary* line voltages that may range from 2200 V to as high as 46,000 V.

The transformer is protected from overloads and faults by fuses or so-called

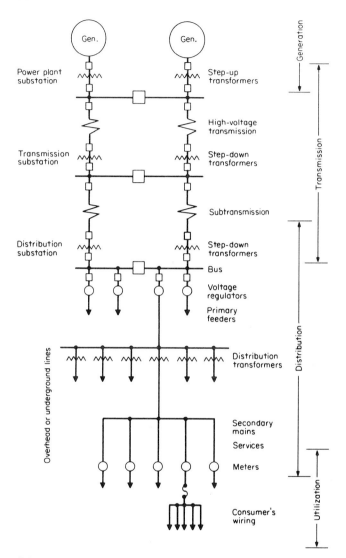

FIG. 1-1 Typical electric system showing operational divisions.
Note overlap of divisions.

weak links on the high-voltage side; the latter also usually include circuit-breaking devices on the low-voltage side. These operate to disconnect the transformer in the event of overloads or faults. The circuit breakers (where they exist) on the secondary, or low-voltage, side operate only if the condition is caused by faults or overloads in the secondary mains, services, or consumers' premises; the primary fuse or weak link in addition operates in the event of a failure within the transformer itself.

If the transformer is situated on an overhead system, it is also protected from lightning or line voltage surges by a surge arrester, which drains the voltage surge to ground before it can do damage to the transformer.

The transformer is connected to the primary circuit, which may be a lateral or spur consisting of one phase of the usual three-phase primary main. This is done usually through a *line* or *sectionalizing fuse,* whose function is to disconnect the lateral from the main in the event of fault or overload in the lateral. The lateral conductors carry the sum of the energy components flowing through each of the transformers, which represent not only the energy used by the consumers connected thereto, but also the energy lost in the lines and transformers to that point.

The three-phase main may consist of several three-phase branches connected together, sometimes through other line or sectionalizing fuses, but sometimes also through switches. Each of the branches may have several single-phase laterals connected to it through line or sectionalizing fuses.

Where single-phase or three-phase overhead lines run for any considerable distance without distribution transformer installations connected to them, surge arresters may be installed on the lines for protection, as described earlier.

Some three-phase laterals may sometimes also be connected to the three-phase main through *circuit reclosers.* The recloser acts to disconnect the lateral from the main should a fault occur on the lateral, much as a line or sectionalizing fuse. However, it acts to reconnect the lateral to the main, reenergizing it one or more times after a time delay in a predetermined sequence before remaining open permanently. This is done so that a fault which may be only of a temporary nature, such as a tree limb falling on the line, will not cause a prolonged interruption of service to the consumers connected to the lateral.

The three-phase mains emanate from a *distribution substation,* supplied from a *bus* in that station. The three-phase mains, usually referred to as a *circuit* or *feeder,* are connected to the bus through a protective circuit breaker and sometimes a voltage regulator. The voltage regulator is usually a modified form of transformer and serves to maintain outgoing voltage within a predetermined band or range on the circuit or feeder as its load varies. It is sometimes placed electrically in the substation circuit so that it regulates the voltage of the entire bus rather than a single outgoing circuit or feeder, and sometimes along the route of a feeder for partial feeder regulation. The circuit breaker in the feeder acts to disconnect that feeder from the bus in the event of overload or fault on the outgoing or distribution feeder.

The substation bus usually supplies several distribution feeders and carries the sum of the energy supplied to each of the distribution feeders connected to it. In turn, the bus is supplied through one or more transformers and associated circuit breaker protection. These substation transformers step down the voltage of their supply circuit, usually called the *subtransmission* system, which operates at voltages usually from 23,000 to 138,000 V.

The subtransmission systems may supply several distribution substations and may act as *tie feeders* between two or more substations that are either of the *bulk power* or *transmission* type or of the distribution type. They may also be tapped to supply some distribution load, usually through a circuit breaker, for a single consumer, generally an industrial plant or a commercial consumer having a substantially large load.

The transmission or bulk power substation serves much the same purposes as a distribution substation, except that, as the name implies, it handles much greater amounts of energy: the sum of the energy individually supplied to the subtransmission lines and associated distribution substations and losses. Voltages at the transmission substations are reduced to outgoing subtransmission line voltages from transmission voltages that may range from 69,000 to upwards of 750,000 V.

The transmission lines usually emanate from another substation associated with a power generating plant. This last substation operates in much the same manner as other substations, but serves to step up to transmission line voltage values the voltages produced by the generators. Because of material and insulation limitations, generator voltages may range from a few thousand volts for older and smaller units to some 20,000 volts for more recent, larger ones. Both buses and transformers in these substations are protected by circuit breakers, surge arresters, and other protective devices.

In all the systems described, conductors should be large enough that the energy loss in them will not be excessive, nor the loss in voltage so great that normal nominal voltage ranges at the consumers' services cannot be maintained.

In some instances, voltage regulators and capacitors are installed at strategic points on overhead primary circuits as a means of compensating for voltage drops or losses, and incidentally help in holding down energy losses in the conductors.

In many of the distribution system arrangements, some of the several elements between the generating plant and the consumer may not be necessary. In a relatively small area, such as a small town, that is served by a power plant situated in or very near the service area, the distribution feeder may emanate directly from the power plant bus, and all other elements may be eliminated, as indicated in Fig. 1-2. This is perhaps one extreme; in many other instances only some of the other elements may not be necessary; e.g., a similar small area somewhat distant from the generating plant may find it necessary to install a distribution substation supplied by a transmission line of appropriate voltage only.

In the case of areas of high load density and rather severe service reliability requirements, the distribution system becomes more complex and more expensive. The several secondary mains to which the consumers' services are connected may all be connected into a mesh or network. The transformers supplying these

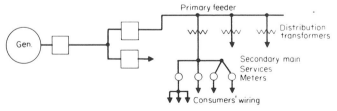

FIG. 1-2 "Abbreviated" electric system.

secondary mains or network are supplied from several different primary feeders, so that if one or more of these feeders is out of service for any reason, the secondary network is supplied from the remaining ones and service to the consumers is not interrupted. To prevent a feeding-back from the energized secondary network through the transformers connected to feeders out of service (thereby energizing the primary and creating unsafe conditions), automatically operated circuit breakers, called *network protectors,* are connected between the secondary network and the secondary of the transformers; these open when the direction of energy flow is reversed.

The two examples cited here are perhaps the two extremes in the design of distribution systems, the first the simplest, the latter the most complex. There are many variations in between these, and the basic ones will be described in their appropriate places.

Only distribution systems, however, will be the subject of further description and discussion in this book. In general, these include the distribution substation, primary feeders, transformers, secondary mains, services, and other elements between the substation and the consumers' points of service.

CHAPTER 2

DISTRIBUTION SYSTEM CONSIDERATIONS

In determining the design of distribution systems, three broad classifications of choices need to be considered:

1. The type of electric system: dc or ac, and if ac, single-phase or polyphase.

2. The type of delivery system: radial, loop, or network. Radial systems include duplicate and throwover systems.

3. The type of construction: overhead or underground.

DESIRED FEATURES

Electrical energy may be distributed over two or more wires. The principal features desired are safety; smooth and even flow of power, as far as is practical; and economy.

Safety

The safety factor usually requires a voltage low enough to be safe when the electric energy is utilized by the ordinary consumer.

Smooth and Even Flow of Power

A steady, uniform, nonfluctuating flow of power is highly desirable, both for lighting and for the operation of motors for power purposes. Although a direct current system fills these requirements admirably, it is limited in the distance over which it can economically supply power at utilization voltage.

Alternating current systems deliver power in a fluctuating manner following the cyclic variations of the voltage generated. Such fluctuations of power are not objectionable for heating, lighting, and small motors, but are not entirely satisfactory for the operation of some devices such as large motors, which must deliver mechanical power steadily and therefore require a steady input of electric power. This may be done by supplying electricity to the motors by two or three circuits, each supplying a portion of the power, whose fluctuations are purposely made not to occur at the same time, thereby decreasing or damping out the effect of the fluctuations. These two or three separate alternating current circuits (each often referred to as a single-phase circuit) are combined into one polyphase (two- or three-phase) circuit. The voltages for polyphase circuits or systems are supplied from polyphase generators.

Economy

The third factor requires the minimum use of conductors for delivery of electric energy. This usually calls for the use of higher voltages where conditions permit and the elimination of some conductors by providing a common return path for two or more circuits.

TYPES OF ELECTRIC SYSTEMS
Direct Current Systems

Direct current systems usually consist of two or three wires. Although such distribution systems are no longer employed, except in very special instances, older ones now exist and will continue to exist for some time. Direct current systems are essentially the same as single-phase ac systems of two or three wires; the same discussion for those systems also applies to dc systems.

Alternating Current Single-Phase Systems

Two-Wire Systems   The simplest and oldest circuit consists of two conductors between which a relatively constant voltage is maintained, with the load connected between the two conductors; refer to Fig. 2-1.

FIG. 2-1 AC single-phase two-wire system.

In almost all cases, one conductor is grounded. The grounding of one conductor, usually called the *neutral*, is basically a safety measure. Should the live conductor come in contact accidentally with the neutral conductor, the voltage of the live conductor will be dissipated throughout a relatively large body of earth and thereby rendered harmless.

In calculating power (I^2R) losses in the conductors, the resistance of the conductors must be considered. In the case of the neutral conductor, because the ground, in parallel with the conductor, reduces the effective resistance, the "return" current will divide between the conductor and ground in inverse proportion to their resistances. Thus the I^2R loss in the neutral conductor will be lower than that in the live conductor; the I^2R loss in the earth may, for practical purposes, be disregarded.

In calculating voltage drop in the circuits, both the resistance and reactance of the two conductors must be considered. (In dc circuits, reactance does not exist during normal flow of current.) This combination of reactance and resistance, known as impedance, is measured in ohms (Ω). Because the current in the grounded neutral conductor may be less than the current in the live conductor, the voltage drop in the neutral conductor may also be less.

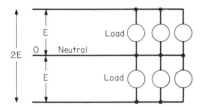

FIG. 2-2 AC single-phase three-wire system.

Three-Wire Systems Essentially the three-wire system is a combination of two two-wire systems with a single wire serving as the neutral of each of the two-wire systems. At a given instant, if one of the live conductors is E volts (say 120 V) "above" the neutral, the other live conductor will be E volts (120 V) "below" the neutral, and the voltage between the two live (or outside) conductors will be $2E$ (240 V). Refer to Fig. 2-2.

If the load is balanced between the two (two-wire) systems, the common neutral conductor carries no current and the system acts as a two-wire system at twice the voltage of the component system; each unit of load (such as a lamp) of one component system is in series with a similar unit of the other system. If the load is not balanced, the neutral conductor carries a current equal to the difference between the currents in the outside conductors. Here again, the neutral conductor is usually connected to ground.

For a balanced system, power loss and voltage drop are determined in the same way as for a two-wire circuit consisting of the outside conductors; the neutral is neglected.

Where the loads on the two portions of the three-wire circuit are unbalanced, voltages at the utilization or receiving ends may be different. These are shown schematically in Fig. 2-3. Let the distance between the dashed lines represent the voltage. There will be a voltage drop, with reference to the neutral, in each of the conductors 1 and 2. The neutral conductor will carry the difference in currents, that is, $I_2 - I_1$, or I_n. This current in the neutral conductor will produce a voltage drop in that conductor, as indicated in Fig. 2-3. The result will be a

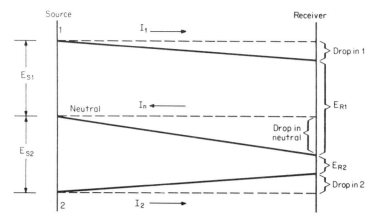

FIG. 2-3 Unbalanced load—single-phase three-wire system.

much larger drop in voltage between conductor 2 and neutral than between conductor 1 and neutral. If the unbalance is so large that I_n is greater than I_1, the receiving end voltage E_{R1} will be greater than the sending end voltage E_{S1}, and there will be an actual rise in voltage across that side.

The limiting case occurs when $I_1 = 0$ and $I_n = I_2$. In that case, all the load is carried on side 2; the rise in voltage on side 1 will be half as much as the drop in voltage on the loaded side 2. However, if an equal load is now added to side 1, the loads in both parts of the circuit will be balanced and I_n will equal 0. The drop in voltage between conductor 2 and the neutral will be reduced to half that obtained with the load on side 2 only, although the load now supplied is doubled.

Voltage drops in the conductors will depend on the currents flowing in them and their impedances. The power loss in each conductor (I^2R) will depend on the current flowing in it and its resistance.

In all of this discussion, the size of the neutral has been assumed to be the same as the live or outside conductors.

Series Systems The series type of circuit is used chiefly for street lighting and, although being rapidly replaced by multiple-circuit lighting, nevertheless still exists in substantial numbers. It consists of a single-conductor loop in which the current is maintained at a constant value, the loads connected in series; see Fig. 2-4. The voltage between the conductors at the source or at any other point depends on the amount of load connected beyond that point. The voltage at the source is equal to the vectorial sum of the voltages across the various loads and the voltage drop in the conductor.

The voltage drop in each section of the conductor depends on the current flowing in it (which is constant in value) and the impedance of that section of the conductor.

The power supplied the circuit equals the sum of the power for the individual units of load and the line losses. Power loss in each section of the conductor will

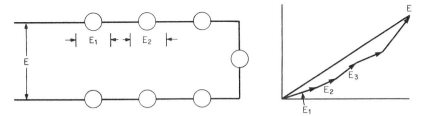

FIG. 2-4 AC single-phase series system and voltage vector diagram.

depend on the current (squared) and the resistance of that section of the conductor.

Alternating Current Two-Phase Systems

Two-phase systems are rapidly becoming obsolete, but a good number of them exist and may continue to exist for some time.

Four-Wire Systems   The four-wire system consists of two single-phase two-wire systems in which the voltage in one system is 90° out of phase with the voltage in the other system, both usually supplied from the same generator. Refer to Fig. 2-5.

In determining the power, power loss, and voltage drops in such a system, the values are calculated as for two separate single-phase two-wire systems.

Three-Wire Systems   The three-wire system is equivalent to a four-wire two-phase system, with one wire (the neutral) made common to both phases; refer to Fig. 2-6. The current in the outside or phase wires is the same as in the four-wire system; the current in the common wire is the vector sum of these currents but opposite in phase. When the load is exactly balanced in the two phases, these currents are equal and 90° out of phase with each other and the resultant neutral current is equal to $\sqrt{2}$ or 1.41 times the phase current.

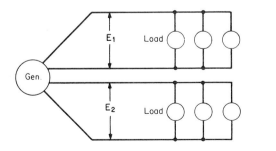 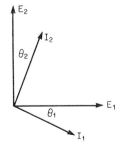

FIG. 2-5 AC two-phase four-wire system and vector diagram.

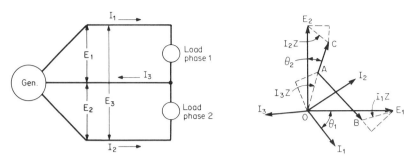

FIG. 2-6 AC two-phase three-wire system and vector diagram.

The voltage between phase wires and common wire is the normal phase voltage, and, neglecting the difference in neutral IR drop, the same as in the four-wire system. The voltage between phase wires is equal to $\sqrt{2}$ or 1.41 times that voltage.

The power delivered is equal to the sum of the powers delivered by the two phases. The power loss is equal to the sum of the power losses in each of the three wires.

The voltage drop is affected by the distortion of the phase relation caused by the larger current in the third or common wire. In Fig. 2-6, if E_1 and E_2 are the phase voltages at the source and I_1 and I_2 the corresponding phase currents (assuming balanced loading), I_3 is the current in the common wire. The voltage (IZ) drops in the two conductors, subtracted vectorially from the source voltages E_1 and E_2, give the resultant voltages at the receiver of AB for phase 1 and AC for phase 2. The voltage drop numerically is equal to $E_1 - AB$ for phase 1, and $E_2 - AC$ for phase 2. It is apparent that these voltage drops are unequal and that the action of the current in the common wire is to distort the relations between the voltages and currents—the effect shown in Fig. 2-6 is exaggerated for illustration.

Five-Wire Systems The five-wire system is equivalent to a two-phase four-wire system with the midpoint of both phases brought out and joined in a fifth wire. The voltage is of the same value from any phase wire to the common neutral, or fifth, wire. The value may be in the nature of 120 V, which is used for lighting and small motor loads, while the voltage between opposite pairs of phase wires, E, may be 240 V, used for larger-power loads. The voltage between adjacent phase wires is $\sqrt{2}$, or 1.41, times 120 V (about 170 V). See Fig. 2-7.

If the load is exactly balanced on all four phase wires, the common or neutral wire carries no current. If it is not balanced, the neutral conductor carries the vector sum of the unbalanced currents in the two phases.

Alternating Current Three-Phase Systems

Four-Wire Systems The three-phase four-wire system is perhaps the most widely used. It is equivalent to three single-phase two-wire systems supplied from the same generator. The voltage of each phase is 120° out of phase with the voltages of the other two phases, but one conductor is used as a common conductor for all of the system. The current I_n in that common or neutral conductor is equal to the vector sum of the currents in the three phases, but opposite in phase, as shown in Fig. 2-8.

If these three currents are nearly equal, the neutral current will be small, since these phase currents are 120° out of phase with each other. The neutral is usually grounded. Single-phase loads may be connected between one phase wire and the neutral, but may also be connected between phase wires if desired. In this latter instance, the voltage is $\sqrt{3}$ or 1.73 times the line-to-neutral voltage E. Three-phase loads may have each of the separate phases connected to the three phase conductors and the neutral, or the separate phases may be connected to the three phase conductors only.

Power delivered is equal to the sum of the powers in each of the three phases. Power loss is equal to the sum of the I^2R losses in all four wires.

The voltage drop in each phase is affected by the distortion of the phase relations due to voltage drop caused by the current in the neutral conductor. This is not so, however, when the neutral conductor is grounded at both the sending and receiving ends, in which case the neutral drop is theoretically zero, the current returning through ground. The voltage drop may be obtained vectorially by applying the impedance drop of each phase to its voltage. The neutral point is shifted from O to A by the voltage drop in the neutral conductor and the resulting voltages at the receiver are shown by E_{1R}, E_{2R}, and E_{3R}. The voltage drops in each phase are numerically equal to the difference in length between E_{1S} and E_{1R}, E_{2S} and E_{2R}, and E_{3S} and E_{3R}. The effects of the distortion due to voltage drop in the neutral conductors are exaggerated in Fig. 2-8 for illustration.

Three-Wire Systems If the load is equally balanced on the three phases of a four-wire system, the neutral carries no current and hence could be removed, making a three-wire system. It is not necessary, however, that the load be exactly balanced on a three-wire system.

Considering balanced loads, on a three-phase three-wire system, a three-phase load may be connected with each phase connected between two phase wires—a

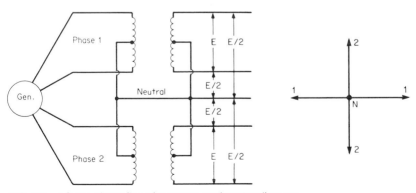

FIG. 2-7 AC two-phase five-wire system and vector diagram.

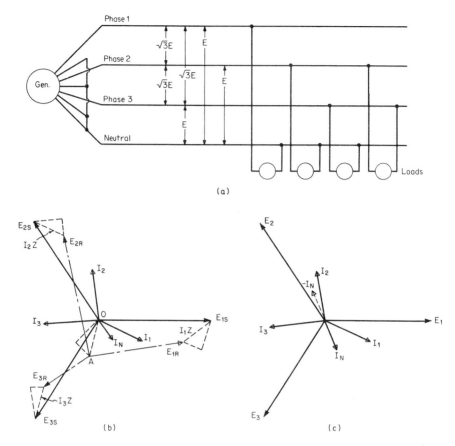

FIG. 2-8 (a) AC three-phase four-wire system; (b) voltage and current vector diagram; (c) current vector diagram.

delta (Δ) connection—or with each phase between one phase wire and a common neutral point—the star or wye (Y) connection, as shown in Fig. 2-9.

The voltage between line conductors is the delta voltage E_Δ, while the line current is the wye current I_Y. The relations in magnitude and phase between the various delta and wye voltages and currents for the same load are shown in Fig. 2-9. For the delta connection, I_Y is equal to the vector difference between the adjacent delta currents; hence:

$$I_Y = \sqrt{3} \text{ (or } 1.73) \text{ times } I$$

and

$$E_\Delta = \sqrt{3} \text{ (or } 1.73) \text{ times } E_Y$$

Power delivered, when balanced loads are considered, is equal to 3 times the power delivered by one phase. Power loss is equal to the sum of the losses in each phase, or when balanced conditions exist, it is 3 times the power loss in any one phase.

The voltage drop in each phase, referred to the wye (Y) voltages, may be determined by adding the impedance drop in one conductor vectorially to E_Y, when balanced loads are considered. The same thing is done in determining voltages where unbalanced loads are considered. If $E_{\Delta S}$ is the voltage between phases at the source and E_{YS} the phase-to-neutral voltage, the drop due to

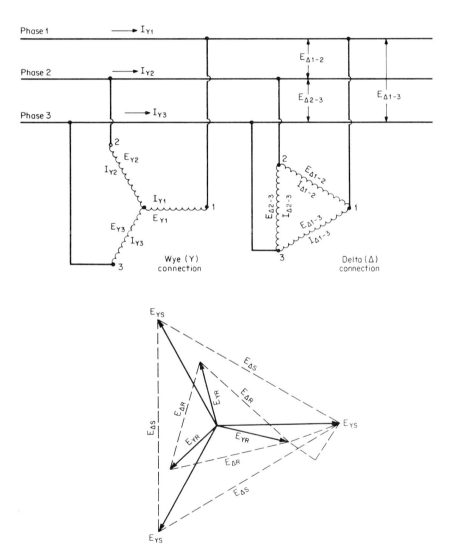

FIG. 2-9 AC three-phase three-wire system and voltage vector diagrams (current vector not shown).

conductor impedance, IZ, is subtracted vectorially from E_{YS} for each of the three phases, and the resulting voltages between phases at the receiving end ($E_{\Delta R}$) are obtained. The effects shown in the vector diagram of Fig. 2-9 are exaggerated for illustration.

Alternating Current Six-Phase Systems

Six-Wire Systems   Six-phase systems consist essentially of two three-phase systems connected so that each phase of one system will be displaced 180° with reference to the same phase of the other system. These may consist of two banks of three transformers connected separately with the polarity of one bank reversed with reference to the second bank; or one bank of transformers may be employed, with the secondary windings divided into two equal parts and both ends of each winding part brought out to separate terminals (for a total of 12 terminals).

The windings may be connected in a double-delta fashion as shown in Fig. 2-10a, or in a double-wye arrangement as shown in Fig. 2-10b. The associated vector diagrams of the voltage relationships are also indicated.

In the double-wye connection, it is not necessary to have the windings brought out to 12 terminals; the neutral connection may be made by connecting together the midtap from each of the three secondary windings.

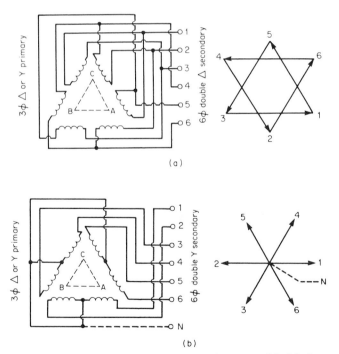

FIG. 2-10   AC six-phase six-wire double-delta system (a), AC six-phase six-wire (and seven-wire) double-wye system (b), and voltage vector diagrams.

Such systems are almost exclusively used in supplying rectifiers or synchronous converters to serve direct current loads; the synchronous converter also aids in improving power factor on the alternating current supply system.

Seven-Wire Systems   A seventh, or neutral, wire may be brought out from the common junction of the double-wye connection, as indicated by the dashed line in Fig. 2-10*b*.

The seven-wire system may be used for distribution purposes, with the neutral connected to other common neutral systems. The disadvantage of the additional conductor is balanced against two major advantages:

1. The ability to serve single-phase loads from a source of higher voltage, i.e., twice the line-to-neutral voltage, compared with 1.73 times the line-to-neutral voltage in a three-phase system.

2. Reduction in overall line losses, as each conductor will carry only one-sixth of the load, compared with one-third in a three-phase system—only half the load per conductor in a three-phase system. The losses, therefore, will be one-quarter those in a three-phase (three- or four-wire balanced) system.

The overall savings in fuel costs for supplying the lesser losses may exceed the increased carrying charges associated with the additional conductors. The improved voltage in supplying three-phase delta (power) loads from such a system also contributes to its acceptability. As fuel and operating costs increase, such systems may find wider application.

Comparison Between Alternating Current Systems

A comparison of efficiencies for the several alternating current systems, assuming the same (balanced) loads, the same voltage between conductors, and the same conductor size is summarized in Table 2-1, which uses a single-phase two-wire circuit as a basis for comparison.

TABLE 2-1   COMPARATIVE EFFICIENCIES OF AC SYSTEMS

Type of ac system		Amount of conductor	Power loss	Voltage drop (approximate)
Single-phase	2-wire	1.0	1.00	1.00
	3-wire	1.5	0.25	0.25
Two-phase	3-wire	1.5	0.50	0.50
	4-wire	2.0	0.25	0.25
	5-wire	2.5	0.25	0.25
Three-phase	3-wire*	1.5	0.167	0.167
	3-wire**	1.5	0.50	0.50
	4-wire*	2.0	0.167	0.167
Six-phase	6-wire	3.0	0.042	0.042
	7-wire	3.5	0.042	0.042

*Wye (*Y*) voltage same as single-phase.

**Delta (Δ) voltage same as single-phase.

TYPES OF DELIVERY SYSTEMS

The delivery of electric energy from the generating plant to the consumer may consist of several more or less distinct parts that are nevertheless somewhat interrelated, described generally in Chap. 1. The part considered "distribution," i.e., from the bulk supply substation to the meter at the consumer's premises, can be conveniently divided into two subdivisions:

1. Primary distribution, which carries the load at higher than utilization voltages from the substation (or other source) to the point where the voltage is stepped down to the value at which the energy is utilized by the consumer.

2. Secondary distribution, which includes that part of the system operating at utilization voltages, up to the meter at the consumer's premises.

Primary Distribution

Primary distribution systems include three basic types:

1. Radial systems, including duplicate and throwover systems

2. Loop systems, including both open and closed loops

3. Primary network systems

Radial Systems The radial-type system is the simplest and the one most commonly used. It comprises separate feeders or circuits "radiating" out of the substation or source, each feeder usually serving a given area. The feeder may be considered as consisting of a main or trunk portion from which there radiate spurs or laterals to which distribution transformers are connected, as illustrated in Fig. 2-11.

The spurs or laterals are usually connected to the primary main through fuses, so that a fault on the lateral will not cause an interruption to the entire

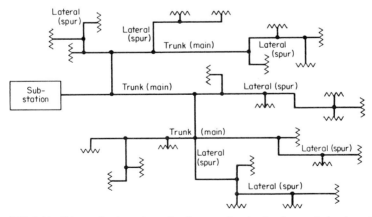

FIG. 2-11 Primary feeder schematic diagram showing trunk or main feeds and laterals or spurs.

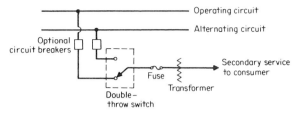

FIG. 2-12 Schematic diagram of alternate feed–throwover arrangement for critical consumers.

feeder. Should the fuse fail to clear the line, or should a fault develop on the feeder main, the circuit breaker back at the substation or source will open and the entire feeder will be deenergized.

To hold down the extent and duration of interruptions, provisions are made to sectionalize the feeder so that unfaulted portions may be reenergized as quickly as practical. To maximize such reenergization, emergency ties to adjacent feeders are incorporated in the design and construction; thus each part of a feeder not in trouble can be tied to an adjacent feeder. Often spare capacity is provided for in the feeders to prevent overload when parts of an adjacent feeder in trouble are connected to them. In many cases, there may be enough diversity between loads on adjacent feeders to require no extra capacity to be installed for these emergencies.

Supply to hospitals, military establishments, and other sensitive consumers may not be capable of tolerating any long interruption. In such cases, a second feeder (or additional feeders) may be provided, sometimes located along a separate route, to provide another, separate alternative source of supply. Switching from the normal to the alternative feeder may be accomplished by a throwover switching arrangement (which may be a circuit breaker) that may be operated manually or automatically. In many cases, two separate circuit breakers, one on each feeder, with electrical interlocks (to prevent connecting a good feeder to the one in trouble), are employed with automatic throwover control by relays. See Fig. 2-12.

Loop Systems Another means of restricting the duration of interruption employs feeders designed as loops, which essentially provide a two-way primary feed for critical consumers. Here, should the supply from one direction fail, the entire load of the feeder may be carried from the other end, but sufficient spare capacity must be provided in the feeder. This type of system may be operated with the loop normally open or with the loop normally closed.

Open Loop In the open-loop system, the several sections of the feeder are connected together through disconnecting devices, with the loads connected to the several sections, and both ends of the feeder connected to the supply. At a predetermined point in the feeder, the disconnecting device is intentionally left open. Essentially, this constitutes two feeders whose ends are separated by a

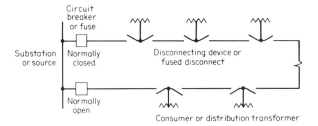

FIG. 2-13 Open-loop circuit schematic diagram.

disconnecting device, which may be a fuse, switch, or circuit breaker. See Fig. 2-13.

In the event of a fault, the section of the primary on which the fault occurs can be disconnected at both its ends and service reestablished to the unfaulted portions by closing the loop at the point where it is normally left open, and reclosing the breaker at the substation (or supply source) on the other, unfaulted portion of the feeder.

Such loops are not normally closed, since a fault would cause the breakers (or fuses) at both ends to open, leaving the entire feeder deenergized and no knowledge of where the fault has occurred. The disconnecting devices between sections are manually operated and may be relatively inexpensive fuses, cutouts, or switches.

Closed Loop   Where a greater degree of reliability is desired, the feeder may be operated as a closed loop. Here, the disconnecting devices are usually the more expensive circuit breakers. The breakers are actuated by relays, which operate to open only the circuit breakers on each end of the faulted section, leaving the remaining portion of the entire feeder energized. In many instances, proper relay operation can only be achieved by means of pilot wires which run from circuit breaker to circuit breaker and are costly to install and maintain; in some instances these pilot wires may be rented telephone circuits. See Fig. 2-14.

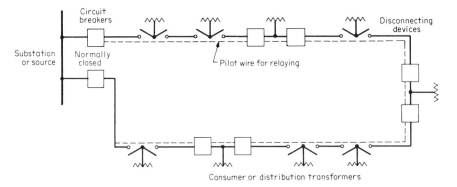

FIG. 2-14 Closed-loop circuit.

   To hold down costs, circuit breakers may be installed only between certain sections of the feeder loop, and ordinary, less expensive disconnecting devices installed between the intermediate sections. A fault will then deenergize several sections of the loop; when the fault is located, the disconnecting devices on both ends of the faulted section may be opened and the unfaulted sections reenergized by closing the proper circuit breakers.

Primary Network Systems   Although economic studies indicated that under some conditions the primary network may be less expensive and more reliable than some variations of the radial system, relatively few primary network systems have been put into actual operation and only a few still remain in service.

   This system is formed by tying together primary mains ordinarily found in radial systems to form a mesh or grid. The grid is supplied by a number of power transformers supplied in turn from subtransmission and transmission lines at higher voltages. A circuit breaker between the transformer and grid, controlled by reverse-current and automatic reclosing relays, protects the primary network from feeding fault current through the transformer when faults occur on the supply subtransmission or transmission lines. Faults on sections of the primaries constituting the grid are isolated by circuit breakers and fuses. See Fig. 2-15.

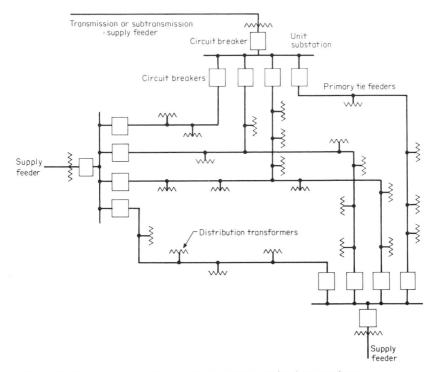

FIG. 2-15   Primary network. Sectionalizing devices on feeders not shown.

This type of system eliminates the conventional substation and long primary trunk feeders, replacing them with a greater number of "unit" substations strategically placed throughout the network. The additional sites necessary are often difficult to obtain. Moreover, difficulty is experienced in maintaining proper operation of the voltage regulators (where they exist) on the primary feeders when interconnected.

Secondary Distribution

Secondary distribution systems operate at relatively low utilization voltages and, like primary systems, involve considerations of service reliability and voltage regulation. The secondary system may be of four general types:

1. An individual transformer for each consumer; i.e., a single service from each transformer.

2. A common secondary main associated with one transformer from which a group of consumers is supplied.

3. A continuous secondary main associated with two or more transformers, connected to the same primary feeder, from which a group of consumers is supplied. This is sometimes known as *banking* of transformer secondaries.

4. A continuous secondary main or grid fed by a number of transformers, connected to two or more primary feeders, from which a large group of consumers is supplied. This is known as a *low-voltage* or *secondary* network.

Each of these types has its application to which it is particularly suited.

Individual Transformer–Single Service   Individual-transformer service is applicable to certain loads that are more or less isolated, such as in rural areas where consumers are far apart and long secondary mains are impractical, or where a particular consumer has an extraordinarily large or unusual load even though situated among a number of ordinary consumers.

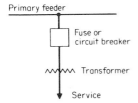

FIG. 2-16 Single-service secondary supply.

In this type of system, the cost of the several transformers and the sum of power losses in the units may be greater (for comparative purposes) than those for one transformer supplying a group of consumers from its associated secondary main. The diversity among consumers' loads and demands permits a

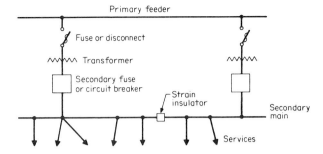

FIG. 2-17 Common-secondary-main supply.

transformer of smaller capacity than the capacity of the sum of the individual transformers to be installed. On the other hand, the cost and losses in the secondary main are obviated, as is also the voltage drop in the main. Where low voltage may be undesirable for a particular consumer, it may be well to apply this type of service to the one consumer. Refer to Fig. 2-16.

Common Secondary Main   Perhaps the most common type of secondary system in use employs a common secondary main. It takes advantage of diversity between consumers' loads and demands, as indicated above. Moreover, the larger transformer can accommodate starting currents of motors with less resulting voltage dip than would be the case with small individual transformers. See Fig. 2-17.

In many instances, the secondary mains installed are more or less continuous, but cut into sections insulated from each other as conditions require. As loads change or increase, the position of these division points may be readily changed, sometimes holding off the need to install additional transformer capacity. Also, additional separate sections can be created and a new transformer installed to serve as load or voltage conditions require.

Banked Secondaries   The secondary system employing banked secondaries is not very commonly used, although such installations exist and are usually limited to overhead systems.

This type of system may be viewed as a single-feeder low-voltage network, and the secondary may be a long section or grid to which the transformers are connected. Fuses or automatic circuit breakers located between the transformer and secondary main serve to clear the transformer from the bank in case of failure of the transformer. Fuses may also be placed in the secondary main between transformer banks. See Fig. 2-18.

Some advantages claimed for this type of system include uninterrupted service, though perhaps with a reduction in voltage, should a transformer fail; better distribution of load among transformers; better normal voltage conditions resulting from such load distribution; an ability to accommodate load increases by changing only one or some of the transformers, or by installing a new transformer at some intermediate location without disturbing the existing arrange-

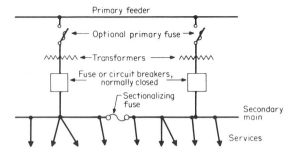

FIG. 2-18 Banked secondary supply.

ment; the possibility that diversity between demands on adjacent transformers will reduce the total transformer load; more capacity available for inrush currents that may cause flicker; and more capacity as well to burn secondary faults clear.

Some disadvantages associated with this type of system are as follows: should one transformer fail, the additional loads imposed on adjacent units may cause them to fail, and in turn their loads would cause still other transformers to fail (this is known as *cascading*); the transformers banked must have very nearly the same impedance and other characteristics, or the loads will not be distributed equitably among them; and sufficient reserve capacity must be provided to carry emergency loads safely, obviating the savings possible from the diversity of the demands on the several transformers.

Banked secondaries, while providing for failure of transformers, do not provide against faults on the primary main or feeder. Further, a hazard on any transformer disconnected for any reason may result from a back feed if the secondary energizes the primary (which may have been considered safe).

Secondary Networks   Secondary networks at present provide the highest degree of service reliability and serve areas of high load density, where revenues justify their cost and where this kind of reliability is imperative. In some instances, a single consumer may be supplied from this type of system by what are known as *spot networks.*

In general, the secondary network is created by connecting together the secondary mains fed from transformers supplied by two or more primary feeders. Automatically operated circuit breakers in the secondary connection between the transformer and the secondary mains, known as *network protectors,* serve to disconnect the transformer from the network when its primary feeder is deenergized; this prevents a back feed from the secondary into the primary feeder. This is especially important for safety when the primary feeder is deenergized from fault or other cause. The circuit breaker or protector is backed up by a fuse so that, should the protector fail to operate, the fuse will blow and disconnect the transformer from the secondary mains. See Fig. 2-19.

The number of primary feeders supplying a network is very important. With only two feeders, only one feeder may be out of service at a time, and there must be sufficient spare transformer capacity available so as not to overload the units remaining in service; therefore this type of network is sometimes referred to as a *single-contingency* network.

Most networks are supplied from three or more primary feeders, where the network can operate with the loss of two feeders and the spare transformer capacity can be proportionately less. These are referred to as *second-contingency* networks.

Secondary mains not only should be so designed that they provide for an equitable division of load between transformers and for good voltage regulation with all transformers in service, but they also must do so when some of the transformers are no longer in service when their primary feeders are deenergized. They must also be able to divide fault current properly among the transformers, and must provide for burning faults clear at any point while interrupting service to a minimum number of consumers; this often limits the size of secondary mains, usually to less than 500 cmil × 10³, so that when additional secondary main capacity is required, two or more smaller size conductors have to be paralleled. In some networks, where insufficient fault current might cause long sections of secondary mains to be destroyed before the fault is burned clear, sections of secondary mains are fused at each end.

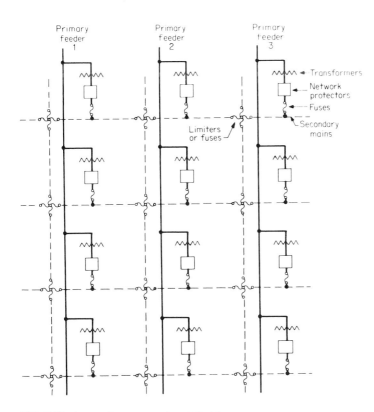

FIG. 2-19 Low-voltage secondary network.

Because these networks may represent very large loads, their size and capacity may have to be limited to such values as can be successfully handled by the generating or other power sources should they become entirely deenergized for any reason. When they are deenergized for any length of time, the inrush currents are very large, as diversity among consumers may be lost, and this may be the limiting factor in restricting the size and capacity of such networks.

Voltages

For all types of service, primary voltages are becoming higher. Original feeder primary voltages of about 1000 V have climbed to nominal 2400, 4160, 7620, 13,800, 23,000, and 46,000 V. Moreover, primary feeders that originally operated as single-phase and two-phase circuits are all now essentially three-phase circuits; even those originally operated as delta ungrounded circuits are now converted to wye systems, with their neutral common to the secondary neutral conductor and grounded.

Secondary voltages have changed from nominal 110/220 V single-phase values to those now operating at 120/240 V single-phase and 120/208 or 120/240 V for three-phase circuits, the 120-V utilization being applied to lighting and small-motor loads while the 208- and 240-V three-phase values are applied to larger-motor loads. More recently, secondary systems have employed utilization voltage values of 277 and 480 V, with fluorescent lighting operating single-phase at 277 V and larger motors operating at a three-phase 480 V. To supply some lighting and small motors single-phase at 120 V, autotransformers of small capacity are employed to step down the 277 V to 120 V.

Secondary voltages and connections will be explored further in discussing transformers and transformer connections.

OVERHEAD VERSUS UNDERGROUND

Although the original distribution system pioneered by Thomas Edison was a direct current low-voltage system installed underground, the widespread expansion of electric systems was based principally on the adoption of alternating current (through the application of transformers) and the very economic overhead type of construction.

While the chief limitation to the adoption of underground systems is economic, there are other reasons that argue against its selection. The necessity for ducts, for manholes, and for cables that require expensive insulation and lead sheaths, short pulls, and a relatively large number of splices, and the special requirements to make equipment waterproof and safe for installation underground all tend to make investment costs several times as great as for overhead systems of comparable characteristics.

Where loads become so great, however, that the number of pole lines and the congestion of conductors on such lines become impractical from safety, operational, and appearance viewpoints, there is no alternative but to place the lines underground. In such areas, traffic conditions are usually so severe that difficulty is experienced in building and maintaining overhead systems; moreover, the heavy traffic itself presents additional hazards from vehicles striking the poles.

While an underground system is not exposed to damage and interruptions from storms, traffic, etc., on the other hand, when trouble does occur, it is very much more difficult and time-consuming to locate and repair than in the overhead system. For this reason, additional provisions and expenditures are made for maintaining service reliability; these include duplicate facilities, throwover schemes, networks, etc. Also, the lesser ability for heat radiation in an underground system does not permit the loading and overloading of conductors and equipment possible with overhead systems.

With plastics taking over the functions of insulation and sheathing in underground cables, and the ability of these materials to be buried directly in the ground, the economic advantage of overhead systems, though still favorable, is markedly reduced. The recent greater emphasis on environment (appearance) also has contributed to a greater pressure for underground installations. Overhead systems will, however, prevail to a very great extent for some time, and will be in almost exclusive use in rural areas.

PART TWO

PLANNING AND DESIGN

CHAPTER 3

LOAD CHARACTERISTICS

In the planning of an electrical distribution system, as in any other enterprise, it is necessary to know three basic things:

1. The quantity of the product or service desired (per unit of time)

2. The quality of the product or service desired

3. The location of the market and the individual consumers

Logically, then, it would be well to begin with the basic building blocks, the individual consumers, and then determine efficient means of supplying their wants, individually and collectively.

CONNECTED LOADS

A good place to start is the tabulation of all electric devices (lamps, appliances, equipment, etc.) that consumers *can* connect to their supply system. The ratings of the devices at specified voltages (and sometimes frequency and temperature) limits are usually contained in the nameplate or other published data accompanying the devices. The devices can be classified into four broad general categories: lighting, power, heating, and electronic. Each of these has different characteristics and requirements.

Lighting Loads

Included under lighting are incandescent and fluorescent lamps, neon lights, and mercury vapor, sodium vapor, and metal halide lights. Nominal voltages specified for lighting are usually 120, 240, and 277 Volts (variations may exist from the base 120-V value, e.g., 115 and 125 V). All operate with dc or single-phase ac; the discussion will be in terms of ac, with comments concerning dc operation where applicable.

Incandescent Lighting   Incandescent lamps operate at essentially unity power factor. Their light output drops considerably at reduced voltage, being some 16 percent less with a 5 percent lowered voltage, and decreasing at a geometrically faster rate from then on. They are also sensitive to sudden rapid voltage variations, producing a noticeable (and annoying) flicker at variations of as little as 3 Volts (on a 120-V base). Street lighting of the incandescent type can be operated in a multiple or a series fashion. The former operates as other lighting in a multiple or parallel circuit, while the light output for the series type depends on the amount of deviation from the standard value of current flowing through it (usually 6.6, 15, or 20 A); it is sensitive to variations of as little as 1 percent in the value of the current. The life of incandescent lamps is considerably reduced at voltages appreciably above normal.

Fluorescent and Neon Lighting   Fluorescent lamps and neon lights operate at power factors of about 50 percent, but usually have corrective capacitors included so that, for planning purposes, they may also be considered to operate at 100 percent or unity power factor. Their light output, per unit input of electrical energy, is considerably greater (25 percent or more) than that of a similarly rated incandescent lamp. The life of fluorescent lamps and neon lights is affected by the number of switching operations they undergo. If fluorescent lamps are used on dc circuits, special auxiliaries and series resistance must be employed; operation is inferior to that on ac, with much less light produced per unit of energy and rated life reduced 20 percent. Neon lights are not usually employed on dc circuits. Fluorescent lamps, neon lights, mercury and sodium vapor, and metal halide lights *may*, if improperly installed or when deteriorating, cause radio and TV interference.

High-Intensity Vapor Lighting   Mercury vapor (high pressure) and sodium vapor (high and low pressure) and metal halide lights operate at power factors of 70 to 80 percent, but also are associated with capacitors to raise the effective value to 100 percent. They are not as susceptible to voltage variations as are incandescent lamps. Their light output and life expectancy are greater than those for fluorescent lamps. They may be employed on dc circuits, but require additional starting auxiliaries. They are generally restricted to applications where large amounts of lighting are desirable, such as on expressways, in large manufacturing areas, or in photographic work; they are somewhat more expensive than other types and have the disadvantage of taking some time after being energized before maximum light output occurs.

Power Loads

Generally included in power loads are motors of all sizes: direct current shunt, compound and series types; alternating current single-phase and polyphase, induction and synchronous types; and universal (series) for both dc and ac

TABLE 3-1 SUMMARY OF MOTOR GENERAL CHARACTERISTICS

Type of motor	Horsepower rating	Speed characteristics	General capability
DC shunt	Up to 200	Constant speed or variable speed by armature or field control	Light or medium starting duty
DC compound	Up to 200	Varying speeds up to 25% from no load to full load	Heavy starting duty, heavy loads for short periods
DC series	Up to 200	Variable speeds	Heavy intermittent starting or heavy loads for short periods
AC 1φ	$^1/_{50}$ or less to 10	Constant speed; usually with some line means for varying speed	Constant speed; no speed control for light or heavy starting duty
AC polyphase (2φ or 3φ) induction	Squirrel-cage to 200; wound rotor to 1000	Speed change by field controls or reduced voltage; available for constant torque and constant hp	Constant speed; light, medium, and heavy starting duty
AC 3φ synchronous	Usually 300 to 5000 +	Constant speed; for power factor control by over- or underexcitation	Constant speed; light, medium, and heavy starting duty and loads

operation. Table 3-1 summarizes the characteristics and general application of these various types of motors.

Single-Phase Fractional-Horsepower Motors The majority of fractional horsepower motors, generally used in appliances of various kinds, are single-phase and operate at power factor values of 50 to 70 percent, but many have corrective capacitors associated with them. When they operate without speed controls or starters, their starting currents may cause lights on the same circuit to flicker; where starts are relatively frequent, as with refrigerators and oil burners, the flicker may be annoying.

Induction Motors Most commercial and industrial ac motors are of the induction type; limited speed control may be obtained in some types by varying the applied voltage. Where accurate speed control is desirable, such as for elevators and printing presses, dc motors are employed, sometimes served from ac sources through motor-generator sets. Induction motors may operate at power factors of 50 to 95 percent but generally operate on the order of 80 to 90 percent; at less than full load, the power factors may drop to 50 to 60 percent. Most large motors for industrial loads (from about 2 hp and larger) are usually three-phase (although many older two-phase motors still exist). Voltage variations of about − 10 percent can be accommodated with little lowering of motor efficiency and power factor values.

Synchronous Motors   Synchronous motors, usually of large sizes, can operate at power factors leading or lagging 100 percent by adjusting their excitation: overexcitement draws leading current, underexcitement lagging current. Often this type of motor is used for power factor correction for the entire installation.

Since larger motors are apt to cause voltages to dip when starting, circuits separate from lighting circuits are provided to eliminate flicker problems; sometimes separate supply transformers are also provided. Also causing similar flicker problems are chemical and electrolytic devices and mechanical devices operated by coils or solenoids.

Heating Loads

The heating category may be conveniently divided into residential (small) and industrial (large) applications.

Residential Heating   Residential heating includes ranges for cooking; hot water heaters; toasters, irons, clothes dryers, and other such appliances; and house heating. These are all resistance loads, varying from a relatively few watts to several kilowatts, most of which operate at 120 V, while the larger ones are served at 240 V; all are single-phase. The power factor of such devices is essentially unity. The resistance of the elements involved is practically constant; hence current will vary directly as the applied voltage. The effect of reduced voltage and accompanying reduced current is merely to cause a corresponding reduction in the heat produced or a slowing down of the operation of the appliance or device. While voltage variation, therefore, is not critical, it is usually kept to small values since very often the smaller devices are connected to the same circuits as are lighting loads, although hot water heaters, ranges, and other larger loads are usually supplied from separate circuits. (Microwave ovens employ high-frequency induction heating and are described below.)

Industrial Heating   Industrial heating may include large space heaters, ovens (baking, heat-treating, enameling, etc.), furnaces (steel, brass, etc.), welders, and high-frequency heating devices. The first two are resistance-type loads and operate much as the smaller residential devices, with operation at 120 or 240 V, single-phase, and at unity power factor. Ovens, however, may be operated almost continuously for reasons of economy, and some may be three-phase units.

Electric Furnaces   Furnaces may draw heavy currents more or less intermittently during part of the heat process and a fairly steady lesser current for the rest; on the whole, the power factor will be fairly high since continuous operation is indicated for economy reasons. The power factor of a furnace load varies with the type of furnace from as low as 60 percent to as high as 95 percent, with the greater number about 75 or 80 percent. Sizes of furnaces vary widely; smaller units with a rating of several hundred kilowatts are single-phase, while the larger, of several thousand kilowatts, are usually three-phase. Voltage regulation, while not critical, should be fairly close because of its possible effect on the material in the furnace.

Welders Welders draw very large currents for very short intermittent periods of time. They operate at a comparatively low voltage of 30 to 50 V, served from a separate transformer having a high current capacity. Larger welders may employ a motor-generator set between the welder and the power system to prevent annoying voltage dips. The power factor of welder loads is relatively low, varying with the load. The timing of the weld is of great importance and may be regulated by electronic timing devices.

High-Frequency Heating High-frequency heating generates heat in materials by high-frequency sources of electric power derived from the normal (60-Hz) power supply. High-frequency heating is of two types: induction and dielectric.

   Induction heating. In induction heating, the material is conducting (metals, etc.) and is placed inside a coil connected to a high-frequency source of power; the high-frequency magnetic field induces in the material high-frequency eddy currents which heat it. Because of the skin effect, the induced currents will tend to crowd near the surface; as the frequency is increased, the depth of the currents induced will decrease, thus providing a method of controlling the depth to which an object is heating.

   Dielectric heating. In dielectric heating, a poor conducting material (plastic, plywood, etc.) is placed between two electrodes connected to a high-frequency source; the arrangement constitutes a capacitor, and an alternating electrostatic field will be set up in the material. (Some slight heating will also be set up from the induction effect described above, depending on the conducting ability of the material.) The alternating field passing uniformly through the material displaces or stresses the molecules, first in one direction and then in the other as the field reverses its polarity. Friction between the molecules occurs and generates heat uniformly throughout the material. Such friction and heat are proportional to the rate of field reversals; hence, the higher the frequency, the faster the heating. Because of heat radiation from the surface, however, the center may be hotter than the outside layers. Residential-type microwave ovens are an application of dielectric heating.

Oscillators Oscillators are used as the source of high-frequency power required for both induction and dielectric heating. This is an electronic application, and its characteristics and requirements are described in the following section.

Electronic Loads

The electronic load category includes radio, television, x-rays, laser equipment, computers, digital time and timing devices, rectifiers, oscillators for high-frequency current production, and many other electronically operated devices. In general, these employ electron tubes or solid-state devices such as transistors, semiconductors, etc. Practically all of these devices operate at voltages lower than the commercial power sources and employ transformers or other devices to obtain their specific voltages of operation. They are all affected by voltage variations.

   Voltage variations may have a marked effect on electron tubes, affecting their current-carrying abilities or emissions as well as their life expectancy. Because of the reduced life of the heater element and higher rate of evaporation of active materials from the cathode surface, the cathode life of electron tubes may be reduced as much as one-half by only a 5 percent rise in cathode voltage. Industrial-type tubes are normally designed to operate with a voltage tolerance of ±5 percent, though closer tolerances are often specified.

While voltage variations also affect the operation of solid-state devices, the effect on their life expectancy is not as serious as in the case of electron tubes. On the other hand, variations in frequency of the power supply have little effect on electron tubes but may have a pronounced effect on solid-state devices.

Both types of devices are very sensitive to voltage dips, and, from the power supply viewpoint, operate at essentially unity power factor. Some applications, such as computers, may require an uninterrupted source of supply, and various schemes are employed to achieve this, including the use of motor-generator sets capable of running on batteries for a limited time; the motor-generator set also eliminates the problems of voltage dips on the commercial power supply.

Except for some rectifier applications, most of these devices operate from single-phase ac supply circuits; large rectifiers may be supplied from three-phase sources.

Oscillators for commercial purposes employ industrial-type electron tubes in conjunction with capacitors and inductances that may be varied to produce the desired high-frequency sources. The regular tolerances in voltage supply from commercial power sources are suitable for this application.

CONSUMER FACTORS

It is obvious that an individual consumer is not apt to be using all of the electrical devices that constitute his or her "connected load" at the same time, or to their full capacity. It would evidently be unnecessary to provide facilities to serve such a total possible load, and much more economical to provide only for a probable load, the load creating the demand on the distribution facilities.

Maximum Demand

The actual load in use by a consumer creates a demand for electric energy that varies from hour to hour over a period of time but reaches its greatest value at some point. This may be called the consumer's instantaneous maximum demand; in practice, however, the maximum demand is taken as that which is sustained over a more definite period of time, usually 15, 30, or 60 min. These are referred to as 15-, 30-, or 60-min integrated demands, respectively.

Demand Factor

The ratio of the maximum demand to the total connected load is called the *demand factor*. It is a convenient form for expressing the relationship between connected load and demand. For example, a consumer may have ten 10-hp motors installed; at any one time, some will not be in use and others will not be fully loaded, so that the actual demand may be only 50 hp; the demand factor is 50 divided by 100, or 50 percent.

The demand factor differs for different types of loads, and by averaging a large number of loads of each type, typical demand factors can be obtained. These values are important in determining the size of facilities to be installed for a particular service; they are extremely useful in making estimates in planning new distribution systems or in expanding existing ones.

Load Factor

The load factor is a characteristic related to the demand factor, expressing the ratio of the average load or demand for a period of time (say a day) to the maximum demand (say 60 min) during that period. For example, a consumer household may have a maximum demand of 2 kW during the evening when many of its lights, the TV, the dishwasher, and other appliances are in use. During the 24-h period, the energy consumed may be 12 kWh; thus the average demand or load is 12 kWh divided by 24 h, or 0.5 kW, and the load factor in this case is 0.5 kW divided by 2 kW, or 25 percent. This provides a means of estimating particular consumers' maximum demand if both their consumption and a typical load factor for their kind of load are known.

Diversity

Consumer load diversity describes the variation in the time of use, or of maximum use, of two or more connected loads. Load diversity is the difference between the sum of the maximum demands of two or more individual consumers' loads and the maximum demand of the combined loads (also called the maximum diversified demand or maximum coincident demand). For example, one consumer's maximum demand may occur in the morning, while another's may occur in the afternoon, and still another's in the early morning hours, as shown in Fig. 3-1.

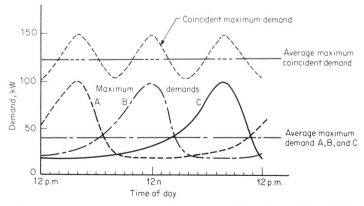

FIG. 3-1 Maximum demands and average maximum demands, coincident and noncoincident.

Diversity Factor

The diversity factor is the ratio of the sum of maximum demands of each of the component loads to the maximum demand of the load as a whole (or the coincident maximum demand). For example, each of the loads mentioned above may have a maximum demand of 100 kW, while the coincident maximum demand on the system supplying the three may be only 150 kW. The diversity factor is then 300 (100 + 100 + 100) divided by 150, or 2, or 200 percent. Such diversity exists between consumers, between transformers, and between feeders, substations, etc. Note that the *demand factor* is defined so that it is always less than 1 or 100 percent, while the *diversity factor* is the reciprocal of the demand factor and is always greater than 1 or 100 percent. This is a most important factor in the economical planning and design of distribution facilities.

Coincidence Factor

The coincidence factor is the ratio of the maximum coincident total demand of a group of consumers to the sum of the maximum demands of each of the consumers.

Utilization Factor

The ratio of the maximum demand of a system to the rated capacity of the system is known as the utilization factor. Both the maximum demand and the rated capacity are expressed in the same units. The factor indicates the degree to which a system is being loaded during the load peak with respect to its capacity. The rated capacity of a system is usually determined by its thermal capacity, but may also be determined by voltage drop limitations, the smaller of the two determining the capacity.

Power Factor

The ratio of power (in watts) to the product of the voltage and current (in volt-amperes) is called the power factor. It is a measure of the relation between current and voltage out of phase with each other brought about by reactance in the circuit (including the device served). Since facilities must be designed to carry the current and provide for losses which vary as the square of the current, and for voltage drops which are approximately proportional to the current, it is necessary that current values be known. The power factor enables loads and losses designated in watts to be converted to amperes. Transformer sizes, wire and cable sizes, fuses, switch ratings, etc., are all based on values of current they must carry safely and economically.

CONSUMER CLASSIFICATION

As aids in planning, consumers may be conveniently classified into certain categories and certain ranges of load densities expressed in kVA per square mile (where this unit is too broad to be useful, watts per square foot for specific occupancies may be used).

Residential

Downtown, apartment buildings, hotels	10 to	50,000 kVA/mi²
Urban, suburban:		
Large homes (plots)	1 to	5,000 kVA/mi²
Small homes	0.5 to	1,000 kVA/mi²
Two-family homes	1 to	5,000 kVA/mi²
Rural, including farm loads	Less than 0.1 to	5 kVA/mi²

Commercial

Stores and shopping centers	10 to	500,000 kVA/mi²
Office buildings	10 to	500,000 kVA/mi²
Service centers, warehouses	10 to	500,000 kVA/mi²
Hospitals, nursing homes	1 to	50,000 kVA/mi²
Schools, churches, clubs, etc.	1 to	500 kVA/mi²
Street and area lighting	1 to	500 kVA/mi²

Industrial

Large manufacturing plants ⎫	Extremely wide variations;
Small manufacturing plants ⎬	consider them as spot
Military bases ⎭	concentrations of loads.

Further classifications may be based on such items as the dependence on electric service because of the critical nature of the consumer's operations, under either normal or emergency conditions; the resultant cost if critical processes are interrupted; or the sensitivity of loads to small voltage deviations.

FLUCTUATION IN DEMAND

There are three main factors that greatly influence the magnitude of maximum demand and the time of its occurrence. The most frequent is the weather as it affects light intensity during daylight hours and temperatures throughout the day and year. The sharpest factor and perhaps that of least duration is special events which result in a temporary slowdown of activities or a greatly increased usage of lighting, radio, and TV and associated increases in water pumping, cooking, and other loads. The largest factor is changes in business conditions accompanied by significant changes in industrial demands and consumption; while much less significant, fluctuations in both residential and commercial consumer demands also follow such changes in business conditions.

The nature, magnitude, and time of these fluctuations are generally unpredictable. Some estimate of them can be gleaned, however, from past experiences, which may vary widely in different areas of the country. Provision for these fluctuations should be taken into account in the planning of distribution systems.

FUTURE REQUIREMENTS

Good engineering requires that probable future growth of loads be considered in planning. This is usually provided for by spare capacity in the present design of the several elements, or by provisions for possible future additions or alterations, or both of these. Load growth is rarely uniform throughout an area, so that growths in various parts of a system will be different from each other and from that of the system as a whole.

Economics

How far present capacity should provide for future load is largely a question of economics: the cost of carrying excess capacity until it is needed versus the cost of replacing smaller units with larger when it becomes necessary. This is a problem of the future worth of present expenditure, which is affected by fluctuations in rates of interest and inflation. Standard sizes of the materials and equipment involved automatically provide for a limited amount of spare capacity for growth, so that any economic analysis can only be approximate. The relatively large proportion of labor to material in the construction of a distribution system or its parts lends itself to the installation of capacity greater than its immediate need. Such spare capacity incidentally provides a cushion for accommodating some of the unforeseen fluctuations in demands described above.

Past Performance

Data from past performances, such as total system loads, substation loads, and feeder loads, can be used as a basis for estimating such growth. The variations from year to year, or from month to month, can furnish a trend for such growth; separate trends can be developed for different parts or areas. Where such data are nonexistent or patently unreliable, estimates can include a fixed percentage growth above the values on which planning is made.

Future Performance

To obtain some idea of what may occur in the future, it may be well to look back a generation or two. Earlier, consumers' appliances could be contained in a relatively short table. To attempt to list all the electrically operated devices, appliances, and gadgets presently to be found in homes and commercial establishments would be an almost endless task. To attempt to foretell what may develop in the future would be an exercise in futility.

The advent of widespread air conditioning and space heating, together with the almost universal use of television, not only substantially changed consumers' maximum demands and consumption, but also materially affected loads, diversity, coincidence and (for larger units) power factors, and utilization factors as well.

While the demand factor may indicate how the connected loads are being used, the utilization factor indicates how the capacity of the supply system is being used. Since the capacity of the supply system is determined by its thermal capability, the increased sustained demand on these facilities will lower their thermal capability, and hence the system capability.

The greater use of electronically operated computers will tend to call for narrower limits of voltage control (regulation and flicker) and a greater degree of service reliability by stiffening the supply distribution system, or through the installation of auxiliary equipment owned and maintained by the consumer or rented as another service by the utility; the choice will be determined by future developments.

VOLTAGE REQUIREMENTS

Electric devices utilizing secondary or low voltage in the United States have been standardized by almost all manufacturers at 120/240 V. While many utilities are following these standards for their systems, there are a significant number operating at 115/230 V and some at 125/250 V (and a small and diminishing number at 110/220 V). For polyphase or three-phase loads, the established ratings are 208 and 416 V for wye-connected systems and 240 and 480 V for delta-connected systems. In a few cities, and some downtown and heavy load centers, network systems supply a single-phase voltage of 277 Volts and a three-phase voltage of 480 V. (A few two-phase 115/230 V systems still exist.)

With distribution systems designed for practical voltage tolerances expressed in volts plus or minus in relation to their normal, "standard," base single-phase voltage of 115, 120, and 125 V, voltages of 110 to 130 V can exist at the terminals of the loads (lamps, appliances, etc.). This is a spread of ± 10 V, or ± 8.3 percent on a 120-V base, to which a flicker voltage drop of 3 V, or 2.5 percent, should be added to allow for motor starts. This would then give a total spread of 23 V, or 19.1 percent, from + 8.3 to − 10.8 percent—approximately within most manufacturers' rated maximum tolerances of ± 10 percent for motors and heating devices. Closer coordination between manufacturers and utilities could do much to improve this situation. Electronic devices are more sensitive to voltage variations, and difficulty may be experienced with their operation at variations of this magnitude.

Some utilities provide for an estimated voltage drop of up to 2 V in the consumer's wiring by specifying normal voltage at the service point 2 V higher than mentioned earlier, i.e., 122 instead of 120 V (117 and 127 V on other bases). Their designs, however, provide for the same high and low voltage limits, but the variations above and below the usual base are unequal, e.g., 128 V high, 122 normal, 114 low. The 23-V spread and − 10.8 percent variation mentioned above will be the same.

Not only does satisfactory operation of lights, appliances, and other devices make for good consumer relations, but the effect of high and low voltages (principally because of unity power factor lighting loads) on both revenue and fuel conservation measures should also be considered.

Higher voltages—660 V, 2400 V, and others—are also employed for larger motors, rectifiers, and some other purposes. Consumers using such large voltages are usually served at primary voltages and meet their own utilization voltage standards.

SERVICE RELIABILITY

Reliability of service generally is interpreted to mean the continuity of service or the lack of interruption to service. For a distribution system, or any of its parts, absolute reliability or continuity of service 100 percent of the time for 100 percent of its consumers is an impossibility, although this goal can be approached. The costs to achieve such goals, even partially, are usually not warranted.

Degree of Service Reliability

As a practical matter, all consumers may not require a uniformly high degree of service reliability. For some consumers, an extremely high degree of service is essential; these may include hospitals, military establishments, some larger theaters, department stores, apartment buildings, hotels, etc., where the safety of the public is concerned; often auxiliary sources of supply are provided to supplement the utility company supply.

For some other types of loads, a high degree of reliability is desirable but not so essential from the public safety viewpoint; smaller apartments and theaters are examples of these, as well as some manufacturing or service processes where interruption may result in substantial monetary losses. To the average residential or commercial consumer, however, a short interruption (and in some cases even an occasional long one) is more of an inconvenience than a hazard or cause for monetary loss.

As a rule, provisions for higher degrees of service reliability involve higher expenditures, for both additional facilities and increased maintenance. The expenditure to provide reliability should bear some proportion to the degree of reliability needed. Various system designs, outlined in Chap. 2, provide for varying degrees of service reliability, from a simple, unsectionalized radial feeder to a low-voltage secondary network supplied from a multiplicity of primary feeders isolated from each other. Each type of service should produce revenues to justify the additional expenditures for achieving the service reliability desired or required; exception may be made for such public services as hospitals and military establishments.

Overhead versus Underground

In this regard, comparisons between overhead and underground systems should be borne in mind. Overhead systems are generally much less costly but are more vulnerable to the hazards of nature (wind, ice, lightning, flood, etc.) and to the actions of people (vehicles hitting poles, kites, etc.); they are, however, easier to maintain, especially as faults can be more easily located and repaired. Underground systems, generally more expensive and less vulnerable to the vagaries of nature and people, nevertheless require longer times for the location and repair of faults that may occur.

Reliability Indices

Service reliability indices are maintained to measure and obtain trends in the performance of a distribution system and its components. Some of these include number of interruptions per consumer served; number of consumers affected per consumer served; number of consumer hours of interruption per consumer served; average duration of interruption (hours) per consumer affected; average number of consumers affected per consumer served; and average duration (hours) per consumer served. Further indices are maintained as to causes and duration of interruptions in the several parts of the system, e.g., on the basis of miles of conductor installed, on the miles of circuit, by voltage classification, or by geographic divisions. Compilation and analyses of these data lend themselves to computer application.

Trends

The trends, over a period of time, not only measure the effectiveness of system designs (and operating procedures), but also point out areas of need for further improvement of service continuity.

CHAPTER 4

ELECTRICAL DESIGN

The electrical design of distribution facilities is based on the loads they are to carry safely and the permissible voltage variations; the final design, however, cannot be divorced from mechanical, economic, and other considerations. Several different designs may serve the same electrical requirements adequately; each, in turn, may be modified by mechanical considerations. The design ultimately selected must reflect economic considerations: specifically, the design that results in the least annual expense in supplying the load or loads in question. This necessarily involves the evaluation of losses, as well as capital, maintenance, and operation expenses. Often, other considerations must also be taken into account including government regulations (at all levels), national and local industry construction and safety codes, taxes, public relations, and some other, intangible requirements.

SERVICES

Rather than design a separate service for each consumer, it is more practical and economical to determine the capacity and construction requirements on a group basis for different types of consumers. The maximum demand for a consumer group is determined by the connected load, to which a demand factor may be applied. The factor can be an estimate based on observation or a logical analysis of the operations of the several devices comprising the load, or it may be obtained from previous experience. Further, service entrance equipment specified by national and local codes (and often installed by the consumer) has minimum ratings based on the number and kinds of circuits installed within the consumer's premises.

Each group represents consumers whose maximum demands fall within cer-

TABLE 4-1 SERVICE DROP SPECIFICATIONS: THREE-WIRE
SINGLE-PHASE 120/240 V SUPPLY

Group	Maximum demand range, kVA	Ampere range	Conductor size Copper	Conductor size Aluminum
A	0 to 2	0–20	no. 6	no. 4
B	2 to 5	20–40	no. 4	no. 2
C	5 to 9	40–70	no. 2	no. 0
D	9 to 20	70–150	no. 00	no. 0000

tain ranges, expressed in volt-amperes or kVA, and for which certain conductor
sizes are specified, listed in Table 4-1. These values are based on a single-phase
three-wire 120/240 V supply via three-conductor self-supporting cable for a 100-
ft length, carrying the maximum amperes listed and producing a 1 percent
voltage drop. Losses at the maximum loads are less than 2 percent. For longer
service drops (sometimes in rural areas), these values may be exceeded.

Three-conductor self-supporting service cables are almost always specified
because of their appearance and ease of installation as compared to older (now
almost obsolete) open-wire-type services; voltage drop for the same load and
length is slightly less than for the open-wire type. The sizes of conductors specified are more than ample to support the mechanical stresses imposed on them,
even in severe weather conditions.

Local conditions, including varying costs of both labor and materials, rates
of growth, and other factors, may substantially change the values shown in Table
4-1. Services for the relatively fewer larger commercial and industrial consumers
served at secondary voltages are usually determined individually.

THE SECONDARY SYSTEM
Transformer-Secondary Combination

The combination of transformers, secondary circuit or main, and the consumers'
services makes up the secondary system. Secondary systems are predominantly
single-phase, except for larger commercial and industrial consumers, who are
supplied from three-phase systems. While the discussion will be limited to single-
phase systems, the principles and methods employed in their design will serve
for other types of secondary systems.

The number of consumers' services and their loads, the voltage drop, the size
of conductors, and the spacing and size of transformers are all variables that
are interdependent. They are factors which must be considered in combination
to arrive at a satisfactory design. There are many theoretical combinations of
these factors that will achieve economical solutions to the problems of design.

For practical purposes, however, these combinations can be reduced to fewer
and more manageable numbers. Certain assumptions can be made safely:

1. The load can be considered uniformly distributed along a secondary whose
   length can be considered fixed. Although not strictly true, this assumption

does represent a majority of conditions, but concentrated or scattered loads must be considered separately.

2. The length of secondary circuit is fixed either by geography or by the type of design; e.g., each city block could be fed by one or more secondary circuits. Refer to Fig. 4-1.

3. In practice, the number and sizes of conductors and transformers are limited, usually to two or three in number, and to certain standard sizes because of manufacturing, purchasing, stocking, and construction economies.

The problem then is to determine the proper combination of conductor or wire, transformer, and transformer spacing for the least annual cost, using the materials available while providing for satisfactory voltage variations, including flicker. Also, the design should consider not only present loads, but the economics of supplying future loads as well.

Conductor Size

It may be well to begin with a determination of the size of conductor. The maximum demand for each of the consumers is known and a coincidence or diversity factor, determined by analyses or from previous experience, applied. Assuming uniformly distributed loads, the loading on each half of the circuit (each direction from the transformer) can be expressed as load density in kilowatts per thousand feet or similar units.

Voltage Drop

For determining voltage drop, the load can be assumed to be concentrated at the midpoint of the secondary main between the transformer and the last consumer, i.e., one-quarter of the length of the conductor from the transformer. The total load connected to one-half the circuit is converted into a coincident maximum demand in amperes.

A maximum tolerable voltage drop (to the last consumer) is assumed and divided by the coincident demand in half the circuit, expressed in amperes. The result will give the maximum permissible resistance of the conductor. On the basis of its length (one-quarter of the circuit), the resistance per unit (1000 ft)

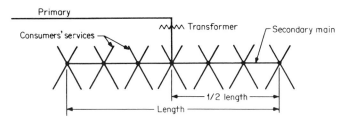

FIG. 4-1   Radial secondary of fixed length and uniformly distributed load.

can be determined. The standard-size conductor whose unit resistance is equal to or less than that calculated can be selected.

This assumes the loads are at or near unity power factor; where this is not so, impedance values based on the spacing between conductors must be used. Also, the drop in one conductor is calculated, which assumes no current in the neutral conductor and a load balanced equally between the two energized or line conductors; where this is not so, voltage drop in the neutral conductor must also be calculated and the greater of the drops in the two line conductors used in selecting the standard-size conductor.

Losses

The next step is to determine the loss in the secondary mains. The value of current and the unit resistance of the conductor are known; for the purposes of determining losses, the full load can be considered to be at one-third the distance from the transformer. This approximate value in watts or kilowatts is multiplied by an estimate of the "equivalent hours" duration to obtain the energy losses in watthours or kilowatthours. This should be multiplied by 4 for the entire length of the two conductors (neglecting the neutral).

It should be noted that while load curves for a particular period (day, month, year) vary with the value of current, corresponding curves for losses vary with the *square* of the current, even though the curves may have a similar configuration. Like the load factor, a *loss factor* is the ratio of average power loss for a certain period of time (day, month, year) to the maximum loss or loss at peak load (for a stipulated time: 15, 30, or 60 min) during the same period. This value can be determined with sufficient accuracy by analysis of a few typical daily load curves for the period involved. This loss factor always lies between the load factor (for long, sustained peak loads) and the square of the load factor (for short, sharp peaks). The loss factor multiplied by 24 equals the daily equivalent hours.

Returning to the energy losses, in watthours or kilowatthours, these are evaluated at the system cost per kilowatthour (which includes not only fuel costs, but carrying charges on equipment, operating costs, and other overheads). This value is compared with the carrying charges (including maintenance costs and appropriate overheads) on the *installed* cost of the conductors. If the two values are reasonably close, the conductor selected is economically satisfactory, according to Kelvin's law.

Kelvin's Law

Kelvin's law is generally expressed as follows: The most economical size of conductor is that for which the annual charge on the investment is equal to the annual cost of energy loss.

If these two values are not reasonably close, another size of conductor may be chosen, or the length of the secondary main (and its connected loads and its coincident maximum demand) may be changed; and either process may be repeated until the values of annual charges and annual cost of energy losses are reasonably close.

EXAMPLE 4-1   REFER TO FIG. 4-1.

Assume that the maximum demand of each consumer is 2 kW and the diversity or coincidence factor is 0.67. The total load on half the length of the conductor is 12 consumers × 2 kW = 24 kW, or 24,000 W; this divided by (approximately) 240 V equals 100 A per conductor, except the neutral, which carries no current (on the assumption that the load is balanced between line conductors). Assume the total length is 1000 ft.

For a voltage spread of ±5 V from the first customer to the last, on a 120-V base, the minimum voltage at the last customer is 115 V; if a 1-V drop in the service is assumed, the voltage required at the last service is 116 V. The maximum voltage at the first consumer is 126 V. The maximum permissible drop in 500 ft of secondary main is 126 − 116 = 10 V.

Assume the total load is applied at one-quarter of the length from the transformer; from Ohm's law,

$$R = \frac{E}{I} = \frac{10 \text{ V}}{100 \text{ A}} = 0.10 \ \Omega, \text{ or } 0.40 \ \Omega \text{ per } 1000 \text{ ft}$$

From Table 9-2 in Chap. 9, the resistance of no. 7 copper conductor is 0.498 Ω per 1000 ft and its weight is 63 lb per 1000 ft.

For energy loss: Assume the total load (for half the circuit) is applied one-third the distance from the transformer (or 500/3 ft) and the current is 100 A:

$$\text{Loss} = I^2R = 100^2 \times \frac{0.498}{1000} \times \frac{500}{3} = 830 \text{ W or } 0.83 \text{ kW}$$

For the total circuit (neglecting the neutral), multiply by 4, which gives 3.32 kW. Assume a loss factor of 0.75 of the square of the coincident load factor of 0.67, or 0.335.

$$0.335 \times 24 \text{ h} \times 365 \text{ days} = 2937 \text{ equiv h}$$
$$\text{Loss} = 3.32 \times 2937 = 9750.8 \text{ kWh/yr}$$

Assume a cost of 5 cents per kilowatthour, or $487.54 annually. Then estimate the cost of eight sections of no. 7 copper wire: 63 lb/1000 ft × 3 at 80 cents per pound, or $151 for material. Assume the labor cost is $100 a section; then eight sections at $100 plus the $151 for material gives a cost of $951. Assume an annual carrying charge of 20 percent, or $190.20. This is less than half the annual cost of the losses.

For the most economical conductor, the cost of losses should approximate the carrying charge of the conductor installed. By examination, the losses can be halved if the resistance of the conductor is halved.

From Table 9-2, no. 4 copper conductor has a resistance of 0.249 Ω per 1000 ft and a weight of 126.4 lb per 1000 ft.

$$\text{Loss} = I^2R = 100^2 \times \frac{0.249}{1000} \times \frac{500}{3} = 415 \text{ W or } 0.415 \text{ kW}$$

For the total circuit (neglecting the neutral), multiply by 4 to give 1.660 kW. The loss is 1.660 times 2937, or 4875 kWh per year; at 5 cents per kWh that is $243.77 a year. Now estimate the cost of eight sections of no. 4 copper wire: 126.4 lb per 1000 ft times 3 at 80 cents a pound amounts to $303.36 for material; the labor cost of $100 a section is about the same; then eight sections at $100 plus the $303.36 gives a cost of $1103.36. At an annual carrying charge of 20 percent that amounts to $220.67 a year. This is comparable to the annual cost of losses, $243.77.

Improvement in voltage drop is directly proportional to the lowering of resistance; hence,

$$\text{10-V drop} \times \frac{0.249 \text{ (for no. 7 Cu)}}{0.498 \text{ (for no. 4 Cu)}} = 5.0 \text{ V}$$

This improvement not only will take care of the 3-V additional flicker drop, but will also improve operation of most of the connected devices (and increase revenue). No. 4 copper is apparently the better choice.

Transformer Size

Having determined the *tentative* size of conductor, the next step is to determine the size of the transformer to be installed. The value of the diversified coincident demand for the loads connected to the secondary main having been determined, the nearest standard-size transformer (in kVA) to the demand (in kW) is tentatively selected. To allow for future growth and not to prejudice the life of the transformer, the size chosen is usually larger than the demand.

The most economical load of a transformer is that for which the annual cost of its copper loss is equal to the annual carrying charges of the transformer *installed* plus the annual cost of the core loss. The core loss can be considered constant regardless of the load carried by the transformer. Values of core loss and transformer resistance, both expressed as percentages at full load, vary with the manufacturer, vintage, size, and other characteristics, and are found in the transformer specifications; core loss is usually a fraction of one percent, while resistance is usually less than 2 percent (reflecting the high efficiency of transformers). Here, too, if the two values are not close, another size of transformer may be chosen, or the secondary circuit may be changed so that two or more transformers supply the load. It may be necessary to review the conductor size and loads for the new resulting circuits.

EXAMPLE 4-2

The diversified maximum demand for 20 consumers at 2.25 kW demand each and a load factor of 0.67 is 30 kW. Assume the loads are at or near unity power factor.

The load can be served from one 25-kVa transformer, if the period of overload is not excessive; the next larger standard-size transformer is 50 kVA. From manufacturer's specifications (typical), the no-load loss for the 25-kVA transformer is 0.5 percent of the full load; for 50 kVA, it is 0.4

percent. Resistance as a percentage of full load for 25 kVA is 1.4 percent; for 50 kVA, it is 1.25 percent.

$$\text{No load loss (25 kVA)} = \frac{0.5}{100} \times 25{,}000 = 125 \text{ W or } 0.125 \text{ kW}$$

0.125 kW × 2937 equiv h = 367 kWh

At 5 cents per kilowatthour that equals $18.35 a year.

$$\text{No load loss (50 kVA)} = \frac{0.4}{100} \times 50{,}000 = 200 \text{ W or } 0.200 \text{ kW}$$

0.200 kW × 2937 equiv h = 587.4 kWh

At 5 cents a kilowatthour that comes to $29.37 a year.

Annual carrying charges: Assume a cost of $8 per kVA for 25 kVA and $6 per kVA for 50 kVA (typical) plus a $100 installation cost. For 25 kVA,

$$25 \times \$8 + \$100 = \$300$$

The carrying charge of 20 percent comes to $60.00. For 50 kVA,

$$50 \times \$6 + \$100 = \$400$$

The carrying charge of 20 percent amounts to $80.00. The total annual carrying charges are:

$$25 \text{ kVA:} \quad \$18.35 + \$60.00 = \$ \ 78.35$$
$$50 \text{ kVA:} \quad \$29.37 + \$80.00 = \$109.37$$

Load loss: Resistance as a percentage of full load is the same as the loss at full load. For 25 kVA:

$$\frac{1.4}{100} \times 25{,}000 = 350 \text{ W or } 0.350 \text{ kW}$$

Adjust for a 30-kW load:

$$\left(\frac{30}{25}\right)^2 \times 350 = 504 \text{ W or } 0.504 \text{ kW}$$

For 50 kVA:

$$\frac{1.25}{100} \times 50{,}000 = 625 \text{ W or } 0.625 \text{ kW}$$

Adjust for a 30-kW load:

$$\left(\frac{30}{50}\right)^2 \times 625 = 225 \text{ W or } 0.225 \text{ kW}$$

Annual cost of losses:

$$25 \text{ kVA:} \quad 0.504 \text{ kW} \times 2937 \text{ equiv h} = 1480 \text{ kWh}$$

At 5 cents a kilowatthour the cost is $74.00.

50 kVA: 0.225 kW × 2937 equiv h = 660.8 kWh

At 5 cents a kilowatthour the cost is $33.04.

To compare the annual carrying charges to the cost of losses, for 25 kVA, the respective costs are $78.35 and $74.00; for 50 kVA, they are $109.37 and $33.04.

While the 25-kVA transformer figures are reasonably close and represent the more economical unit, they do so at a probable cost of shortening of life and no provision for future growth. Despite the greater difference, the 50-kVA unit probably should be preferred. As an exercise, different secondary-transformer combinations might be investigated—say, two secondary circuits with two 15-kVA or two 25-kVA transformers, or three secondary circuits with three 15-kVA units.

It is obvious that any secondary-transformer configuration represents a compromise. Much depends on the relative costs of material and labor, which may vary widely from time to time and from place to place. Further, other considerations may play a great part in the final determination; e.g., conductor sizes may change to meet mechanical requirements.

Future Growth

To provide for future growth, loads are adjusted upward by a percentage estimated to represent probable increase over a specified period of time. Facilities to serve these increased loads are designed in the same manner described. The difference in investment costs for each design is evaluated in terms of the future worth of the present increment of cost of the additional facilities provided for growth. This is compared to the cost of installing the facilities at the future time. If it is less, it is desirable to provide for the future load at the time of initial installation. If not, provision for future load growth should be dropped, or scaled down to values and timing that will justify some value of additional cost.

To accommodate the load growth, the transformer and conductors can be replaced with larger ones, or more popularly, the secondary circuit can be divided into two or more parts without changing conductors; a suitably sized transformer is then added to the newly formed secondary circuits. Comparison of costs and annual carrying charges dictate the method selected.

Networks

The analysis described pertains to radial secondary circuits. Where networks are involved, the same principles and methods can be applied by assuming the network to be divided into a number of adjacent radial-type circuits, as shown in Fig. 4-2; no appreciable error is introduced.

The general principles and methods applied to overhead single-phase radial-type secondary circuits may be applied to underground circuits and three-phase three- or four-wire circuits by proper adjustment of terms to fit the cases. With

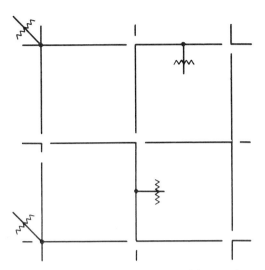

FIG. 4-2 Division of network into radial secondary circuits for design analysis.

underground circuits, the lesser current-carrying capacity of a size of conductor, without overheating, must be taken into account. In network design, the ability to burn clear the conductors in the cable under fault or short-circuit conditions should also be ascertained. These additional considerations may be taken into account after the economic studies are made.

Rural Systems

Where consumers are scattered, such as in rural areas on in the case of three-phase consumers in an area supplied essentially at single phase, the load may be served either by extending the secondary from one transformer or bank of transformers, or by installing a separate transformer or transformers to serve those consumers. Annual carrying charges, including costs of losses, should be compared in selecting the method of supply.

There are many other problems in the design of secondary systems, but they lend themselves to the application of the same basic principles and methods, with proper consideration given to their particular requirements.

THE PRIMARY SYSTEM

The primary system comprises the facilities that deliver power from the distribution substation to the distribution transformers. These take the form of one or more distribution feeders or circuits emanating from the substation, each supplying a portion of the entire load served from that substation. The feeders are made up of mains (or trunks) from which branches (or laterals, or spurs) are provided to supply the several transformers serving loads within the feeder's designated area.

Feeder Mains

The feeder mains are usually three-phase three- or four-wire circuits, and the branches are predominantly single-phase, although they may consist of two or three phases of the three-phase circuit if the loads carried on them are large or require polyphase supply.

Like the secondary circuit, the design of the primary feeder is based on the maximum voltage variation permissible at the farthest consumer. This depends on the size, type, and location of the loads to be supplied, the size of the conductors, and the operating voltage, which may also be limited by local codes and regulations.

Conductor Size

The size of the conductors for the "main" portion of the feeder is usually larger than that of the branches. While a conductor's size may be reduced as it proceeds farther from the substation because of the smaller load it is normally required to carry, this is seldom done. The size of the conductor of the main near the substation is often carried all the way to its extremities; indeed, it is sometimes made even larger than normal operating conditions would dictate. This not only provides for rapid growth, in which it may be found desirable to divide the load so that the direction of supply may be reversed, but also provides spare capacity to carry all or part of the load of adjacent feeders under contingency conditions. Moreover, the larger conductor size may substantially reduce the voltage variation on this portion of the circuit, permitting greater freedom in the design of the branches. The sizes of wire for both main and branches, as in the secondary system previously discussed, will depend on the voltage variation or regulation desired and on economy, which includes evaluation of losses in the conductors.

Sectionalizing

Provision for moderating the effects of faults on the circuit usually takes the form of fuses and switches. Each of the single-phase branches is connected to the main through a fuse; a fault on the branch will blow the fuse and isolate the fault, leaving the remainder of the circuit intact. A fault on the three-phase main will affect the entire circuit; the size of the conductors may be such that the fault current will be beyond the capability of being safely interrupted by a fuse, and the circuit breaker at the substation is called upon to handle the fault current and disconnect the faulted circuit from the substation bus (which may also supply other circuits). Switches are installed in the main of the feeder, enabling the main to be sectionalized, isolating the fault between two switches or other sectionalizing devices. The unfaulted portion back to the substation is reenergized by closing the circuit breaker at the substation; the unfaulted portion beyond the fault is energized from adjacent sources; the portion containing the fault will remain deenergized until the fault condition is repaired and the circuit restored to normal operation. Where the feeder main may consist of two or more parts, circuit breakers in the form of "reclosers" may be installed on each of the parts. Refer to Fig. 4-3.

Reclosers

Reclosers are designed to open when a fault occurs on that part of the main in which they are connected; a timing device, however, enables them to reclose a predetermined number of times for short durations. If the fault is of a temporary nature, such as wires swaying together or a tree limb falling on them, the recloser will remain closed and service will be restored; should the fault persist, the recloser will remain open and disconnect that part of the main from the circuit.

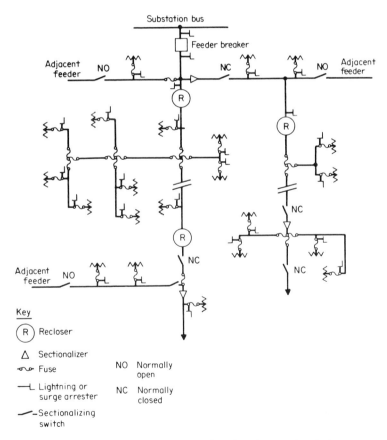

FIG. 4-3 Radial primary feeder showing location of protective devices.

Transformer Fuses

Similarly, a distribution transformer may be connected to the primary main or branch through a fuse. A fault or overload on the transformer or its associated secondary circuit will cause the fuse to blow and disconnect the faulted section from the remainder of the primary circuit.

Load Balancing

On polyphase portions of the feeder, on both main and branches, loads are balanced between phases as closely as practical by connecting transformers and single-phase branches to alternate phases of the circuit; this provides a more uniform balancing of loads along the line (contributing to better load and voltage conditions) than would balancing in large blocks of loads. An approximate method multiplies each load by its distance from the substation; the sum of these, uniformly distributed, should be about the same for each phase.

Operating Voltage

The selection of the primary operating voltage is probably the factor having the greatest influence on the design of the primary system. It has a direct effect on the length of the feeder and its loading, the substation supplying the feeders and the number of feeders, the number of consumers affected by an outage, and on maintenance and operating practices (which, in turn, affect annual carrying charges). Several voltage levels have evolved into "standard" nominal values of primary voltages: 2400, 4160, 7620, 13,200, 23,000, 34,500, 46,000, and 69,000 V.

Delta and Wye Circuits

Many of the older systems employed delta circuits with phase-to-phase voltages approximating 2400 V; as loads increased, it was found economical to convert these into wye circuits with phase-to-phase voltages approximating 4160 V, but with phase-to-neutral voltages remaining at the 2400-V level, permitting the use of the same transformers, insulators, and other single-phase equipment. The wye circuit necessitated a fourth, neutral conductor grounded in many places; later, a single conductor common to both the primary and secondary systems was employed safely, effecting greater economies.

Delta to Wye Conversion

As loads grew and load densities increased, resort was had to higher voltages, making use of subtransmission circuits, a great many of which were delta circuits operating at phase-to-phase voltages of approximately 13,200 V; the wye voltage or phase-to-ground voltage of this level is 7620 V. For the same economy reasons the 13,200-V phase-to-phase delta subtransmission supply circuits to distribution substations were converted to 13,200-V phase-to-neutral wye circuits having phase-to-phase voltages of 23,000 V. Distribution circuits at these higher voltages required fewer substations, whose acquisition in the more developed areas became increasingly difficult. Circuits at these higher voltages also found employment in rural areas where distances between consumers were greater and load diversities lower.

This process continued with the development of distribution circuits operating at 34,500 and 46,000 V from subtransmission lines operating at these voltages. Advantage is taken of taps on transformers supplying these circuits, sometimes as much as 10 percent, in adapting these feeders to distribution requirements. Other voltages, outside the ranges mentioned above, may be found, e.g., 3000, 6600, 8800, 11,000, and 27,000 V.

Advantages

The principal advantages to such conversions from delta to wye systems are:

1. The wye system affords greater feeder capacity and usually improved voltage regulation.

2. Existing transformers, insulators, and other material can be used; in most cases, spacing between conductors is left unchanged.

3. Single-phase branches need not have any work done on them.

4. Existing secondary neutral conductors can be used as the fourth and neutral conductor in establishing the wye circuit. Where a new neutral conductor is required, it can be installed safely and readily in the secondary position on the pole with no conflict with the higher-voltage energized conductors.

5. The entire circuit need not be converted to the higher voltage at one time, but can be converted piecemeal over a period of time; a portion of the three-phase delta circuit can be maintained from a relatively small step-down transformer (pole-mounted) connected to the new supply three-phase wye circuit.

6. Transformers and other equipment at the substation can be rearranged and reutilized, like those on distribution lines.

7. Where the neutral is grounded at the substation and at many points along the feeder, the voltage stresses on the insulation of the lines, transformers, and other devices are limited to the lowest possible value; should an accidental ground occur on any phase, it will be cleared as the circuit breaker opens.

8. Important savings can be realized in the equipment installed on wye systems: transformers need only one high-voltage bushing; only one cutout and lightning arrester are required (if a completely self-protected transformer is used the cutout can be eliminated); and the single high-voltage line conductor may be mounted on one pin at the top of the pole, eliminating the need for a cross arm (and contributing to a neater appearance of the line).

Disadvantages

There are some disadvantages to the conversions from delta to wye:

1. The load and voltage advantages of the higher voltage apply only on the three-phase main and not on the single-phase branches, as they continue to operate at the existing voltage.

2. A ground on a phase conductor constitutes a short circuit, which will de-energize at least that portion of the circuit. On delta circuits, normally operated ungrounded, one or more accidental grounds on the same phase of the circuit will not cause any interruption to service. (The occurrence of a ground on another phase, however, will create a short circuit between phases, possibly connected together through long lengths of conductors; if the impedance of the intervening conductors between grounds is large, the fault current flowing to ground may not be sufficient to open the circuit breaker at the substation, and much damage can ensue until its magnitude either causes the circuit breaker to open or the conductors burn themselves clear at some point. A delta circuit may be hazardous, as a worker, unaware of a ground that may exist at a point farther away, may come in contact with an ungrounded phase wire.)

3. Because of the grounded nature of the wye system, greater care, reflected in greater maintenance costs (e.g., greater and more frequent tree trimming), may be required to achieve the same degree of reliability as in a delta circuit.

4. The higher voltage and the many grounds in a wye circuit may cause greater interference to communications circuits that parallel the power circuits.

5. Some local regulations and codes may require greater safety factors in the construction of facilities operating at the nominally higher voltages.

Higher-Voltage Circuits

When the need is indicated for a still higher-voltage distribution circuit, major reconstruction and a complete replacement of transformers and other devices is usually necessary. The new higher-voltage circuit is generally designed for immediate wye operation, omitting the intermediate delta operation. In addition to the greater construction costs, additional maintenance and operating costs must be considered in determining the economics of going to higher voltages. Beyond about 15 kV, handling such lines and equipment requires either "live line" tools and methods or the deenergizing of lines and equipment. This latter condition may require additional sectionalizing facilities, including a greater number of extensions between feeders to enable loads to be transferred from the circuit to be deenergized.

The greater load-carrying ability of the higher-voltage primary circuits tends to have them serve larger areas and a greater number of consumers, so that an interruption to an entire circuit will have a greater effect on the area served. Rapid sectionalizing and reenergizing means are therefore more necessary and must be considered in evaluating the service reliability factor in economic studies.

Voltage Drop and Losses

Sizes of conductors of primary circuits are also based on acceptable voltage drop and losses in the conductor and the cost of the facilities; the mechanical requirement may be the decisive factor. The principles and methods given for secondary circuits also apply here.

Branches of the primary circuit may supply from one to a great many transformers. Where only one transformer is involved, voltage drops and losses may be calculated as a concentrated load at the end of the line. Where the branch is relatively long and serves a few transformers widely spaced, these values may be derived from a circuit considered to have a distributed load. Where the length is short, or where a larger, more closely situated number of transformers exist, the circuit may be considered as supplying a uniformly distributed load; the total loads of these transformers can be assumed to be concentrated at a point half the length of the branch (from the tap-off at the main to the last transformer) in calculating the maximum voltage drop, and at one-third the distance (from the tap-off at the main) for calculating losses in the entire length of the branch. For single-phase circuits, the characteristics of the neutral conductor should also be considered. For polyphase branches, each phase and the transformers connected to it may be considered separately; the loads on the separate phases may be considered balanced and the neutral ignored.

Voltage and loss calculations for the three-phase main portion of the feeder may be considered to be concentrated at the tap-off point of the main; these, together with the transformers connected to the main, can be considered as a uniformly distributed load on the main. In some instances, the main may proceed from the substation for a certain length before serving any branches or transformers. In this case, the main can be considered in two parts. The portion to which branches and transformers are connected may be considered to have uniformly distributed load, with voltage and losses calculated accordingly. The untapped portion of the main (from the substation to the first load connected to it) may be considered to be a line with the entire load (the uniformly distributed load mentioned earlier) concentrated at its end (where the first load is connected). The loads may be assumed to be balanced and the neutral neglected. The total voltage drop is the sum of the drops in the two portions of the main; the total losses in the feeder main are also the sum of those in the two portions.

In considering the total annual cost of the primary line for comparison with the annual cost of the losses in it, in addition to the cost of the conductors in place, the cost of poles, insulators, switches, etc., must also be included as well as the annual costs of operation and maintenance.

Voltage drops and energy losses are reduced substantially as the applied voltage values increase. For primary circuits, particularly those operating at higher voltages, these values are considerably less than for comparable secondary quantities.

As indicated earlier for secondary systems, the most economical size of conductor for a proposed load (present, future, and contingency) may be determined by an analysis of the annual carrying charges for the system considered and the annual cost of energy losses in the conductor.

Conductor Size

A conductor size, though as near as possible to that indicated by the economic analysis, may still be subject to other considerations. The permissible voltage drop in the several parts of the circuit will determine the minimum size of conductor; if this size is greater than the indicated economical size, economy is disregarded; if smaller, the economical size should be chosen.

The choice of conductor size, however, will not only be limited to those which will carry the load with satisfactory voltage variations, but the size chosen must also be mechanically able to support itself even under unfavorable weather conditions, if overhead, and to withstand installation cable stresses if underground. As a rule, for overhead systems, conductors smaller than no. 6 AWG medium- to hard-drawn copper are not recommended, because of strength limitations, nor are those larger than no. 4/0, because of the difficulty in handling. For underground systems, soft-drawn copper may be used because of its ease in handling; no. 6 or no. 8 is the minimum for reasons of strength as well as load and voltage limitations, and no. 4/0 and 350,000-cmil are the largest sizes that may burn themselves clear under short-circuit conditions; where a conductor larger than these sizes is required, two smaller-size conductors in parallel may be substituted.

As indicated earlier for secondary systems, the sizes of conductors employed for primary (and secondary) circuits should be standardized for any one system, and limited to relatively few in number. Such standardization simplifies, and adds to economy in, their manufacture, purchasing, stocking, and handling in the field.

While the discussion applies principally to radial-type systems, it is also applicable to primary network systems, the network being divided into a number of adjacent radial-type circuits; the analysis will be very similar to that indicated for secondary networks.

VOLTAGE REGULATORS

Where the most economical size of conductor results in voltage drops or regulation greater than permissible, alternatives may be considered. These may include the installation of larger-size conductors, or a voltage regulator, or both, economics indicating the selection. Here the economic comparison is based on the annual carrying charges of the conductor installed together with those for the regulator—the energy losses in the regulator and its operating and maintenance costs.

Sizes

Regulator sizes specify the percentage of regulation in definite steps—e.g., 5 percent, 7½ percent, 10 percent, etc.—and hence the size of conductor that will give satisfactory regulation with each size of regulator is determined and the total annual costs for each alternative are compared. These are also compared with the annual costs for conductors that would prove satisfactory without a regulator. The alternative with the least total annual cost is the one preferred.

Controls

The regulator does not reduce the voltage variation along the feeder with which it is associated. It does reduce the voltage spread at the point of supply to that feeder, or a portion of the feeder. Refer to Fig. 4-32.

The regulator can be applied at the substation to reduce the supply-voltage spread on individual feeders or on the bus supplying a number of feeders. Unless the feeders are of about the same length and have the same kinds and magnitudes of loads, individual feeder regulators are generally preferred.

Where feeder voltages drop below permissible limits, voltage regulators may be inserted in the primary circuit to correct the condition. They should be located at the point on the feeder where, under full load, the voltage falls below the permissible limit; they are usually located some distance before this point in order to provide for some future increase in the loading of the feeder. Voltage regulators may be of either the induction type or of the tap-changing-under-load (TCUL) type; these are described in Chap. 12. They may be either single-phase or three-phase units.

Voltage-Regulating Relays

Regulators are usually controlled automatically, though they may be manually operated in association with a voltmeter. In older units (many of which still exist), the element for automatic control is essentially a *contact-making voltmeter*, which makes a contact to cause the regulator to raise the voltage when the voltmeter reads the minimum permissible outgoing voltage, and another contact to lower the voltage when the voltmeter reads the maximum permissible outgoing voltage. In newer units, electronic (solid-state) relays accomplish this function without any moving parts.

Line-Drop Compensators

Where it is desired to regulate or maintain the voltage band at some distance from the source of the distribution feeder (e.g., at the first consumer or at some other point farther out on the feeder), a *line-drop compensator* is used with the contact-making voltmeter. The line-drop compensator is an electrical miniature of the line to the point where the regulation is desired. See Fig. 4-4. Resistance and reactance values of the line are calculated and a resistance and reactance proportional to these values are set on the compensator; the line current, through a current transformer, flows through the compensator, producing a voltage drop proportional to that current. This drop is subtracted from the line voltage at the regulator terminals, thus applying at the contact-making voltmeter a voltage (varying with the load) representing the voltage at the point of compensation on the feeder. Refer to Fig. 4-32.

The point of compensation should be selected so that the consumer farthest from the regulator will have at least the lowest permissible voltage under the heaviest load while the consumer nearest the regulator will have the highest permissible voltage under light-load conditions.

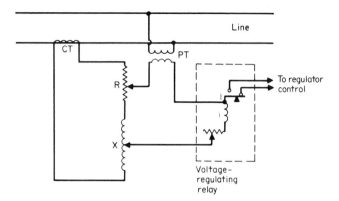

FIG. 4-4 Schematic diagram of line-drop compensator and voltage-regulating relay.

Networks

Where the regulators (at the substation) control the voltage on feeders supplying a secondary network, steps must be taken to prevent the regulators from becoming "unstable," i.e., some moving to their maximum increase position while others on adjacent feeders move to their minimum positions; this condition can reverse itself and be continuous, not only creating periodic voltage variations that might be annoying, but creating troublesome circulating currents. This is especially true for three-phase regulators that cause a phase displacement. Mechanical interconnections, *in-phase* regulators, and *phase shifters* are sometimes used to prevent this instability. Where two feeders only are involved, stability can be maintained by using the current from one line in the compensator for the other.

On some feeders, a lowering in voltage may be necessary under periods of light load or where other means of raising voltage are employed, such as taps, boosters, and capacitors.

TAPS

Where voltage improvement can be obtained by some fixed amount which will not cause voltages to exist outside permissible limits during both light and heavy load conditions, taps can be changed on the distribution transformers on certain portions of the feeder.

For example, assuming an evenly distributed load on a feeder, the taps on the transformers in the first third of the feeder from the substation can be changed to lower the secondary voltage a fixed amount; the taps on the second or center third of the feeder may be left on their normal setting; and those on the farthest third of the feeder may be changed to raise the secondary voltage a fixed amount.

The taps on the transformers merely change their ratios of transformation. If the normal ratio is (say) 20 to 1 to give a secondary voltage of 120 V, tap changes on those nearest the substation would result in a 21 to 1 ratio and a voltage drop of approximately 6 V, which, if subtracted from a high permissible voltage of 126 V, will still leave a voltage of 120 V; or put another way, the tap change allows the highest permissible voltage at the substation to be raised 6 V without exceeding the permissible high-voltage limit at the first consumer. Similarly, on those farthest from the substation, tap changes can result in a 19 to 1 ratio and a voltage increase of 6 V, allowing additional voltage drops in the feeder up to 6 V before the permissible low voltage at the last consumer is not met.

BOOSTERS

An increase or decrease in the primary voltage can also be obtained by the installation of a transformer in the line to provide a fixed voltage drop. A distribution transformer, connected as an autotransformer, may be used to boost or buck the feeder voltage at the point of its installation. The percentage of boost or buck will depend on the ratio of the primary and secondary coils, including the tap used, of the transformer selected. The capacity of the unit is determined by the current-carrying capacity of the secondary coil, through which the entire line current will flow. (Refer to Fig. 4-6k.)

The use of a distribution of normal design in this way is usually done in an emergency. It is an unsafe method, as the secondary is connected directly to the primary. For safety reasons, special attention should be given when connecting or disconnecting such units.

CAPACITORS

Voltage regulation can also be improved by the application of shunt capacitors: at the substation, out on the primary feeder, or both. The current drawn by a capacitor has a leading power factor characteristic and will cause a voltage rise from the location of the capacitor back to the current source. The voltage rise will be equal to the reactance of the circuit (back to the source) multiplied by the capacitor current (taking into account their vector relationship). The rise in voltage is independent of the load on the circuit and is greatest at the location of the capacitors and decreases to the source.

Capacitors provide a constant increase in the level of voltage at the location of the capacitor that is the same under any load condition of the feeder, from light to heavy loading. If capacitors are installed so that they may be switched on during heavy load periods and off at light load periods, voltage regulation can be improved. If a bank of capacitors is so arranged that some of its units can be switched on and off separately, voltage regulation can be improved even further.

Primary Feeder

When they are installed out on a primary feeder, the capacity of the capacitors (in kVA) and the location on the feeder where they are to be installed depends

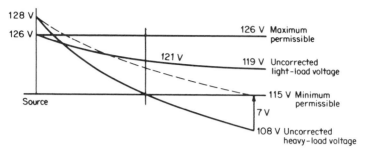

FIG. 4-5 Voltage improvement using capacitors.

on the manner in which the loads are distributed on the feeder, the power factor of the loads, the feeder conductor size and spacing between conductors, and the voltage conditions along the feeder. Like the line voltage regulator, capacitors should be installed approximately at the point where the voltage at heavy load is at the minimum permissible level (with some consideration given to load growth). The conditions under light load will determine what portion of the capacitance installed may be fixed and what may be switched.

EXAMPLE 4-3   Refer to Fig. 4-5.

Assume a 7200-V primary, single-phase. The minimum voltage at the last consumer is 108 V (uncorrected), and the minimum voltage at the capacitor point is 115 V (under heavy load). Thus the minimum correction required is 7 Volts.

The maximum voltage at the last consumer is 127 V (light-load-corrected), and the maximum voltage at the capacitor point is 128 V (also light-load-corrected). Assume the reactance from the source to the capacitor point is 0.277 Ω. Then

$$I_C = \frac{E}{X_L} = \frac{8}{0.277} = 28.8 \text{ A} \qquad \text{heavy load}$$

and

$$\frac{2}{0.277} = 7.22 \text{ A} \qquad \text{light load}$$

$$1 \text{ kVA} = \frac{1000}{7200} = 0.14 \text{ A/kVA}$$

To determine the capacity required to raise the voltage 8 V at heavy load:

$$\text{Capacitor kVA} = \frac{28.8}{0.14} = 206 \text{ kVA}$$

Say eight 25-kVA units.

At light load, the 128 V at the capacitor must be lowered a minimum of 2 V, and the 127 V at the last consumer must be lowered a minimum of 1 V.

To determine the capacity to be disconnected to lower the voltage 2 V at light load:

$$\text{Capacitor kVA} = \frac{7.22}{0.14} = 51.6 \text{ kVA}$$

Say two 25-kVA units.

Substations

Capacitors may also be installed at substations on the bus supplying the outgoing distribution feeders. They are usually installed in relatively large-capacity banks, and it is usually necessary to switch off portions of them at periods of light load to prevent excessively high outgoing voltage. The voltage drops along the feeders supplied from this substation bus remain the same as do their power factors, since the relationship between the voltage and current flowing through each of the feeders supplying their loads is unaffected by the capacitors added to the substation bus. The voltage level of each of the entire feeders is raised depending on the capacitance added at the substation, but the voltage spread on each feeder remains the same. In many instances, the principal reason for the capacitors at the substation bus is not necessarily to control the bus voltage, but, by counteracting the effect of induction (or reactance), to reduce the current to that necessary to supply the load at approximately unity power factor, thereby permitting larger loads to be supplied by the same transmission and substation facilities.

Series Capacitors

Capacitors can also be installed in series with primary feeders to reduce voltage drop, but they are rarely employed in this fashion. Where shunt capacitors, connected in parallel with the load, correct the component of the current due to the inductive reactance of the circuit, series capacitors compensate for the reactive voltage drop in the feeder.

A capacitor in series in a primary feeder serving a lagging-power factor load will cause a rise in voltage as the load increases. The power factor of the load through the series capacitor and feeder must be lagging if the voltage drop is to decrease appreciably. The voltage on the load side of the series capacitor is raised above the source side, acting to improve the voltage regulation of the feeder. Since the voltage rise or drop is produced instantaneously with the variations in the load, the series capacitor response as a voltage regulator is faster and smoother than the induction or TCUL-type regulator; moreover, no contact-making voltmeter and load compensator are required for its operation.

During fault conditions, however, the large fault current passing through the series capacitor can develop excessive voltage across the capacitor, sufficient to cause its destruction. It is essential, therefore, that it be taken out of service as quickly as possible. A resistor and air gap are connected between the terminals of the series capacitor. When the voltage becomes sufficiently high, the gap breaks down and permits the capacitor to be short-circuited through the resistor; the resistor dampens out any oscillatory discharge current so the gap can break down and restrike repetitively without damaging the capacitor. Auxiliary relays operate to short-circuit and bypass the capacitor if the fault persists.

Because of the potential hazard, series capacitors as voltage regulators are usually restricted to supplying single large consumers where flicker may result from frequent motor starts or from electric welders, furnaces, and similar devices that may cause rapid and repetitive load fluctuations.

REACTORS

Primary

Where relatively high-voltage primary feeders (23-kV and above) operate in metallic sheathed cables and are rather long, the capacitance effect of the cable may cause undesirable voltage rises along the feeder. Reactors connected between the primary conductors and the neutral or ground are inserted in the feeder at appropriate points to hold customer voltages within permissible limits; shunt reactors act in a similar fashion as shunt capacitors.

Secondary

Where two or more transformers supply a common load, the transformers may not share the load equitably. This may be due to differing secondary voltages at the transformers' terminals either because the primary feeder voltages are different, or because the transformers have different impedances. Reactors inserted in the secondary leads of one or more of the transformers are installed in an effort to equalize the voltages and make the transformers share the load equitably. This phenomenon is often evident in low-voltage secondary networks, and especially in "spot networks." In this latter case, the terminals of the reactor coil of one transformer are interconnected to those of another transformer with the leads reversed. Hence, the voltage drop in each of the reactances is added to or subtracted from the several secondary voltages, tending to balance the load among the several supply transformers.

TRANSFORMERS

Transformers play a central part in the design of distribution systems; they reduce the high voltage of the primary to the low utilization voltage of the secondary. As with other elements of the distribution circuit, the energy losses and the drop in voltage due to the current flowing through them to supply loads are factors in the selection of the size and location of transformers.

Losses

Energy losses in a transformer are generally of two kinds:

1. No-load loss (also known as iron or core loss) results from the magnetizing or exciting current flowing in the primary coil regardless of the load carried.

Its value of about 0.5 percent at rated full load may vary substantially at voltages above or below rated values. Although small as a power loss, it goes on constantly, accumulating into significant annual energy losses (in kilowatthours).

2. Full-load losses (earlier known as copper losses) result from the load current passing through the resistance of both primary and secondary coils. This I^2R loss varies with the square of the current carried and therefore depends on the shape of the load curve. Since the current flowing in a circuit is inversely proportional to the voltage, the copper loss is inversely proportional to the square of the voltage; hence, for the same-size transformer, the losses in the primary coil are substantially less as the voltage ratings increase.

No-load and full-load losses for the various sizes of transformers vary with different manufacturers and are usually specified by them in some percentage of normal voltage and full-load ratings. No-load losses may be expressed in watts or as a percentage of the full rated load in watts.

Impedance—Resistance and Reactance

Copper losses, as well as voltage regulation, require that resistance and reactance values (and their vector sum, impedance) of the transformer be known. These three values represent both primary and secondary coils of the transformer. They are usually specified by the manufacturer as a percentage related to the percentage voltage drop. That percentage gives a value in volts when applied to either primary or secondary voltage; from that voltage and the full rated current the values in ohms may be derived.

The percentage impedance given for a transformer represents (and is equivalent to) the percentage drop from normal rated primary voltage that would occur when full rated load current flows in the secondary; thus the percentage impedance can be used to determine the impedances (in ohms) of the primary and secondary as follows. In reference to the primary,

$$Z_p = \frac{\% \, Z_p}{100} \frac{E_p}{I_p}$$

and in reference to the secondary,

$$Z_s = \frac{\% \, Z_s}{100} \frac{E_s}{I_s}$$

where I_p, E_p, I_s, and E_s are all full rated load-current values. Since $E_p = nE_s$, where n is the transformer turn ratio,

$$\frac{Z_p}{Z_s} = \left(\frac{E_p}{E_s}\right)^2 = n^2$$

The relationship between resistance R, reactance X, and impedance Z,

$$Z^2 = R^2 + X^2$$

applies whether the quantities are in percentages ($\% \, Z$, $\% \, R$, $\% \, X$), in ohms referred to the primary (Z_p, R_p, X_p), or in ohms referred to the secondard (Z_s,

R_s, X_s). Since the percentage impedance and the percentage resistance for a given size of transformer are specified by the manufacturers, the percentage reactance may be computed:

$$\% \ X = \sqrt{(\% \ Z)^2 - (\% \ R)^2}$$

and the percentage resistance and percentage reactance can be reduced to ohms, referred to either the primary or the secondary side of the transformer:

$$R_p = \frac{\% \ R}{100} \frac{E_p^2}{\text{transformer kVA} \times 1000} \qquad \Omega$$

$$R_s = \frac{\% \ R}{100} \frac{E_s^2}{\text{transformer kVA} \times 1000} = \frac{R_p}{n^2} \qquad \Omega$$

$$X_p = \frac{\% \ X}{100} \frac{E_p}{\text{transformer kVA} \times 1000} \qquad \Omega$$

$$X_s = \frac{\% \ X}{100} \frac{E_s^2}{\text{transformer kVA} \times 1000} = \frac{X_p}{n^2} \qquad \Omega$$

The copper loss W, in watts at full load, is caused by the current passing through the resistance of the windings. Thus,

$$W = R_p I_p^2 = R_s I_s^2$$

Since the manufacturer specifies the copper loss, the equivalent resistance can be computed:

$$R_p = \frac{W}{I_p^2} \text{ and } R_s = \frac{W}{I_s^2} \text{ and } \frac{R_p}{R_s} = \frac{I_s^2}{I_p^2} = n^2$$

Voltage Drop

The determination of voltage drop through the transformer employs values of impedance, resistance, and reactance, as indicated in the previous discussion of primary and secondary systems. The drop must be referred to either the primary or the secondary side:

$$\text{Voltage drop (primary)} = I_p(R_p \cos \theta + X \sin \theta)$$

where I_p is the load current and $\cos \theta$ the power factor.

$$\text{Voltage drop (secondary)} = I_s(R_s \cos \theta + X \sin \theta)$$

$$\frac{\% \text{ voltage drop}}{100} = \frac{\text{voltage drop (primary)}}{\text{rated primary voltage}} = \frac{\text{voltage drop (secondary)}}{\text{rated secondary voltage}}$$

These same phase-to-neutral values of Z, R, and X can also be employed in polyphase circuits. Since phase to phase voltage (for a three-phase circuit) is $\sqrt{3}$ times the phase-to-neutral value and the voltage value in the equation is squared, the ratio between phase-to-phase and phase-to-neutral characteristics is 3 to 1. If the transformer kVA value is the three-phase total kVA and E_p is the phase-to-phase voltage,

Phase-to-phase:

$$R_p = \frac{\% R}{100} \frac{E_p^2 \times 3}{3\phi \text{ kVA} \times 1000} \quad \Omega$$

$$X_p = \frac{\% X}{100} \frac{E_p^2}{3\phi \text{ kVA} \times 1000} \quad \Omega$$

$$Z_p = \frac{\% Z}{100} \frac{E_p^2}{3\phi \text{ kVA} \times 1000} \quad \Omega$$

Phase-to-neutral:

$$R_p = \frac{\% R}{100} \frac{E_p^2}{3\phi \text{ kVA} \times 1000} \quad \Omega$$

X_p and Z_p will use the same expressions as for single-phase values, but with total three-phase transformer capacity in kVA.

Transformers connected in parallel or in banks should have impedances as nearly the same as possible, within a fraction of a percent. Transformers not having essentially the same impedance when in parallel will not divide the load in proportion to their ratings, and when in wye or delta banks they will cause circulating current to flow that reduces the capacity of the transformers, increases losses, and results in excessive heating.

Transformers supplying radial secondary circuits, whether single-phase or in banks, have a reactance of from 3 to 5 percent, which results in reasonable voltage regulation. Transformers supplying low-voltage secondary networks have a higher reactance, sometimes as high as 10 percent, to ensure a better load division among the transformers, particularly under contingency conditions when one (or more) of the supply feeders may be out of service.

Transformer Connections

Transformer connections were already described in Chap. 2, together with some vector diagrams of current and voltage relationships; Fig. 4-6a through k, in which connections involving transformer polarity are shown, also portray those most apt to be found on distribution systems.

Single-Phase   The standard single-phase distribution transformer is generally designed with the secondary coil in two parts, which may be connected in parallel for two-wire 120-V operation, or in series for three-wire 120/240 V operation. The latter is the most commonly used connection for single-phase distribution systems. The load is balanced between two 120-V circuits; with perfect balance, no current flows in the center or neutral wire. Refer to Fig. 4-6a.

Three-Phase   For three-phase systems, the wye-connected secondary can serve single-phase loads at 120 V for each phase; when the load is balanced, the neutral will carry no current. This connection can also supply three-phase power loads at 208 V between phases, and it is best adapted for use on secondary networks. It does have the disadvantage of a lowered three-phase (208 V) voltage supply

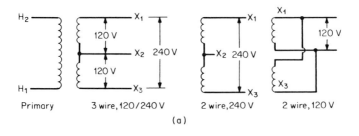

Primary   3 wire, 120/240 V   2 wire, 240 V   2 wire, 120 V

(a)

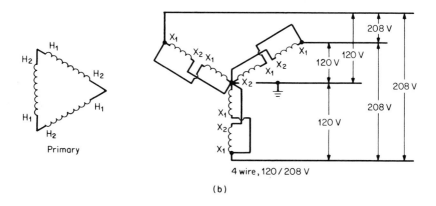

Primary

4 wire, 120/208 V

(b)

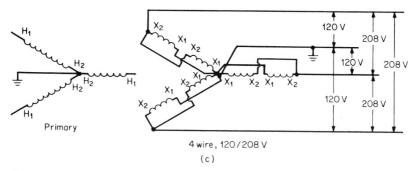

Primary

4 wire, 120/208 V

(c)

FIG. 4-6 (a) Single-phase two- and three-wire secondary connections. (b) Three-phase delta-to-wye four-wire secondary connection. (c) Three-phase wye-to-wye four-wire secondary connection. (d) Three-phase open delta three- and four-wire secondary connections. (e) Three-phase open wye–open delta three- and four-wire secondary connections. (f) Three-phase delta-delta three- and four-wire secondary connections. (g) Three-phase wye-delta three- and four-wire secondary connections.

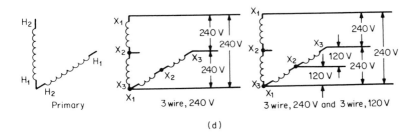

Primary   3 wire, 240 V   3 wire, 240 V and 3 wire, 120 V

(d)

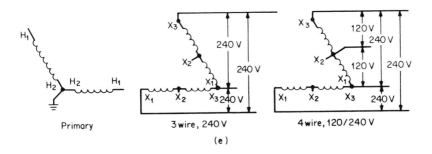

Primary   3 wire, 240 V   4 wire, 120/240 V

(e)

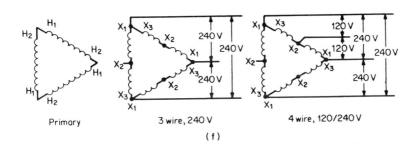

Primary   3 wire, 240 V   4 wire, 120/240 V

(f)

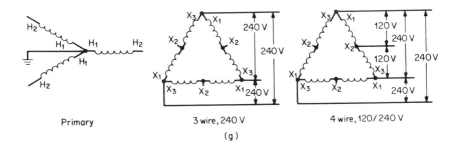

Primary   3 wire, 240 V   4 wire, 120/240 V

(g)

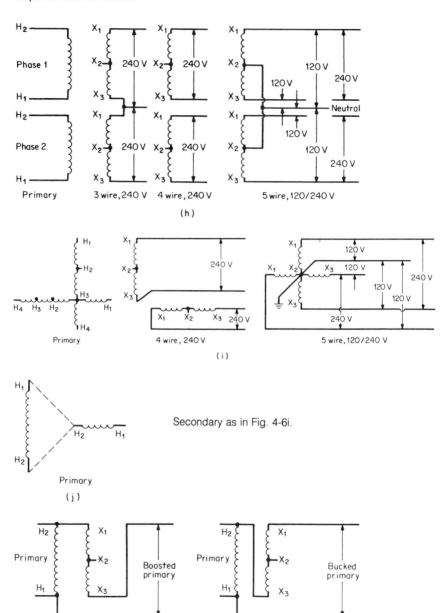

FIG. 4-6 (*h*) Two-phase three-, four-, and five-wire secondary connections. (*i*) Three-phase/two-phase (or vice versa) four- and five-wire secondary connections (also known as the Scott connection). (*j*) Three-phase/two-phase (or vice versa) four- and five-wire secondary connection. Secondary connections same as for Fig. 4-6*i*. (*k*) Single-phase boost-buck primary connections.

to three-phase motors with standard ratings of 240 V; the 32-V difference, or 13.3 percent, below the rating may affect the operation of the motors. To remedy the situation, the secondary voltage is often raised to 125 V, yielding about 217 V between phases or about only 10 percent less than the standard 240-V rating, more likely to be within the design tolerances for satisfactory operation. See Fig. 4-6*b* and *c*.

The primary supply to this four-wire wye secondary connection can be either delta- or wye-connected; in the latter case, the wye is usually grounded to prevent voltage unbalances from unbalanced secondary loads from distorting phase relationships. Often, further economy is achieved if both the primary and secondary circuits employ a common neutral conductor.

Small amounts of three-phase power loads may be supplied on a chiefly single-phase system by a small-diameter-conductor extension of another phase and the installation of a small-capacity single-phase transformer in an open-wye or open-delta bank (on the primary side) with the principal single-phase transformer. The secondary of this second single-phase transformer is connected in an open-delta configuration with the secondary of the principal transformer, providing a small three-phase delta power supply to the small three-phase requirement. Because of the phase relationship of the voltage and current, however, only 86 percent of the capacity of this second, small single-phase transformer can be utilized. This is an economical method of supplying a small, isolated three-phase load in the midst of an area supplied from single-phase facilities. See Fig. 4-6*d* and *e*.

Two-Phase Although two-phase systems are virtually extinct, there are still some two-phase power loads in existence. These may be supplied from three-phase delta or wye systems through proper connections of two single-phase transformers. Two such connections are shown in Fig. 4-6*h*, as are those for three-phase to two-phase (and vice versa) in Fig. 4-6*i* and *j*.

Boost-Buck Earlier, reference was made to the use of single-phase transformers to boost or buck the line voltage of a primary feeder. Here, the primary and secondary of the transformer are connected in series, essentially operating as an autotransformer. The incoming primary coil is connected across the primary circuit, while the outgoing primary is connected between a common terminal of the primary of the transformer and the terminal of the secondary coil; the voltage of the secondary coil is either added to boost the primary voltage or subtracted to buck it. The capacity of the secondary coil limits the primary current that may flow through it. These connections are also shown in Fig. 4-6*k*.

Three-Phase Units

Connections for three-phase transformers and lead markings are shown in the IEEE classification of polarities illustrated in Table 4-2.

TABLE 4-2 IEEE CLASSIFICATION OF POLARITIES OF THREE-PHASE TRANSFORMERS

Three-phase transformers without taps

Group 1:
angular
displacement
0°

Group 2:
angular
displacement
180°

  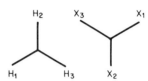

Group 3:
angular
displacement
30°

Three-phase transformers with taps

Group 3:
angular
displacement
30°

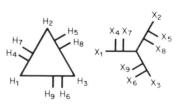

Courtesy General Electric Company

Autotransformers

Under certain conditions, when the ratio of transformation desired is low, usually not greater than about 5 to 1, and electrical isolation between primary and secondary circuits is not essential, the autotransformer has some advantages.

The autotransformer consists of one winding, a part of which may serve as both primary and secondary. In a two-winding transformer, all of the energy is transformed by magnetic action. In the autotransformer, a portion only is transformed magnetically and the remainder flows conductively through a part of its windings. Since only a portion of the energy is transferred, the autotransformer can be smaller than a two-winding unit; comparable costs of the unit and its installation are less. Also, the losses from currents flowing through it are lessened, resulting in greater efficiency and improved voltage regulation.

A schematic diagram is shown in Fig. 4-7. Voltage, current, and turn relationships are indicated. The same ratio of transformation exists as in a two-winding transformer:

$$\frac{N_p}{N_s} = \frac{E_p}{E_s} = \frac{I_s}{I_p}$$

The figure shows a step-down arrangement of the voltage from E_p to E_s; reversing E_s and E_p will give a step-up arrangement. The current in the coil from a to b is the sum of the exciting current I_p and the nontransformed, relatively large conductive current I_{ss}. This part of the coil, a to b, must be a sufficiently large conductor to carry this load, whereas the portion from b to c carries only the magnetizing current and hence can employ smaller-size conductors.

The percentage of volt-amperes transferred and the percentage of voltage transformed are the same (using the high voltage as a base). For example, if the autotransformer lowers the voltage 5 percent, it actually transforms only 5 percent of the volt-amperes supplied the load. Since the size of the autotransformer depends on only the volt-amperes transformed, the size of the unit can be only 5 percent of the load; if the load supplied is 100 kVA, the size of the autotransformer required would be 5 kVA.

As the ratio of transformation increases, the autotransformer becomes less and less economical until, at about 5 to 1, its advantages no longer apply. A

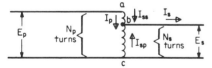

FIG. 4-7 Autotransformer.

larger and larger part of the coil would require a larger conductor and heavier insulation, since both portions are connected electrically.

The electric connection between the incoming and outgoing circuits is a disadvantage, as a disturbance on one side affects the other. For example, a ground on either circuit is a ground on both, and a ground on the high-voltage side may impose a high voltage on the low side and on the loads connected thereto. The low-voltage side may be insulated to withstand the higher voltage, but the connected loads may not be so protected.

The autotransformer generally has a comparatively lower impedance, which may cause greater fault currents to flow through it during a contingency. The autotransformer, therefore, must be built very ruggedly to withstand the greater mechanical stresses produced, or else external impedances should be connected in the circuit to limit the magnitude of the fault currents, or both should be done in some combination.

Autotransformers may be used on both single-phase and polyphase circuits, as indicated previously. Voltage regulators of both the induction and TCUL types are autotransformers in principle.

Ratings and Temperature

The rating of any piece of electrical equipment is limited by the maximum permissible temperature in any of its components. For transformers, an additional consideration is the permissible voltage drop through the unit.

The maximum temperature generally accepted is that beyond which the insulation is apt to be damaged. Standards set by engineering and manufacturing groups specify an allowable temperature rise of 55°C above an ambient of 40°C, based on the average temperature of the windings; allowing a 10°C difference between "hot spot" and average temperatures in the windings, a maximum temperature of 105°C is indicated. This value is well below the temperature at which insulation fails, providing a large factor of safety.

Transformers are rated in volt-amperes (or kVA) rather than watts. Since the characteristics of the circuit and its loads affect the power factor of the power being transformed, a poor power factor can cause a large current flow in the coils of the transformer, producing losses and heat, with relatively little actual power delivered. The rating that takes into account the current flow and the voltage applied is the volt-ampere rating.

The rating is based on the current the transformer will carry *continuously* without exceeding the temperature rise limitations. In selecting a transformer to accommodate a load, other factors besides the maximum value of the load must be taken into consideration. Load duration and cyclic variations; the variations in ambient temperatures, especially because of latitudes and seasons; weather patterns of rain, snow, and ice; the age and condition of the transformer and its components—these are all factors that influence how much a transformer may be loaded with respect to its rating. Moreover, since transformers are not tailored to fit the load but are manufactured in standard sizes, there is normally a margin of transformer capacity available for supplying loads above the rating of the transformer for short periods of time.

Transformer Sizes

Standard sizes of distribution transformers change from time to time as economics and situations change. Kilovolt-ampere capacities presently in greatest use include:

Single-phase units: 10, 25, 37½, 50, 100, 167, 250, 333, and 500 kVA. Older units that still exist in service include 1, 1½, 3, 5, 7½, 15, 75, 150, and 200 kVA.

Three-phase units: 75, 150, 300, 500, 1000, and 3000 kVA. Other, older units still in service include 5, 7½, 10, 15, 25, 50, 100, 200, and 450 kVA.

Voltage Ratings

Standards of voltage ratings, on both the primary and secondary sides, as well as the numerical and percentage voltage variations above and below nominal voltage ratings, are specified in the selection of taps included in the primary winding of the transformer; these, too, are subject to revision from time to time in response to changing requirements.

SUBSTATIONS
Location versus Distribution Voltage

Perhaps the first consideration regarding a distribution substation is its location. In general, it should be situated as close to the load center to be served as practical. This implies that all loads can be served without undue voltage regulation, including future loads that can be expected in a reasonable period of time. The difficulty in obtaining substation sites is an important factor in selecting the distribution voltage, both in original designs and in later conversions.

The higher the distribution voltage, the farther apart substations may be located, but they also become larger in capacity and in the number of customers served. Thus, the problem of the number and location of distribution substations involves not only the study of transmission and subtransmission designs, but more emphasis on service reliability and consideration of additional costs that may be justified. The subjects of sectionalizing, field-installed voltage regulators and reclosers, capacitors, and ties to adjacent sources are discussed elsewhere, but are pertinent to the problem.

Supply Feeders and Circuit Breaker Requirements

The number and sources of supply subtransmission feeders to the distribution substation will depend not only on the load to be served, but also on the degree of service reliability sought. Some rural substations may be supplied from only one subtransmission feeder, while substations serving urban and suburban areas have a minimum of two supply feeders and may have several more. Each additional incoming feeder, however, adds to the bus and switching requirements, including auxiliary devices for their protection, all of which add to costs.

Circuit Breaker Arrangements

Some basic arrangements of incoming high-voltage circuit breakers and transformers are shown in Fig. 4-8. Each scheme progressively adds to the reliability of service to the substation and the loads it supplies. For example, in scheme *a*,

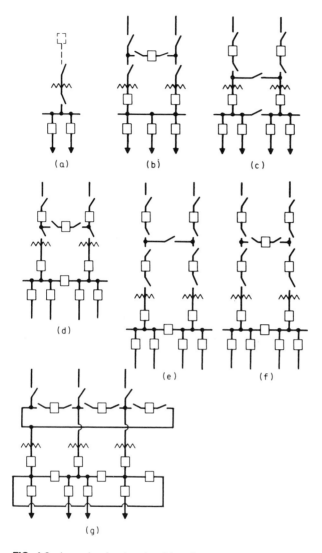

FIG. 4-8 Incoming feeder circuit breaker arrangements.

a failure on the transmission line or substation transformer or bus will trip the breaker back at the transmission source, and service may not be restored until the fault is *found and repaired;* in scheme *b,* such failures will trip the circuit breaker, but service can be restored as soon as the fault is *isolated;* in schemes *c, d, e, f,* and *g* (the last incorporating a ring bus), failures on the incoming transmission lines, transformers, or high-voltage circuit breaker will not interrupt (except for a short time or momentarily) the supply to the bus serving distribution feeders. Since the cost of high-voltage circuit breakers, together with their accessories, is often as great as or greater than the cost of the transformers with which they may be associated, it is essential that the cost of additional circuit breakers not outweigh the protective advantages gained. It may prove desirable that a minimum number of circuit breakers be installed initially and others added as deemed necessary for any improvement in service reliability that time, increments of load, and customers' requirements may indicate.

Interrupting Duty

The circuit breakers must not only interrupt the normal load current, but must be mechanically able to withstand the forces resulting from the large magnetic fields created by the fault current flowing through them. Since the field will depend on the magnitude of the fault current, which in turn also depends on the voltage of the circuit, the stresses that must be accommodated depend on both of these values. A circuit breaker, therefore, is rated not only on its applied voltage and normal current-carrying capacity, but on its interrupting ability, expressed in volt-amperes (or kVA or MVA); for example, 100-A, 35-kV, or 50,000-kVA interrupting "duty" or capability.

Insulation Coordination—BIL

Circuit breakers and other equipment are subject to high-voltage surges resulting from lightning or switching operations, and the insulation of their energized parts must be capable of withstanding them. Lightning or surge arresters are installed on the conductors and buses of each phase as close to the circuit breakers as practical, with the intent of draining off the voltage surge to ground before it reaches the breaker.

To provide adequate insulation economically and to restrict and localize possible damage to the circuit breaker, the insulation provided for the several parts is coordinated. Internal parts are insulated as equally as practical, but their insulation is generally stronger than that of the bushings, which in turn is stronger than that of the "discharge" point of the associated arrester. Thus, a surge not drained to ground by the arrester will next tend to flash over at the bushings, outside the tank, where damage would be confined, comparatively light, and easier to repair. In general, the insulation of the weakest point in the circuit breaker should be weaker by such a margin as to ensure it will break down before the insulation of the principal equipment it is protecting.

The coordination of insulation requires the establishment of a basic insulation level (BIL) above which the insulation of the component parts of the system should be maintained, and below which lightning or surge arresters and other protective devices operate. This is discussed further in connection with protective devices.

Substation transformers also have their insulation coordinated with that of associated circuit breakers, buses, and other devices.

Capacitors

As mentioned earlier, banks of capacitors may be connected to the high-voltage incoming bus in connection with voltage regulation and increasing the capacity or capability of the substation to supply load. All or portions of these banks may be switched on and off to provide flexibility in maintaining voltage regulation and power factors. This is done with one or more circuit breakers, and arresters or other protective devices as indicated.

Transformers

Substation transformers may consist of three-phase units or banks of three single-phase units. The size of these individual installations may range from 150 kVA (three-phase) in small rural stations to upwards of 25,000 kVA at larger urban and suburban substations. Their impedances are generally low, restricting unregulated voltage variations at the bus to a few percent, except where fault current levels are high. In this case, transformer impedances are increased to limit fault current duty to design limits.

The impedances of the transformer banks in a station should match each other as closely as practical to have the banks share the load as equally as practical.

The transformers may be connected in a delta or wye pattern, on both the incoming high-voltage (subtransmission) side and the outgoing low-voltage (primary circuit) side. The transformers are ordinarily of the two-winding standard type, operating much as the distribution transformers.

For many reasons, including the random and nonuniform movement of the molecules in the core of the transformer, the alternating magnetic field that is set up may be distorted, producing serrated sine waves on both sides of the transformer. These serrations can be broken down into a series of harmonics or waves with frequencies of 3, 5, 7, etc., times the basic frequency (usually 60 cycles per second). If the transformers have a ground on either side, the harmonics or fluctuations flow to ground and the original sine wave essentially remains undistorted. If the windings are connected in delta fashion, these fluctuations circulate around the delta, filtering out the harmonics and eliminating them from the sine wave formed in the windings; however, they do cause some unnecessary heating.

Where the transformer windings are connected in a wye arrangement *without* a ground or neutral back to the source, the harmonics may be particularly bothersome. To overcome these, each of the single-phase transformations (singly or within a three-phase unit) is provided with a third, small-capacity winding; the three such windings are connected in delta (even though the main primary and secondary windings are connected in wye). The delta thus formed allows the harmonics to circulate within it, producing a little heat but essentially filtering them out, so that the sine wave produced on both the high and low sides of the transformer will be a more pure sine wave.

Low-Side Bus Arrangements

The low sides of the transformers are connected to their buses usually through circuit breakers. Several configurations are shown in Fig. 4-9. Some provision is

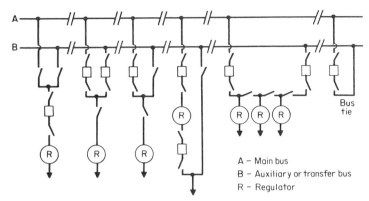

FIG. 4-9 Arrangement of distribution feeder buses at substations (see also Fig. 4-8).

usually made for permitting circuit breakers, switches, regulators, and other devices to be taken out of service for maintenance or for other reasons without causing an interruption to the outgoing distribution feeders. Each of the outgoing distribution feeders is usually equipped with its own circuit breaker. The relays operating these, as well as the transformer high-side circuit breakers, and the capacitors (if any) are coordinated so that only the proper circuit breaker will operate to clear a fault that may occur on some portion of the system; this is considered in more detail on pages 92 to 96.

Voltage Regulators

Each distribution feeder may have its voltage individually regulated, employing three single-phase regulators or one three-phase regulator. If all of the distribution feeders have approximately the same load cycles and voltage regulation (even if corrected by capacitors, field regulators, or other means out on the feeder) the bus to which they are connected may be regulated in place of individual feeder regulators. While this calls for a certain amount of compromise, it may prove economical in many instances.

Mobile Substations

Substations are often designed for three single-phase transformers so that, where they are connected in delta on the incoming side, they can operate in open delta in the event of failure of one of the units. In some instances, a spare single-phase transformer is installed at the substation so that, in the event of failure of one of the transformers, a replacement can be made readily.

   With the advent of lighter transformers and improved transportation equipment, it has proven practical to mount a three-phase transformer and associated switching and surge arresters on a trailer especially designed for that purpose. Such a mobile substation can be readily transported to a substation where a failure has occurred. The terminal arrangements of both the mobile substation and the fixed substation are so designed that often service can be restored more quickly than by reconnecting the spare unit (which no longer need be provided).

The mobile substation not only can be effective where the failure may involve more than one transformer, but can service a number of substations in a more economical fashion than the installation of spare transformers at many, if not all, substations. Further, it may also be installed as a separate, temporary substation, picking up portions of the load of one or more substations whose facilities may be overloaded.

PROTECTIVE DEVICES

For the distribution system to function satisfactorily, faults on any part of it must be isolated or disconnected from the rest of the system as quickly as possible; indeed, if possible, they should be prevented from happening. The principal devices to accomplish this include fuses, automatic sectionalizers, reclosers, circuit breakers, and lightning or surge arresters. Success, however, depends on their coordination so that their operations do not conflict with each other. Figure 4-3 indicates where these devices are connected on the system.

Fuses

Time-Current Characteristic   A fuse consists basically of a metallic element that melts when "excessive" current flows through it. The magnitude of the excessive current will vary inversely with its duration. This time-current characteristic is determined not only by the type of metal used and its dimensions (including its configuration), but also on the type of its enclosure and holder. The latter not only affect the melting time, but, in addition, affect the arc clearing time. The *clearing time* of the fuse, then, is the sum of the melting time and the arc clearing time. Refer to Figs. 4-10 and 4-11. Note that for curve *b* in Fig. 4-11, the clearing time for a certain value of current is less than for curve *a;* the fuse with the characteristic *b* is therefore referred to as a "fast" fuse, compared with the fuse of curve *a.* Refer to Fig. 4-11.

Fuses are rated in terms of voltage, normal current-carrying ability, and interruption characteristics usually shown by time-current curves. Each curve actually represents a band between a minimum and a maximum clearing time

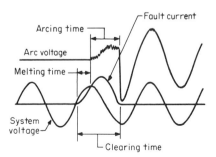

FIG. 4-10 Oscillogram of link melting and fault current interruption. *(Courtesy McGraw Edison Co.)*

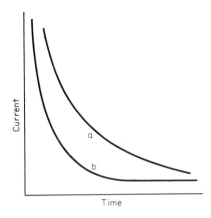

FIG. 4-11 Typical time-current characteristic for fuses.

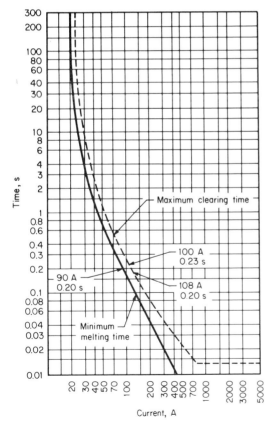

FIG. 4-12 Typical time-current characteristic for a 10-K fuse link. *(Courtesy McGraw Edison Co.)*

for a particular fuse; the difference between them is a predetermined percentage adjustment made to allow for manufacturing tolerances and to ensure an adequate clearing time. A set of such curves is developed for each of the different ratings and types of fuses; see Figs. 4-12 to 4-14.

Fuse Coordination   The number, rating, and type of the interrupting devices shown in Fig. 4-3 depend on the system voltage, normal current, maximum fault current, the sections and equipment connected to them, and other local conditions. The devices are usually located at branch intersections and at other key points. When two or more such devices are employed in a circuit, they will be coordinated so that only the faulted portion will be deenergized. In Fig. 4-28 fuse D must clear before sectionalizer C, and C must clear before recloser B. Likewise, fuse G must clear before F, F before E, and both E and B before A. At the transformer locations, fuse M must clear before D, and N before G. All

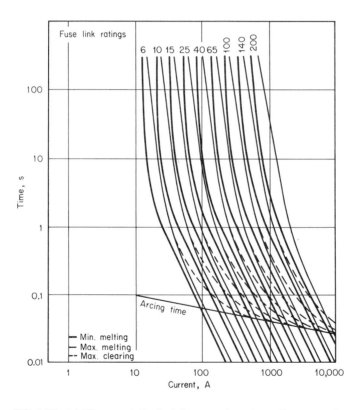

FIG. 4-13 (a) "Representative" minimum and maximum time-current characteristic curves for EEI-NEMA type K (fast) fuse links. *(Courtesy Westinghouse Electric Co.)*

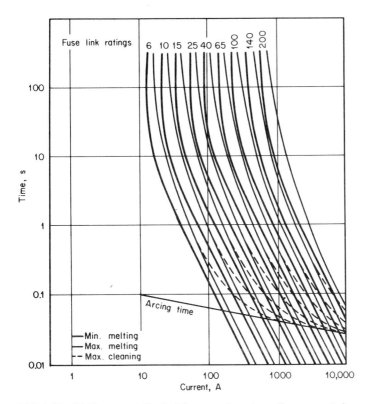

FIG. 4-13 (b) "Representative" minimum and maximum time-current characteristic curves for EEI-NEMA type T (slow) fuse links. *(Courtesy Westinghouse Electric Co.)*

of these devices must be coordinated; i.e., their ratings should provide for carrying normal load currents and for responding correctly to a fault.

Fault current will flow from the source to the fault through the various devices in its path. The magnitude of this fault current will depend on the impedance (resistance for dc circuits) between the source and the point of fault, or roughly, on the distance between them. When a fault is distant from the source, the impedance of this part of the circuit is high and the fault current is low; when the fault is close to the source, the fault current is high.

At the coordinating point farthest from the source, therefore, the fuse will have the lowest rating consistent with the maximum normal load at this point; at the other coordinating points along the path of the current the fuses will have increased ratings as they are closer to the source. These are indicated in Fig. 4-15 and Table 4-3. The characteristics of these fuses must also coordinate with those of other protective devices in the same path and with those of the circuit breaker at the source.

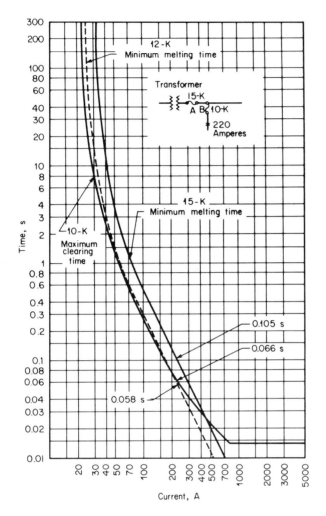

FIG. 4-14 Coordination by time-current curves. *(Courtesy McGraw Edison Co.)*

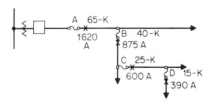

FIG. 4-15 Example of fuse-coordinated system.

TABLE 4-3 FUSE-COORDINATED SYSTEM

Position	Fault current, A	Protective link	Maximum clearing time M, s	Minimum melting time N, s	Clearing factor	
					Ratio CT/MT, M(pos)/N(pos)	Percent
A	1620	65-K	0.078	—	—	—
B	875	40-K	0.066	0.120	0.078-A/0.120-B	65.0
C	600	25-K	0.061	0.093	0.066-B/0.093-C	71.0
D	390	15-K	—	0.084	0.061-C/0.084-D	72.6

Courtesy McGraw Edison Co.

Repeater Fuses

Line fuses are sometimes installed in groups of two or three (per phase), known as repeater fuses, having a time delay between each two fuse units. When a fault occurs, the first fuse will blow and the second fuse will be mechanically placed in the circuit by the opening of the first; if the fault persists, the second fuse will blow; if there is a third fuse, the process is repeated. If the fault is permanent, all of the fuses will blow and the faulted part of the circuit will be deenergized. New fuses must be installed to restore the line to normal.

Where capacitors are applied to feeders for power factor correction, fuses chosen to protect the line from the bank (and vice versa) must also coordinate with sectionalizing and other devices in the circuit back to the source.

Transformer Fuses

Fuses on the primary side of distribution transformers serve to disconnect the transformer from the circuit not only in the event of a fault in the transformer or on the secondary, but also when the normal load on the transformer becomes so high that failure is imminent. Fuses on the secondary side protect the transformer from faults or overloads on the secondary circuit it serves.

The characteristics of a primary fuse are a compromise between protection from a fault and protection from overload, yet the fuse also has to coordinate with other fuses on the line. One attempt at a solution is the completely self-protected (CSP) transformer, in which the primary fuse, with characteristics based only on protection against fault, is situated within the transformer tank (and, to differentiate, is called a *link*) while overload protection is accomplished by low-voltage circuit breakers (instead of fuses) on the secondary side of the transformer that are also situated within the tank. The circuit breakers, once open, however, must be reclosed manually.

Fuses are provided on the line side of the protectors on low-voltage secondary networks. These are backup protection in the event the protector fails to open during back feed from the network into the primary when it is faulted or deliberately grounded.

Secondary fuses, known as *limiters,* are also provided at the juncture of secondary mains to isolate faulted sections of the secondary mains and to prevent the spread of burning in conductors (usually in cables) where sufficient fault current does not exist to burn them clear in a small portion of the mains.

Automatic Line Sectionalizers

Automatic line sectionalizers are connected on the distribution feeder in series with line and sectionalizing fuses; they are also in series with and electrically farther from the source than reclosers or circuit breakers with reclosing cycles. These devices are decreasing in usage, but many exist on distribution systems.

When a fault occurs on the circuit beyond the sectionalizer, the fault current initiates a *fault-counting relay* that is coordinated with the characteristics of the fuses and other devices. Each time the circuit is deenergized (from reclosers or circuit breakers), the relay moves toward the trip position; just before the final operation that will lock out the recloser or circuit breaker if the fault persists, the sectionalizer will trip (while no fault current is flowing) and open the circuit at that point, removing the fault and permitting the circuit breaker or recloser to close and reset into its normal position; service is thus restored to the rest of the circuit up to the location of the sectionalizer. If the fault is of a temporary nature and is cleared before the reclosing devices complete their operations, the sectionalizer will reset to its normal position after the circuit is reenergized.

Sectionalizers are rated on continuous current-carrying capacity, minimum tripping and counting current, and maximum momentary fault current, as well as for maximum system voltage, load-break current, and impulse voltage or basic insulation level (BIL).

More than one sectionalizer can be connected in series with a reclosing device. The sectionalizer nearest the reclosing device can be set to operate after (say) three operations while the more remote one is set for (say) two such operations.

Sectionalizers are relatively low-cost devices; they are not required to interrupt fault current although fault current flows through them. They may be operated manually and are considered the same as load-break switches.

Reclosers

Reclosers are essentially circuit breakers of lower capacity, both as to normal current and interrupting duty. They are usually installed on major branches of distribution feeders in series with other sectionalizing devices; they perform the same function as repeater fuses connected in the circuit or circuit breakers at the substation.

Reclosers are designed to remain open, or "locked out," after a selected sequence of tripping operations. A fault will trip the recloser; if the fault is temporary in nature and no longer exists, the next tripping operation does not take place and the recloser returns to its normally closed position, ready for another incident. If the fault persists, the recloser will close and the operation will be repeated until the recloser locks out. The reclosers are usually set for three automatic reclosing operations before locking out; the first operation is usually "instantaneous," i.e., occurring as quickly as the breaker contacts can open with no time delay; the second and third operations have time delays inserted, that for the second tripping smaller than that for the third; a fourth tripping will result in the recloser's remaining open until it is automatically or manually restored to normal, ready for the next incident.

Reclosers can operate on one or more time-current characteristic curves, as shown in Fig. 4-16. The reclosing characteristics of the recloser for each operation are coordinated with those of the fuses at the coordinating points in the circuit and with those of the relays controlling the circuit breaker at the substation. These are illustrated in Fig. 4-17. The first and basic curve is a tripping setting representing the minimum clearing time of the recloser; the other curves are determined by deliberate time delays introduced by making minor changes in the hydraulic and mechanical linkage system.

Reclosers may be single-phase units or three-phase units. The latter usually consist of three single-phase units mechanically interlinked for a common lockout operation, and are installed in a common tank. Figure 4-18a and b are typical oscillograms of single-phase and three-phase faults interrupted by these units.

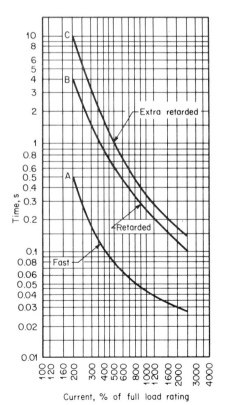

FIG. 4-16 Time-current curves of standard-duty reclosers. *(Courtesy McGraw Edison Co.)*

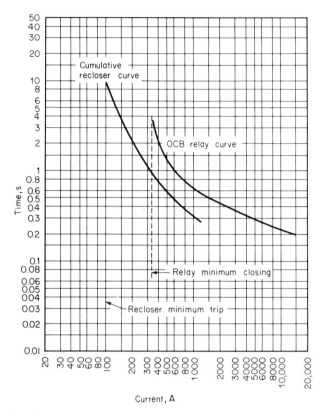

FIG. 4-17 Coordination of OCB relay and recloser. *(Courtesy McGraw Edison Co.)*

Circuit Breakers—Relays

Where the fault current is beyond the ability of a fuse or recloser to interrupt it safely, or where repeated operation within a short period of time makes it more economical, a circuit breaker is used. The ability of circuit breakers has been touched upon earlier; their time-current characteristics, however, are dependent on the protective relays associated with them and must be coordinated with those of down-line reclosers, fuses, and other protective devices.

Overcurrent Relays

Overcurrent relays close their contacts to actuate the circuit that causes the circuit breaker to open or close when the current flowing in them reaches a predetermined value.

Instantaneous Without time delay deliberately added, the relay will close its contacts "instantaneously," i.e., in a relatively short time, in the nature of 0.5 to

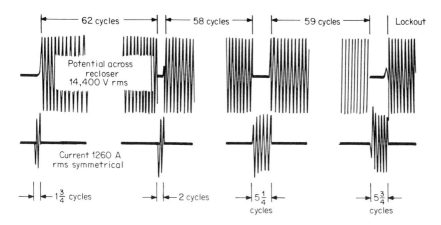

Interruption of 1260 A at 14,400 V, power factor of 38 percent, by B22 standard-duty recloser

A. First opening. Total clearing time, 1¾ cycles. Arcing time at contacts, ½ cycle. Reclosing time, 62 cycles.

B. Second opening. Total Clearing time, 2 cycles. Arcing time at contacts, ½ cycle. Reclosing time, 58 cycles.

C. Third opening. Total clearing time, 5¼ cycles. Arcing time at contacts, ½ cycle. Reclosing time, 59 cycles.

D. Fourth opening. Total clearing time, 5¾ cycles. Arcing time at contacts, ½ cycle. Recloser locks out.

(a)

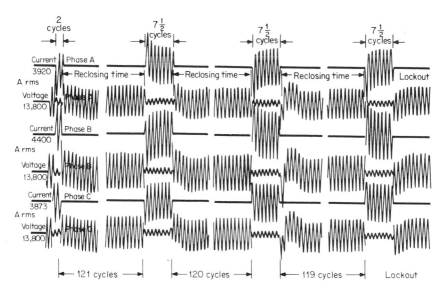

Interruption of three-phase fault current of 4000 A at 13,800 V, power factor of 12.5 percent, by C13 three-phase heavy- duty recloser.

A. First opening. Total clearing time, phase A, 1½ cycles; phase B, 1½ cycles; phase C, 2 cycles. Arcing time, all phases, approximately 1/2 cycle. Reclosing time, 121 cycles.

B. Second opening. Total clearing time, all three phases, 7½ cycles. Arcing time, all phases, approximately ½ cycle. Reclosing time, 120 cycles.

C. Third opening. Total clearing time, phases A and C, 7½ cycles; phase B, 8 cycles. Arcing time, all phases, approximately ½ cycle. Reclosing time, 119 cycles.

D. Fourth opening. Total clearing time, all three phases, 7½ cycles. Arcing time, all phases, approximately ½ cycle. Recloser locks out.

(b)

FIG. 4-18 (a) Typical oscillogram of recloser operation. (b) Typical oscillogram of three-phase fault interruption by heavy-duty recloser. *(Courtesy McGraw Edison Co.)*

perhaps 20 cycles. To prevent frequent operation of the breaker from transient, nonpersistent conditions, undesirably high settings may be applied to the relay. The time-current characteristic of this type of relay is shown in curve *a* in Fig. 4-19*a*.

Inverse Time The operation of the relay may be made to vary approximately inversely with the magnitude of the current. The current setting may be varied and time delay introduced by varying the restraint on the movable element of the relay; these are indicated in curves *b* and *c* in Fig. 4-19*a*. Greater selectivity between relays and fuses in the circuit may thus be obtained.

Definite Time A definite time delay can be introduced before the relay begins to operate, allowing greater selectivity to be achieved. This feature is often added to the inverse-time characteristic beyond a certain value of current after which the relay operation is completed after the fixed time delay. This inverse definite minimum time feature is employed in most overcurrent relay applications. It is shown in curve *d* of Fig. 4-19*a*, in which the flat portion of the characteristic results in only a small relay time increase for small values of fault current. Refer to Fig. 4-19*b*.

The distribution circuit may be sectionalized with reclosers, automatic sectionalizers, and fuses, at which points faults may be isolated without affecting

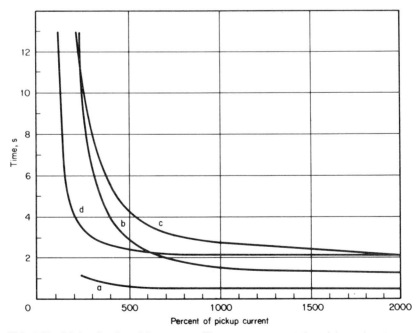

FIG. 4-19 (a) A collection of time curves. These are representative of the various types of time curves which are used on overcurrent relays.

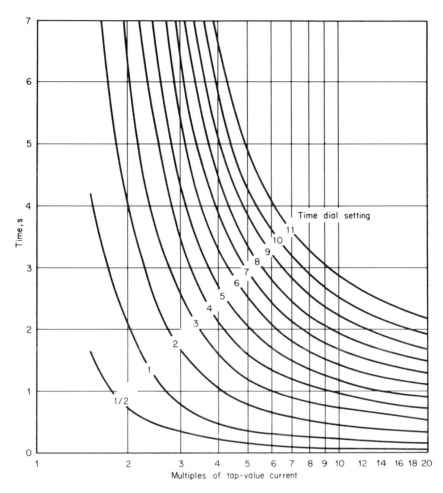

FIG. 4-19 (b) Type CO-8 overcurrent relay time-current curves, 50–60 cycles. (Courtesy Westinghouse Electric Co.)

the entire circuit; fuses are also provided on the primary side of distribution transformers. The definite time characteristic of the relay associated with the circuit breakers at the substation is coordinated with the characteristics of reclosers and fuses on the distribution circuit, as shown in Fig. 4-20.

Directional Relays

Directional relays are essentially overcurrent relays to which an element similar to a wattmeter is added, both sets of contacts being in series. The overcurrent element will operate to close its contacts regardless of the direction of flow of power in the line; the wattmeter element will tend to turn in one direction under

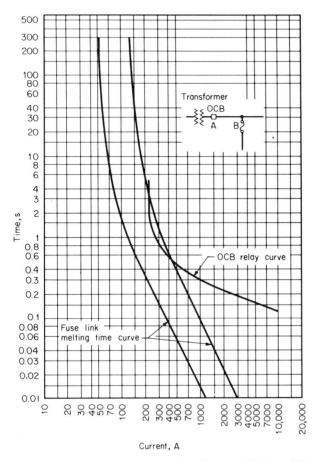

FIG. 4-20 OCB-fuse link coordination. *(Courtesy McGraw Edison Co.)*

normal flow of power and in the reverse direction when power flows in the opposite direction. Hence, both sets of contacts must be closed and power flowing in a given direction before the relay will operate. Both elements may be combined into one so that only a single set of contacts is required.

This type of relay is used in primary or secondary network operations to open the protectors to prevent current from the network from energizing the high side of the transformers and their supply feeder during contingencies. Refer to Fig. 4-21.

Differential Relays

Differential relays operate on the difference between the current entering the line or equipment being protected and the current leaving it. As long as the incoming current and the outgoing current are essentially equal, the relay will

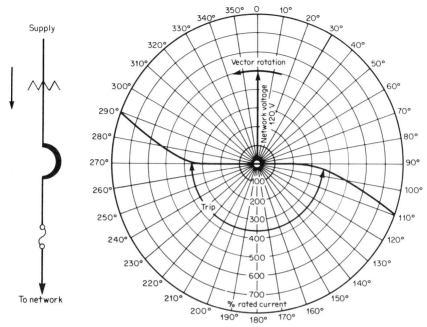

FIG. 4-21 Typical tripping characteristic of the CN-33 network master relay under balanced three-phase conditions. *(Courtesy Westinghouse Electric Co.)*

not operate. A fault within the line or equipment, however, will disturb this equilibrium, and the relay will operate to trip the supply circuit breaker or breakers on both sides of the line or equipment being protected. This type of relay is used to protect buses, transformers, and regulators at the substation. Since the voltages at which these operate may be high, current transformers installed on both sides of the equipment, with proper ratios in the case of transformers, supply the currents to the relay. Refer to Fig. 4-22.

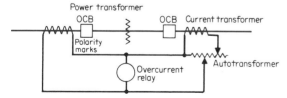

FIG. 4-22 One-line diagram of current-differential protection.

Surge or Lightning Arresters

The function of a surge or lightning arrester is to limit the voltage stresses on the insulation of the equipment being protected by permitting surges in voltage to drain to ground before damage occurs. The surges in voltage generally are caused by lightning (either by direct stroke or by induction from a nearby stroke) or by switching.

Arresters consist of two basic components: a spark gap and a nonlinear resistance element (for a valve type) or an expulsion chamber (for an expulsion type). When a surge occurs, the spark gap breaks down or sparks over, and permits current to flow through the resistance (or chamber) element to ground. Since the arrester at this point presents a low-impedance path, a large current, referred to as *60-cycle follow current*, flows through the arrester. The nonlinear resistance, at the higher voltages, will tend to restrict this current and eventually cause it to cease to flow; here, the magnitude of the follow current is independent of the system capacity. The expulsion chamber will confine the arc, build up pressures that eventually blow out the arc, and cause the follow current to cease to flow; here, the follow current is a function of the system capacity and the expulsion chamber must be suitably designed. After each such operation, the arrester must be capable of repeating this operating cycle.

Insulation Coordination It must be kept in mind that while the arrester is operating, the surge voltage is also "attacking" the insulation of the line or equipment it is protecting; the arrester, however, drains the high voltage to ground, reducing its magnitude, *before* sufficient time has elapsed to damage the insulation of the line or equipment.

Insulation characteristics, therefore, can be expressed as functions of voltage and the time it is impressed. This is usually shown as a volt-time curve, known

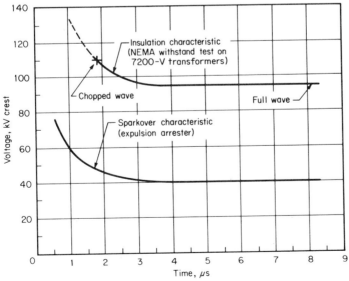

FIG. 4-23 Insulation coordination. *(Courtesy McGraw Edison Co.)*

as the *impulse level*, and represents the voltage and its duration the equipment can withstand.

The arrester also has a volt-time curve that indicates the voltage and time at which the spark gap begins to break down and permit the passage of the surge to ground.

The insulation characteristic of the line or equipment being protected must be at a higher voltage level than the volt-time characteristic of the arrester protecting it; indeed, a sufficient voltage differential must be provided to ensure safe and positive protection. Figure 4-23 illustrates typical curves and their relationship. While the impulse level of the line or equipment must be high enough that the arrester provides adequate protection, it should be as low as practical to hold down insulation costs.

Basic Insulation Level (BIL) The coordination of insulation requires the establishment of a minimum level above which are the components of a system and below which are the protected devices associated with those components. A joint committee of electrical engineers, utilities, and manufacturers adopted basic insulation levels which define the impulse voltages capable of being withstood by insulation of various insulation classes: "Basic impulse insulation levels are reference levels expressed in impulse crest voltages with a standard wave not longer than 1.5 by 40 microseconds. Apparatus insulation as demonstrated by suitable tests shall be equal to or greater than the basic insulation level."

The standard 1.5- by 40-μs wave selected simulates lightning surges, which are more prevalent than switching surges, and are more readily reproduced in the laboratory. The wave, shown in Fig. 4-24, reaches its maximum or crest value at 1.5 μs and at 40 μs reaches one-half its maximum value on the "wave tail." The steep rising portion of the wave is called the wave front, and the rate of rise (in kilowatts per microsecond) determines the slope or steepness of the wave front.

Recommended insulation levels for equipment of various voltage classes are listed in Table 4-4. These levels and the sparkover characteristic of arresters are determined by impulse tests.

The sparkover-time characteristics of the valve- and the expulsion-type arrester differ in that the characteristic curves for expulsion types are not as flat as those for the valve types: the two are shown in Fig. 4-25*a* and *b*, respectively.

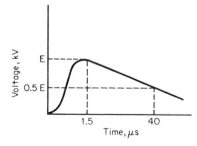

FIG. 4-24 1.5- by 40-μs voltage wave *(Courtesy McGraw Edison Co.)*

TABLE 4-4 TYPICAL BASIC INSULATION
LEVELS*

	Basic insulation level, kV (standard 1.5- × 40-µs wave)	
Voltage class, kV	Distribution class	Power class (station, transmission lines)
1.2	30	45
2.5	45	60
5.0	60	75
8.7	75	95
15	95	110
23	110	150
34.5	150	200
46	200	250
69	250	350

*For current industry recommended values, refer to the
latest revision of the National Electric Safety Code.

The flat curves of the valve types indicate their rapid response, especially to
lightning voltages even during steep wave-front surges. Although expulsion
arresters do not have sparkover voltages as low as the valve types, they do provide
adequate protection for distribution systems; consideration must be given, however, to the maximum fault current at the expulsion arrester and its effect on
coordination with overcurrent protection devices on the system. The curves are
based on a standard 1.5- by 40-µs test wave.

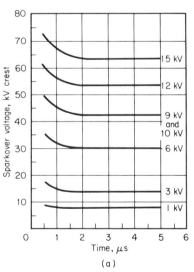
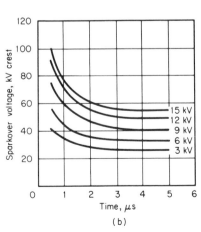

FIG. 4-25 (a) Sparkover characteristics of distribution valve arresters. (b) Sparkover characteristics of expulsion arresters. (Courtesy McGraw Edison Co.)

Arrester Connection   Arresters should be placed as close to the equipment to be protected, and the lengths of the connections to the line and to the ground should be kept as short, as possible. That is because these connections offer relatively high-impedance paths to voltage surges, so that large currents flowing through them could cause a voltage drop in them which, added to the surge voltage, could impose additional stress on the insulation of the equipment being protected. Moreover, on longer lines, such surges can be "reflected," essentially doubling the value of the surge voltage.

Short leads and minimum distance between the arrester and equipment protected are desirable for all arrester applications. Further, if the equipment being protected has a ground, that ground and the arrester ground should be interconnected to relieve any potential stress that may develop from the voltage drop across the ground impedance.

Arresters should be connected to the primary side of distribution transformers and to capacitors, underground risers, and other equipment; at certain points on long primary lines; and at reclosers in substations. One arrester should be connected to each phase. For station circuit breakers, transformers, outdoor regulators, and reclosers situated on primary lines, arresters should preferably be connected to both incoming and outgoing sides of such equipment. Voltage ratings of arresters should take cognizance of whether the systems are delta or wye, grounded or ungrounded, and of the voltage distortions resulting from an accidental ground on one phase.

FAULT-CURRENT CALCULATION

For the proper coordination of protective devices on a distribution system, it is essential that the magnitude be known of the fault current which they may be called upon to handle. For dc systems, the calculation is a relatively simple application of Ohm's law; for ac systems, the procedures are more complex, but for most problems, practical solutions permit simplified procedures.

For ac systems, four general types of faults can be considered: three phases short-circuited together (with or without a ground), phase to phase to ground, phase to phase, and single phase to ground.

The following simplified equations yield symmetric current values that are sufficiently precise for purposes of coordination.

1. Three-phase fault current:

$$I_{3\phi} = \frac{0.58V}{Z_A + Z_B} \quad \text{A}$$

2. Phase-to-phase-to-ground fault current:

$$I_{\phi-\phi-G} = \frac{V\sqrt{(Z_A + Z_B)^2 + (Z_A + Z_C)^2 + (Z_A + Z_B)(Z_A + Z_C)}}{(Z_A + Z_B)\,[2(Z_A + Z_C) + (Z_A + Z_B)]} \quad \text{A}$$

3. Phase-to-phase fault current:

$$I_{\phi-\phi} = 0.87 I_{3\phi} \quad \text{A}$$

4. Single-phase-to-ground fault current:

$$I_{\phi-G} = \frac{1.73V}{3Z_A + 2Z_B + Z_C} \quad A$$

where V = normal phase-to-phase voltage of feeder
 Z_A = impedance of substation transformer, Ω

Z_B and Z_C = impedances of feeder, Ω

These equations are for 60-cycle overhead radial distribution feeders and assume that an infinite source capacity is capable of maintaining normal voltage at the high side with the fault on the low side; that the impedances making up the total impedance of the faulted circuit are added arithmetically (impedances must be at the same voltage base); that transformers have 6 percent impedance (but other impedances can be calculated on their percentage ratio to 6 percent; and that conventional spacing exists between conductors on poles (though large variations cause only small changes in impedance values that can be disregarded without appreciable error).

EXAMPLE 4-4

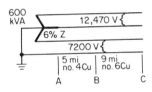

From manufacturers' tables, $Z_A = 15.57\ \Omega$ and $Z_B = 0$. For a fault at point A, at the substation,

1. For a three-phase fault:

$$I_{3\phi} = \frac{0.58 \times 12,470}{15.57 + 0} = 465\ A$$

2. For a phase-to-phase-to-ground fault,

$$I_{\phi-\phi-G} = \frac{12,470\sqrt{15.57^2 + 15.57^2 + 15.57^2}}{15.57[2(15.57) + 15.57]}$$

$$= \frac{12,470\ \sqrt{3}}{3(15.57)} = 463\ A$$

3. For a phase-to-phase fault,

$$I_{\phi-\phi} = 0.87 \times I_{3\phi}$$

$$= 0.87 \times 465 = 404\ A$$

4. For a phase-to-gound fault,

$$I_{\phi-G} = \frac{1.73 \times 12,470}{3 \times 15.57} = 463 \text{ A}$$

For a fault at point B, 5 mi from the substation,

1. For a three-phase fault, with $Z_B = 7.85$ (from manufacturers' tables),

$$I_{3\phi} = \frac{0.58 \times 12,470}{15.57 + 7.85} = 309 \text{ A}$$

2. For a phase-to-phase-to-ground fault, with $Z_C = 15.8 \ \Omega$ (from manufacturers' tables),

$$I_{\phi-\phi-G} = \frac{12,470\sqrt{23.42^2 + 31.37^2 + (23.42)(31.37)}}{23.42[2(31.37) + 23.42]}$$

$$= \frac{12,470 \times 47.5}{2017.9} = 293 \text{ A}$$

3. For a phase-to-phase fault,

$$I_{\phi-\phi} = 0.87 \times 309 = 269 \text{ A}$$

4. For a phase-to-ground fault,

$$I_{\phi-G} = \frac{1.73 \times 12,470}{3(15.57) + 2(7.86) + 15.8} = 276 \text{ A}$$

For a fault at point C, 14 mi from the substation,

1. For a three-phase fault, with $Z_B = 7.85 + 20.60 = 28.45$ (from manufacturers' tables),

$$I_{3\phi} = \frac{0.58 \times 12,470}{15.57 + 28.45} = 164 \text{ A}$$

2. For a phase-to-phase-to-ground fault, with $Z_C = 15.80 + 35.28 = 51.08$ (from table), $Z_A + Z_B = 15.57 + 28.45 = 44.02 \ \Omega$, and $Z_A + Z_C = 15.57 + 51.08 = 66.65 \ \Omega$,

$$I_{\phi-\phi-G} = \frac{12,470\sqrt{44.02^2 + 66.65^2 + (44.02)(66.65)}}{44.02[2(66.65) + 44.02]}$$

$$= \frac{12,470 \sqrt{9316}}{7810} = 154 \text{ A}$$

3. For a phase-to-phase fault,

$$I_{\phi-\phi} = 0.87 \times 164 = 143 \text{ A}$$

4. For a phase-to-ground fault,

$$I_{\phi-G} = \frac{1.73 \times 12,470}{3(15.57) + 2(28.45) + 51.1} = 139 \text{ A}$$

EXAMPLE 4-5

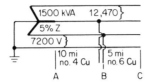

Since the branch in this case has two phases and a neutral, no three-phase fault can occur. $V = 12,470$ V. The impedance values, based on manufacturers' tables, are as follows. $Z_A = 6.28$ Ω for 6 percent impedance. For 5 percent impedance, $Z_A = 5/6 \times 6.28 = 5.2$ Ω. $Z_B = 15.7 + 11.45 = 27.15$ Ω. $Z_C = 31.6 + 19.60 = 51.20$ Ω. $Z_A + Z_B = 32.35$ Ω, and $Z_A + Z_C = 56.40$ Ω.

For a fault at point C,

1. For a phase-to-phase-to-ground fault,

$$I_{\phi-\phi-G} = \frac{12,470\sqrt{32.35^2 + 56.40^2 + (32.35)(56.40)}}{32.35[2(56.40) + 32.35]}$$

$$= \frac{12,470\sqrt{6052}}{4696} = 207 \text{ A}$$

2. For a phase-to-phase fault, since the three-phase fault current would be 224 A,

$$I_{\phi-\phi} = 0.87 \times 224$$

$$= 195 \text{ A}$$

3. For a phase-to-ground fault,

$$I_{\phi-G} = \frac{1.73 \times 12,470}{3(5.2) + 2(27.15) + 51.20} = 178 \text{ A}$$

EXAMPLE 4-6

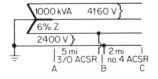

Since the branch here is single-phase, only a phase-to-ground fault can occur. $V = 4160$ V. From manufacturers' tables, $Z_A = 1.037$ Ω, $Z_B = 1.78 + 4.52 = 6.3$ Ω, and $Z_C = 4.16 + 6.90 = 11.1$ Ω.

For a fault at point C,

$$I_{\phi-G} = \frac{1.73 \times 4160}{3(1.037) + 2(6.3) + 11.1}$$

$$= 268 \text{ A}$$

EXAMPLE 4-7

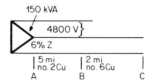

150 kVA
4800 V
6% Z
5 mi no. 2 Cu
2 mi no. 6 Cu
A   B   C

With two branch lines, only a phase-to-phase fault can occur. $V = 4800$ V. From manufacturers' tables, $Z_A = 9.22 \ \Omega$ and $Z_B = 5.55 + 4.52 = 10.07 \ \Omega$.

For a fault at point C,

$$I_{\phi-\phi} = \frac{0.87 \times 0.58 \times 4800}{9.22 + 10.07}$$

$$= 125 \text{ A}$$

Actual Fault Current

The actual fault current may be broken into two components, ac and dc. The simplified equations give values of the ac, or steady-state, component, which remains constant throughout the duration of the fault. The value of this current follows Ohm's law and is equal to the voltage divided by the impedance of the circuit from the source to the point of fault.

As shown in Fig. 4-26a, the ac component is symmetric about its axis and is known as the symmetric current. The dc, or transient, component occurs at the instant of the fault and drops rapidly to zero. Its magnitude depends on the time the fault is initiated; on a system carrying load, its maximum value is equal to the ac component at the beginning of the fault less the load current at that time. Standard measurement of the two components is taken at one-half cycle after the start of the fault. The total (root mean square) fault current is the square root of the sum of the squares of the two components; this composite rms current is not symmetric, and is known as the asymmetric current.

The magnitude of the total fault current during the transient period, therefore, depends on the type of fault and the time of its initiation. After the transient period, the magnitude of the fault current depends only on the type of fault.

The time of fault initiation is measured angularly along the voltage wave, i.e., as a number of degrees from a known point, such as peak voltage or voltage zero. Since there is usually a phase angle between voltage and current, a fault will occur at a different point on the current wave.

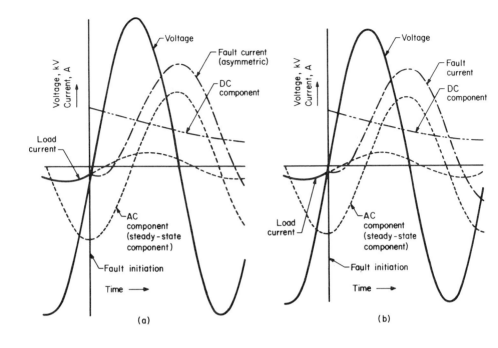

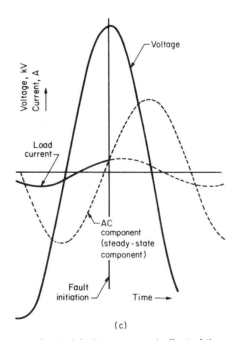

FIG. 4-26 Components of actual fault current and effect of time of fault initiation upon magnitude of rms fault current. Fault initiation at (a) ac component peak, (b) voltage zero, and (c) instant when ac component and load current are equal (no dc component exists). *(Courtesy McGraw Edison Co.)*

Figure 4-26a through c shows the effect of the time of fault initiation. If it occurs at the ac component peak, maximum transient current occurs. This dc component is equal to the difference between the instantaneous values of the ac fault current and the ac load current. Maximum transient current occurs at the ac component peak, but maximum rms fault current occurs at voltage zero. The transient component is zero when the fault occurs at the time the instantaneous values of the load current and steady-state current are equal. Between the maximum and zero transient points, a transient component will exist in which the rms fault current is less than the rms current when the fault occurs at voltage zero. Since protective devices operate during transient periods, they are designed to interrupt the maximum possible fault current.

Factors of Asymmetry

As the reactance to resistance ratio (X/R) and the power factor of a faulted circuit change, the magnitude of the asymmetric current will vary with respect to the symmetric current. As this ratio increases and the power factor decreases, the asymmetric current will increase. A ratio, known as the factor of symmetry, of the asymmetric to the symmetric current can be found for different X/R ratios and corresponding power factors. As shown in Fig. 4-27, a specific factor exists for each X/R ratio and corresponding power factor.

When the symmetric current and the X/R ratio are known, the maximum asymmetric current can be found. This is useful in the design of fuse cutouts which are rated asymmetrically.

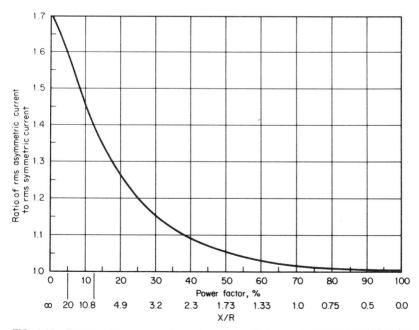

FIG. 4-27 Factors of asymmetry for various power factor percentages and X/R ratios. *(Courtesy McGraw Edison Co.)*

Symmetrical Components

The simplified equations for calculating fault currents are derived from the symmetrical equations:

1. $I_{3\phi} = \dfrac{E_{\phi-N}}{Z_1}$

2. $I_{\phi-\phi-G} = \dfrac{E_{\phi-N}a^2\,(Z_0 + Z_2) - aZ_0 - Z_2}{Z_1Z_0 + Z_1Z_2 + Z_0Z_2}$

3. $I_{\phi-\phi} = \dfrac{3E_{\phi-N}}{Z_1 + Z_2}$

4. $I_{\phi-G} = \dfrac{3E_{\phi-N}}{Z_1 + Z_2 + Z_0}$

where Z_1 is the positive-sequence impedance, Z_2 is the negative-sequence component, Z_0 is the zero-sequence component, and each one represents the summation of like components throughout the faulted circuit. Also, a is the unit vector $e^{j120} = -0.5 + j0.866$, that is, an operator which indicates that the vector to which it is attached has been rotated through 120° in a positive or clockwise direction; a^2 indicates rotation through 240°. Further, theoretically, the three impedance components for a transformer are equal; that is, $Z_1 = Z_2 = Z_0$; also the positive- and negative-sequence components for conductors are equal; that is, $Z_1 = Z_2$. The three transformer components can be set equal to Z_A, the two conductor components equal to Z_B, and the conductor zero-sequence component equal to Z_C.

1. Substituting in the basic equations above:

$$E_{\phi-N} = \dfrac{E_{\phi-\phi}}{\sqrt{3}}$$

$Z_1 = Z_A$ (positive transformer component)

$\quad\ + Z_B$ (positive conductor component)

$$I_{3\phi} = \dfrac{E_{\phi-\phi}/\sqrt{3}}{Z_A + Z_B} = \dfrac{0.58E_{\phi-\phi}}{Z_A + Z_B}$$

2. Eliminating the factor a by trigonometry,

$$
\begin{aligned}
I_{\phi-\phi-G} &= \dfrac{\sqrt{3}(E_{\phi-\phi}/\sqrt{3})\,\sqrt{Z_0^2 + Z_0Z_2 + Z_2^2}}{Z_1Z_0 + Z_1Z_2 + Z_0Z_2} \\[2ex]
&= \dfrac{E_{\phi-\phi}\sqrt{(Z_A + Z_C)^2 + (Z_A + Z_C)(Z_A + Z_B) + (Z_A + Z_B)^2}}{(Z_A + Z_B)(Z_A + Z_C) + (Z_A + Z_B)(Z_A + Z_B) + (Z_A + Z_C)(Z_A + Z_B)} \\[2ex]
&= \dfrac{E_{\phi-\phi}\sqrt{(Z_A + Z_B)^2 + (Z_A + Z_C)^2 + (Z_A + Z_B)(Z_A + Z_C)}}{(Z_A + Z_B)[2(Z_A + Z_C) + (Z_A + Z_B)]}
\end{aligned}
$$

3. $I_{\phi-\phi} = \dfrac{3(E_{\phi-\phi}/\sqrt{3})}{(Z_A + Z_B) + (Z_A + Z_B)} = \dfrac{\sqrt{3}E_{\phi-\phi}}{2(Z_A + Z_B)}$

$= \dfrac{1.73}{2}\dfrac{E_{\phi-\phi}}{Z_A + Z_B} = 0.87I_{3\phi}$

4. $I_{\phi-G} = \dfrac{3(E_{\phi-\phi}/\sqrt{3})}{(Z_A + Z_B) + (Z_A + Z_B) + (Z_A + Z_C)}$

$= \dfrac{1.73E_{\phi-\phi}}{3Z_A + 2Z_B + Z_C}$

EXAMPLE 4-8   System coordination (refer to Fig. 4-28)

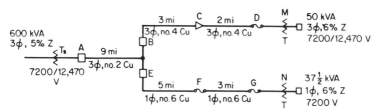

FIG. 4-28   One-line diagram of illustrative typical system.

The impedance of transformer T_s is 15.57 Ω (from the manufacturer's table for 6% Z); Z_A for a 5% Z transformer is 5/6 of 15.57 or 12.98 Ω. The impedance of the lines (from manufacturers' tables):

Line section	Wire size	Length, mi	Impedance, Ω*	
			Z_B	Z_C
3φ AB	no. 2 Cu	9	10.17	23.23
3φ BC	no. 4 Cu	3	4.71	9.48
3φ CD	no. 4 Cu	2	3.14	6.32
1φ EF	no. 6 Cu	5	11.45	19.60
1φ FG	no. 6 Cu	3	6.87	11.76

*Z_B is the positive- or negative-sequence value. Z_C is the zero-sequence value.

Total impedance from substation to load side of protective device:

Protective device	Line voltage	Impedance, Ω			
				Lines*	
		Transformer Z_A	Section	Z_B	Z_C
A	12,470	12.98	—	0.00	0.00
B	12,470	12.98	AB	10.17	23.13
C	12,470	12.98	AC	14.88	32.61
D	12,470	12.98	AD	18.02	38.93
E	12,470	12.98	AE	10.17	23.13
F	12,470	12.98	AF	21.62	42.73
G	12,470	12.98	AG	28.49	54.49

*Z_B is the positive- (or negative-) sequence value; Z_C is the zero-sequence value.

Available fault currents on load side of protective device and selection:

Protective device	Fault currents, A				Protective device selected (EEI-NEMA standards)
	3ϕ	ϕ-ϕ-G	ϕ-ϕ	ϕ-G	
A	557	557	485	557	Circuit breaker and overcurrent relay, or 35-A standard-duty recloser
B	312	293	272	262	15-A standard-duty recloser
C	259	241	225	212	28-A 3-shot sectionalizer*
D	233	217	203	189	8 (A) T link fuse
E	—	—	—	262	15-A 1ϕ standard-duty recloser
F	—	—	—	173	10 (A) K link fuse
G	—	—	—	143	6 (A) K link fuse

*Recloser B has one fast operation when a fault occurs on the load side of D; the sectionalizer counts one operation, but the link fuse at D does not melt. During the second time-delayed operation of recloser B, the link fuse at D melts and clears the fault. The sectionalizer counts this as the second operation; with the fault removed, both recloser and sectionalizer reset and the line is energized to D.

STREET LIGHTING

Street lighting may be supplied from either series or multiple circuits. Series circuits and lamps have been extensively used in the past and many such systems are still in operation. Multiple supply, however, has been used almost exclusively in later installations, especially with the development of economical light-sensitive control devices, which has made practical the connection of such lighting to the same secondary mains that also supply other types of customers.

Multiple Circuits

Multiple street lighting is usually served from secondary mains at normal utilization voltages. The secondary may be a separate circuit serving only street lighting loads, or part of the secondary circuit supplying other customers. The former may have controls exclusive to the circuit, while the latter may have individual controls for individual lamps or groups of lamps. As mentioned earlier, such circuits may be controlled from a variety of relays, some actuated from series street lighting circuits. Photocells, activated by the light intensity of the ambient, and connected between the secondary supply and the streetlight, have proven economical and desirable, as they simplify controls, eliminating additional wires and improving the appearance of overhead lines.

The principal advantage of multiple supply is the use of the secondary distribution systems, with only minor effects on transformer load and voltage conditions; additional advantages include greater safety because of the low voltages involved as well as lower handling and installation costs, as the equipment used is essentially that in use on distribution circuits. In most instances, separate lightning or surge arresters are not required.

Series Circuits

As the name denotes, lamps in the series type of circuit are connected in series and supplied from a constant-current transformer; the usual current output rating is 6.6 A, with the outgoing voltage varying with the load connected. The light output for different-sized lamps at this fixed current rating depends on the length of the filament; for larger-size lamps operating at 15 or 20 A, the filaments are also of heavier cross section in order to carry the greater current safely. Series lamps often are rated in *lumens,* the unit of light flux, while the wattage may vary with the size of the lamp. Lamp efficiency is expressed in lumens per watt, with 600-lm lamps operating at 6.6 A requiring about 43 W while 6000-lm lamps require approximately 320 W. Light efficiency is further increased with reflectors and refractors.

Advantages of series circuits include high efficiency for a widely distributed load, the ability to use a high-voltage and low-current supply that permits long lengths of relatively small-size conductors to be used with low loss and small voltage drop, and the ability to keep the variation in light intensity at a minimum because of the constant current value at each lamp. Disadvantages mainly center around the need for special transformers and control devices, as well as the need for separate lightning or surge arresters at the transformer, the switches, and certain points along an extended circuit.

Constant-Current Transformers

To obtain a constant current output from an essentially constant voltage supply, the two coils of the transformer are so arranged that the distance between them may be varied. The primary coil, receiving power at a constant voltage, is fixed in position. The secondary coil is movable along the common core; its position

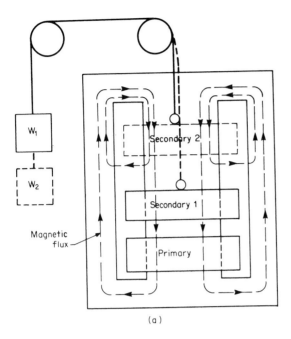

(a)

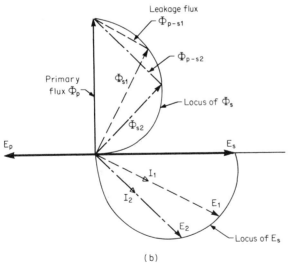

(b)

FIG. 4-29 (a) Movable-coil constant-current transformer. (b) Vector diagram of relation between primary and secondary voltage, current, and magnetic flux.

depends on the load. The secondary coil is balanced by a weight that reacts proportionally with the repulsion forces to maintain its output at the rated value.

When the two coils are close together, the magnetic field or flux produced by the primary interlinks maximally with the secondary coil, inducing a maximum secondary voltage. The coils act as magnets and tend to repel each other. As the load on the secondary decreases, the current in the secondary will tend to increase, increasing the repulsion between the coils. The secondary coil is raised, decreasing the magnetic field or flux linking the two coils, thereby decreasing the secondary voltage, and the current is dropped to its rated value. Refer to Fig. 4-29a and b.

The majority of series street lighting circuits are operated without grounds. When an accidental ground occurs, the circuit is generally not affected. When a second ground occurs, the lamps between the two grounds will go out or burn dimly, giving ready indication of the location of the grounds. In some instances, however, an intentional ground is placed at mid-circuit. This not only reduces the normal stresses on the insulation of the several units in the circuit to one-half the voltage of the constant-current transformer, but serves to help locate the point of accidental ground, as indicated above.

Constant-current transformers may be installed at the substation, in underground vaults, or mounted on poles.

Series Lamps

Series lamps are usually rated at 6.6 A. Larger lamps, generally 4000 lm or larger, are supplied at 15 or 20 A through transformers that supply an individual lamp (referred to as IL transformers) or transformers that supply several lamps (referred to as SL transformers).

Lamp failure in a series circuit causes a momentary opening of the circuit and a high voltage at the terminals of the lamp. A *film disk cutout*, connected between the terminals of the lamp, normally acting as an insulator, breaks down

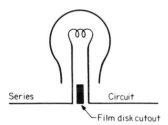

FIG. 4-30 Film disk cutout schematic diagram.

and short-circuits the failed lamp. Refer to Fig. 4-30. Thus circuit continuity and service to the rest of the circuit are restored. The slight reduction in load causes the constant-current transformer to reduce the total voltage on the circuit, maintaining the current at the rated constant value.

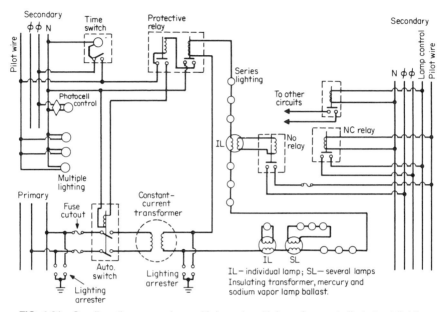

FIG. 4-31 One-line diagram series-multiple and multiple-series controlled street lighting.

Control Devices

Series circuits may be controlled from the substation manually or through time switches that automatically turn them on and off at predetermined times; one such switch operates in accordance with the hours of sunset and sunrise, and is called an *astronomical time switch*.

Control circuits employing a variety of relays and schemes that "cascade," one control circuit controlling one or more additional control circuits, are shown in Fig. 4-31. Note that series circuits can be controlled from multiple circuits supplying only street lights, or from light-sensitive relays supplied from multiple circuits.

PRACTICAL BASIS OF DESIGN

In distribution system studies, particularly where growth and expansion are contemplated, several plans are usually taken under consideration. For purposes of comparison, design voltage-drop limits are determined beforehand, and these same limits are used in all of the studies.

A peak-load voltage profile is developed and plotted in terms of equivalent secondary voltages, as shown in Fig. 4-32. Comparable layouts are made to provide adequate voltages and suitable equipment loadings with the design loads. A light-load voltage profile is assumed to fall within permissible limits, although regulator changes at the substation may be relied upon to maintain maximum permissible voltage at the first customer.

The utilization voltage spread of 15 V, the estimated service and customer wiring voltage drops, and contact-making-voltmeter (voltage-regulating relay) bandwidth setting are derived from industry standards accepted for the design of utilization devices. Reference to Fig. 4-32 shows that there is a spread of 10 V to "spend" in the design of the distribution system between the primary of the first transformer and the end of the secondary of the last transformer on the circuit. In general, industry experience indicates that a feasible and economical allocation of this 10 V is to use a voltage drop of 3 to 4 V in the primary between the first and last transformer and 6 to 7 V in the transformer-secondary combination.

For example, an economical means of providing for load growth, particularly where obtaining substation sites is a problem, is the conversion of 2400/4160 V distribution systems to 7620/13,200 V operation. Conversion may be most easily accomplished by first converting the areas most remote from existing 4160-V substations. The 4160-V substation facilities can thus be kept operating at approximately their capacity although the area served will become smaller as the load density increases. All 4160-V equipment, including distribution transformers, can be used continuously and economic loss from premature retirement of such equipment can be avoided. New equipment for 13,200-V service will take care of load growth on the 13,200 V system.

In comparing an existing 4160-V system with a proposed 13,200-V system, voltage drops and energy losses are calculated, and circuits are designed to keep voltage drops within the limits expressed above—and shown in Fig. 4-32—and to keep energy losses to a minimum.

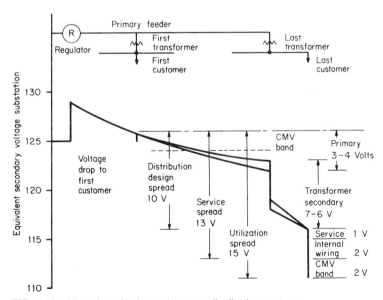

FIG. 4-32 Allocation of voltage drops on distribution systems.

TABLE 4-5A UNIT VOLTAGE DROPS FOR PRESENT 4160-V SYSTEM
(10 kVA, 100 ft, *L/N* voltage drop, 2360/4080 receiver volts)

Wire size	R(25°C)	X(35-in spacing)	Power factor 0.85			Power factor 0.90		
			R cos θ + X sin θ	3φ unit voltage drops	1φ unit voltage drops	R cos θ + X sin θ	3φ unit voltage drops	1φ unit voltage drops
8	0.657	0.151	0.638	0.901	5.40	0.657	0.929	5.57
6	0.413	0.145	0.428	0.605	3.63	0.435	0.615	3.69
4	0.260	0.140	0.295	0.417	2.50	0.295	0.417	2.50
3	0.209	0.135	0.249	0.352	2.11	0.247	0.349	2.09
2	0.167	0.133	0.212	0.300	1.80	0.208	0.294	1.76
1/0	0.105	0.128	0.157	0.222	1.33	0.150	0.212	1.27
2/0	0.0833	0.125	0.138	0.195	1.17	0.130	0.184	1.10
3/0	0.0663	0.120	0.120	0.170	1.02	0.112	0.158	0.950
4/0	0.0527	0.119	0.107	0.151	0.906	0.0994	0.140	0.841
3c, 1/0	0.105	0.031	0.106	0.150	0.899	0.108	0.152	0.915
3c, 300,000 cmil	0.0372	0.0270	0.0458	0.0647	0.388	0.0453	0.0640	0.384
2c,6	0.413	0.034	0.370	—	3.13	0.387	—	3.28

Note: For the two phase wires and the neutral of a three-phase system, use 0.5 times the single-phase unit voltage drops. For the two-phase 4-wire system, use 0.5 times the single-phase unit voltage drops. The above voltage drops are line-to-neutral for a wye-connected system.

TABLE 4-5B UNIT VOLTAGE DROPS: CALCULATIONS FOR PROPOSED CONDITIONS
(10 kVa, 1000 ft, *L/N* voltage drop, 122-V equivalent receiver, 0.90 power factor)

Wire size	R(25°C)	X (54-in spacing)	R cos θ + X sin θ	4160-V system		13,200-V system	
				3φ unit voltage drops	*1φ unit voltage drops*	*3φ unit voltage drops*	*1φ unit voltage drops*
8	0.657	0.160	0.661	0.903	5.42	0.284	1.71
6	0.413	0.155	0.440	0.600	3.60	0.189	1.14
4	0.260	0.150	0.300	0.410	2.46	0.129	0.774
3	0.209	0.145	0.251	0.343	2.06	0.108	0.648
2	0.167	0.143	0.212	0.290	1.74	0.0912	0.548
1/0	0.105	0.138	0.155	0.212	1.27	0.0666	0.400
2/0	0.0833	0.135	0.134	0.183	1.10	0.0576	0.346
3/0	0.0663	0.130	0.116	0.158	0.950	0.0498	0.299
4/0	0.0527	0.128	0.103	0.141	0.845	0.0443	0.266
3c, 1/0	0.105	0.031	0.108	0.148	—	0.0463	—
3c, 300,000 cmil	0.0372	0.0270	0.0453	0.0619	—	0.0195	—
2c, 6	0.413	0.034	0.387	—	3.17	—	0.998

Note: The above voltage drops are line-to-neutral values.

The computation of voltage drops and losses on an extensive distribution circuit is a rather long and tedious process. A method for making calculations easily and providing a record of basic data employs a system of "unit voltage drops"; a self-explanatory data form also indicates the procedure for making the computations.

For convenience, the unit selected for the voltage-drop calculations may be 10,000 kVA·ft, that is, 10 kVA at a distance of 1000 ft or 100 kVA at a distance of 100 ft, etc. Power factors and equivalent spacings used for calculations are indicated in Tables 4-5A and 4-5B for the two voltages under consideration. Circuit loads are assumed distributed in proportion to the connected transformers except where relatively large individual loads are known, in which case equivalent transformer capacity is estimated on the basis of known demands.

Included with the calculations is a sketch of the existing and proposed circuit with station numbers for each branch, transformer or group of transformers, or wire size change; these correspond to numbers on the calculation sheet. Factors for each wire size to convert from percent voltage drop to percent kilowatt loss are shown in Table 4-5C; losses in some smaller branches are

TABLE 4-5C RELATION OF PERCENT VOLTAGE DROP AND PERCENT KILOWATT LOSS

3 phase, unit length, load concentrated at end:

$$\text{Factor} = R \cos \theta + X \sin \theta; \quad \theta = \text{power factor angle}$$

$$\frac{\text{kVA} \times \text{factor}}{(\text{kV})^2 \times 10} = \% \text{ voltage drop} = \% V$$

$$\frac{(\text{kVA})^2 \times R}{(\text{kV})^2 \times 10 \times \text{kVA} \times \text{PF}} = \frac{\text{kVA} \times R}{(\text{kV})^2 \times \text{PF} \times 10} = \% \text{ kW loss} = \% L$$

where PF = power factor and R = resistance of one conductor.

$$\frac{\% L}{\% V} = \frac{\text{kVA} \times R}{(\text{kV})^2 \times \text{PF} \times 10} \qquad \frac{9(\text{kV})^2 \times 10}{\text{kVA} \times \text{factor}} = \frac{R}{\text{PF} \times \text{factor}}$$

For single-phase, the same ratio applies.

Conductor size	R (25°C)	Factor (R cos θ + X sin θ)	% L/% V
6	0.413	0.440	1.04
3	0.209	0.251	0.925
2	0.167	0.212	0.875
1/0	0.105	0.155	0.752
2/0	0.0833	0.134	0.680
4/0	0.0527	0.103	0.569

Assumed power factor = 0.90.

TABLE 4-5D SAMPLE PRIMARY CIRCUIT DATA AND CALCULATIONS

Circuit No. x, substation ABC, nominal circuit voltage 13,200, lamp voltage 120. Average delivered primary voltage 7750 (line-to-neutral for wye circuits), load condition peak.

1. Total transformer kVA:	1614
2. Maximum circuit load:	2760 kVA
3. Capacity factor (item 2 ÷ item 1):	1.71
4. Basic voltage drop, sub to last transformer:	193.5
5. Basic voltage drop, sub to first transformer:	104.1
6. Basic voltage drop, first to last transformer:	89.4
7. Calculated voltage drop, first to last transformer (item 3 × item 6):	152.9
8. Percent voltage drop, first to last transformer (item 7 ÷ average delivered primary voltage, × 100):	1.9
9. Percent voltage drop, sub to first transformer (item 5 ÷ average delivered primary voltage, × 100):	2.3
10. Percent voltage drop, sub to last transformer:	4.2

	Distance, ft			Transformer kVA		$\dfrac{kVA \cdot ft}{10,000} = \dfrac{B \times F}{10,000}$	Unit voltage drop	Basic voltage drop			
Station	Between stations	From sub- station	Conductor size	At station	Subtotal from last station			From previous station (G × H)	From sub- station (subtotal col. I)		Percent kV loss
A	B	C	D	E	F	G	H	I	J		K
1	11,200	11,200	3ϕ, 2/0	115	1614	1807.7	0.0576	104.1	104.1		1.56
2	900	12,100	3ϕ, 1/0	205	1499	134.9	0.0666	8.9	113.0		
3A	1,600	13,700	3ϕ, 1/0	376	1294	207.0	0.0666	13.8	126.8		
4	200	13,900	3ϕ, 1/0	8	918	18.4	0.0666	1.2	128.0		
5	600	14,500	3ϕ, 1/0	50	810	48.6	0.0666	3.2	131.2		
6	1,200	15,700	3ϕ, 1/0	68	860	103.2	0.0666	6.9	138.1		0.56
7	1,700	17,400	3ϕ, 2	15	792	134.6	0.0912	12.3	150.4		
8	300	17,700	3ϕ, 2	38	777	23.3	0.0912	2.1	152.5		
9B	500	18,200	3ϕ, 2	118	739	36.9	0.0912	3.4	155.9		
10C	400	18,600	3ϕ, 2	275	621	24.8	0.0912	2.3	158.2		0.38

TABLE 4-5D SAMPLE PRIMARY CIRCUIT DATA AND CALCULATIONS *(Continued)*

| | Distance, ft | | | Transformer kVA | | $\frac{kVA \cdot ft}{10000} = \frac{B \times F}{10000}$ | | Basic voltage drop | | |
| Station | Between stations | From sub-station | Conductor size | At station | Subtotal from last station | | Unit voltage drop | From previous station $(G \times H)$ | From sub-station (subtotal col. I) | Percent kV loss |
A	B	C	D	E	F	G	H	I	J	K
11	200	19,200	3φ, 6	15	346	6.9	0.189	1.3	159.5	
12	400	19,200	3φ, 6	8	331	13.2	0.189	2.5	162.0	
13	200	19,400	3φ, 6	15	323	6.5	0.189	1.2	163.2	
14	500	19,900	3φ, 6	43	308	15.4	0.189	2.9	166.1	
15	400	20,300	3φ, 6	53	265	10.6	0.189	2.0	168.1	
16	400	20,700	3φ, 6	40	212	8.5	0.189	1.6	169.7	
17	400	21,100	3φ, 6	15	172	6.9	0.189	1.3	171.0	
18	1,000	22,100	3φ, 6	10	157	15.7	0.189	2.9	173.9	
19	400	22,500	3φ, 6	5	147	5.9	0.189	1.1	175.0	
20	1,300	23,800	3φ, 6	13	142	18.5	0.189	3.5	178.5	
21	900	24,700	3φ, 6	26	129	11.6	0.189	2.2	180.7	
22	1,000	25,700	3φ, 2	3	103	10.3	0.0912	0.9	181.6	
23	400	26,100	3φ, 2	3	100	4.0	0.0912	0.4	182.0	
24	300	26,400	3φ, 2	66	97	2.9	0.0912	0.3	182.3	
25	1,100	27,500	1φ, 6	5	31	3.4	1.14	3.9	186.2	
26	1,200	28,700	1φ, 6	8	26	3.1	1.14	3.5	189.7	
27	1,400	30,100	1φ, 6	10	18	2.5	1.14	2.8	192.5	
28	1,100	31,200	1φ, 6	8	8	0.9	1.14	1.0	193.5	0.81
3A	—	13,700	—	—	376	—	—	—	126.8	
A1	300	14,000	3φ, 6	101	376	11.3	0.189	2.1	128.9	
A2	600	14,600	3φ, 6	146	275	16.5	0.189	3.1	132.0	
A3	400	15,000	3φ, 6	10	129	5.2	0.189	0.9	132.9	
A4	1,000	16,000	3φ, 6	8	119	11.9	0.189	2.2	135.1	

A5	400	16,400	2φ, 6	15	111	4.4	0.57	2.5	137.6
A6	200	16,600	2φ, 6	18	96	1.9	0.57	1.1	138.7
A7	1,000	17,600	2φ, 6	68	78	7.8	0.57	4.4	143.1
A8	1,500	19,100	1φ, 6	5	10	1.5	1.14	1.7	144.8
A9	300	19,400	1φ, 6	5	5	0.2	1.14	0.2	145.0	0.42
9B	—	18,200	—	35	118	—	—	—	155.9
B1	1,200	19,400	1φ, 6	10	83	9.9	1.14	11.3	167.2
B2	900	20,300	1φ, 6	40	73	6.6	1.14	7.5	174.7
B3	300	20,600	1φ, 6	12	33	0.9	1.14	1.0	175.7
B4	400	21,000	1φ, 6	5	21	0.8	1.14	0.9	176.6
B5	500	21,500	1φ, 6	8	16	0.8	1.14	0.9	177.5
B6	700	22,200	1φ, 6	8	6	0.6	1.14	0.7	178.2	0.51
10C	—	18,600	—	78	275	—	—	—	158.2
C1	700	19,300	3φ, 6	22	197	13.8	0.189	2.6	160.8
C2	300	19,600	3φ, 6	8	175	5.3	0.189	1.0	161.8
C3	500	20,100	3φ, 6	8	167	8.3	0.189	1.6	163.4
C4	400	20,500	3φ, 6	28	159	6.4	0.189	1.2	164.6
C5	300	20,800	3φ, 6	10	131	3.9	0.189	0.7	165.3
C6	400	21,200	3φ, 6	8	121	4.8	0.189	0.9	166.2
C7	400	21,600	3φ, 6	10	113	4.5	0.189	0.9	167.1
C8	500	22,100	3φ, 6	10	103	5.1	0.189	0.9	168.0
C9	900	23,000	3φ, 2	13	93	8.4	0.0912	0.8	168.8
C10	400	23,400	3φ, 2	40	80	3.2	0.0912	0.3	169.1
C11	200	23,600	3φ, 2	25	40	0.8	0.0912	0.1	169.2
C12	700	24,300	3φ, 2	5	15	1.1	0.0912	0.1	169.3
C13	1,000	25,300	3φ, 2	—	10	1.0	0.0912	0.1	169.4
C14	200	25,500	1φ, 6	10	10	0.2	0.0912	0.0	169.4	0.26
										4.50

Total percent loss

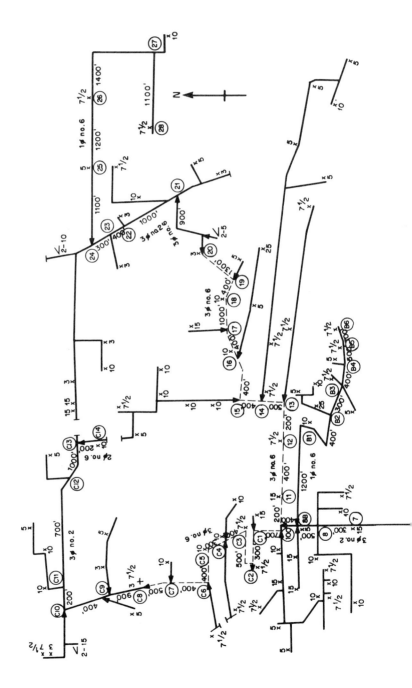

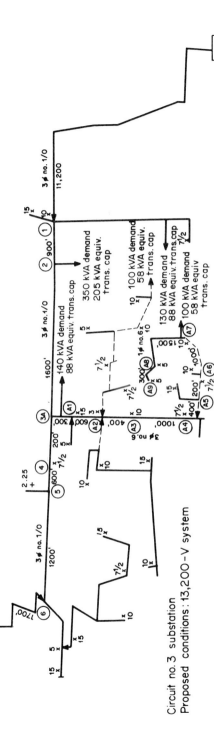

Circuit no. 3 substation
Proposed conditions : 13,200-V system

FIG. 4-33 Proposed higher-voltage circuit.

123

neglected to simplify calculations, but results are sufficiently accurate for purposes of comparison.

In making comparisons, evaluation of losses should include both a demand component reflecting system investment and an energy component. Also, costs of conversion reflecting both additional capital investment and annual operating and maintenance expenses should be taken into account. All of these values are different for each utility. The data sheet and circuit configuration for only the proposed circuit are included in this example; refer to Table 4-5D and Fig. 4-33.

CHAPTER 5

MECHANICAL DESIGN: OVERHEAD

CRITERIA

The mechanical design of the distribution system, and its several parts, must not only be adequate to sustain the normal stresses and strains, but must safely sustain them during abnormal conditions brought about by the vagaries of nature and people. While design criteria for overhead systems are substantially different from those for underground systems, in both instances prudent design takes into account economic and other nontechnical considerations.

For overhead systems, the supports for the conductors and equipment must withstand the forces imposed on them, while the conductors themselves must be sufficiently strong to support their own weight and the forces imposed on them.

National Electric Safety Code (NESC)

Minimum design criteria are suggested in the National Electric Safety Code (NESC) issued by the American National Standards Institute (ANSI). The NESC has received wide acceptance by utilities and other industries in this country and elsewhere.

In general, the code specifies:

1. Clearances between conductors and surrounding structures for different operating voltages and under different local conditions

2. Strength of materials and safety factors used in proposed structures

3. Perhaps the most basic, the probable loading imposed on the conductors and structures based on climatic conditions, approximately defined by geographic areas

TABLE 5-1 WIND AND ICE LOADINGS ON OVERHEAD SYSTEMS

Type of loading	Radial thickness of ice		Wind load on projected area of conductors		Temperature	
	in	*cm*	*lb/ft²*	*kg/m²*	*°F*	*°C*
Heavy	0.5	1.27	4	20	0	− 18.0
Medium	0.25	0.63	4	20	+ 15	− 9.4
Light	0.0	0.0	9	44	+ 30	− 1.1

For current industry recommended values, refer to the latest revision of the National Electric Safety Code.

The NESC divides the country into heavy, medium, and light load areas. The heavy loading area comprises roughly the northeast quarter of the "lower 48" states, and Alaska; the medium loading area comprises the northwest quarter plus a strip across the middle of the country; the light loading area comprises California and all the southern part of the country to a depth of some 300 to 400 mi, and Hawaii. The degrees of loading are indicated in Table 5-1, and the geographic areas for the continental United States in the map shown in Fig. 5-1. The values and areas of demarcation are approximate and should be subject to other practical considerations, including probable deviations based on actual experience, local codes and regulations, and other environmental requirements.

POLES
Stresses
The forces acting on a pole stem from the vertical loading occasioned by the weight it has to carry and from the horizontal loadings applied near the top of the pole. These latter are exerted by the conductors as a result of uneven spans,

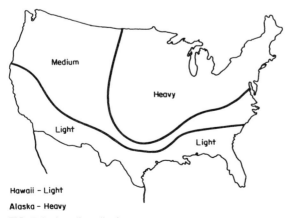

FIG. 5-1 Loading districts.

of offsets and bends in the lines, and of the pressure of wind blowing against them. Both vertical and horizontal loadings include the effects of ice collecting radially about the conductors.

The vertical force on the pole is the dead weight of the conductors with their coatings of ice, cross arms, insulators, and associated hardware. This vertical force exerts a compressive stress that may be considered uniformly distributed over the cross section of the pole. This loading, however, is almost always over-shadowed by the requirements of the horizontal loadings, and is usually not given further attention. Even a very light pole can safely carry the dead weight of a multicircuit, large-conductor line.

EXAMPLE 5-1

Assume relatively long spans of 200 ft, six primary conductors on two cross arms, and one heavy neutral (for two three-phase feeders) and three secondary conductors on a rack; moreover, the cross arms are two sets of double arms, with insulators and associated pins and hardware.

Assume no. 4/0 copper or 397,500-cmil ACSR (250 cmil $\times$ 10^3 copper equivalent), the largest usual distribution conductor, with a radial half-inch coating of ice (at 57 lb/ft³).* Assume a light pole, class 5 pine, with a minimum (top) circumference of 19 in (disregard the effect of the larger, lower cross-sectional areas because of taper; i.e., assume a cylindrical column).

From wire manufacturers' tables, no. 4/0 copper wire has a diameter of 0.528 in and a weight of 640.5 lb per 1000 ft; and 397,500-cmil ACSR has a diameter of 0.806 in and a weight of 620.6 lb per 1000 ft. The total weight, including ice, for 200 ft (100 ft on each side of pole) for the copper conductors is 2120 lb plus 500 lb allowed for four cross arms, insulators, etc., or 2620 lb. For the ACSR conductors, it is 2840 lb, plus the same 500 lb, or 3340 lb.

The cross-sectional area at the top of a class 5 pole is

$$A = \pi r^2 = \pi \left(\frac{19}{2\pi}\right)^2 = \frac{90.25}{3.14} = 28.7 \text{ in}^2$$

This may be rounded off to 25 in².

Conservatively, the dead weight of the copper conductors is 2750 lb divided by 25 in², or 110 lb/in²; that of the ACSR conductors is 3500 lb divided by 25 in², or 140 lb/in².

---

*Traditionally, the wire manufacturing industry has used the abbreviations cm and mcm for circular mils and thousands of circular mils. However, as the metric system becomes more familiar in the United States, the abbreviation cm is more likely to be mistaken for centimeters and the prefix m- for a multiplier of 10^{-3}. Therefore, throughout this book the more current U.S. abbreviation cmil will be used for circular mils; for thousands, either the number will be spelled out in full, or the multiplier 10^3 will be given, whichever is appropriate.

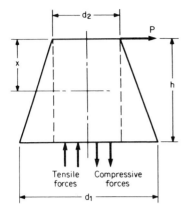

FIG. 5-2 Diagram of stresses on pole.

The maximum permissible compressive stress for wood ranges from about 300 lb/in² for western red cedar to 600 lb/in² for southern long-leaf yellow pine. For the horizontal loading, the pole can be considered a cantilever beam anchored at one end with a load applied at the other. The bending moment produces stresses in the wood, with the maximum fiber stress occurring at the edge of the cross section farthest from the neutral axis; the stresses are compressive on the side on which the load is pulling and tensile on the opposite side. Refer to Fig. 5-2.

Bending Moment The bending moment M is equal to the horizontal force applied P, multiplied by its perpendicular distance to the point where failure may occur, h, usually taken at the ground line.

$$M = Ph$$

Maximum Fiber Stress The maximum fiber stress f at any cross section is

$$f = M\frac{c}{I}$$

where c is the distance from the extreme fibers of cross section to the neutral axis and I is the moment of inertia of the cross section. For a circular cross section, where d is the diameter,

$$c = \tfrac{1}{2}d$$

$$I = \frac{\pi d^4}{64} = 0.0491d^4$$

so that

$$f = \frac{M}{0.0982d^3}$$

and

$$\frac{I}{c} = \frac{\pi d^3}{32} = 0.0982d^3 \qquad \text{(called the section modulus)}$$

Wind Pressure   In arriving at M, if P is the wind pressure on the length of the conductor, including its coating of ice, and h the distance or height from the ground at which the circular cross section is to be determined, the total moment for the several conductors that may be supported is the sum of the values of Ph for all the conductors.

To this must be added the wind pressure on the pole itself. Here, the longitudinal cross section of the pole may be broken down into a rectangle and a triangle, as indicated in Fig. 5-2.

The pressure on the rectangle (P_R) will be

$$P_R = P_1 d_2 h$$

where P_1 is the unit pressure in pounds per square inch. Its moment M_R about the base will be

$$M_R = P_1 d_2 \frac{h^2}{2}$$

The pressure on the triangle (P_T) will be

$$P_T = P_1 (d_1 - d_2) \frac{h}{2}$$

and its moment about the base will be

$$M_T = P_1 (d_1 - d_2) \frac{h^2}{6}$$

and the total pressure on the pole (P_p) will be

$$P_p = P_1 h \frac{d_1 + d_2}{2}$$

and the total moment on the pole (M_p) will be

$$M_p = P_1 h^2 \left(\frac{d_2}{3} + \frac{d_1}{6} \right)$$

The total moment the pole must accommodate will be the sum of the moments for the several conductors and the moment on the pole itself.

EXAMPLE 5-2

Assume a 40-ft pole (required for clearance) set 6 ft in the ground with three no. 4/0 stranded copper conductors on a cross arm with the conductors level at the top of the pole, and 150-ft balanced spans, in a heavy loading area (see Table 5-1).

The moment due to wind on the conductors when ice-coated is as follows. No. 4/0 stranded copper wire has a diameter of 0.528 in; allow 1 in of ice, for a unit diameter of 1.528 in of area to the wind.

$$M = 3 \frac{1.528 \text{ in}}{12 \text{ in/ft}}(150 \text{ ft} \times 4 \text{ lb/ft}^2)(40 - 6 \text{ ft})$$

$$= 7792.8 \text{ ft·lb or } 93,513.6 \text{ in·lb}$$

The moment on the pole itself due to wind (for an estimated diameter at the top of 8 in, and 12 in at the bottom) is as follows:

$$M = \frac{4}{144}(34 \times 12)^2 \times \left(\frac{8}{3} + \frac{12}{6}\right) = 21,580 \text{ in·lb or } 1798 \text{ ft·lb}$$

Say 1800 ft·lb. The total for both is 115,093.6 in·lb; say 115,000 in·lb; and

$$f = \frac{115,000}{0.0982 \times 12^3} = 678 \text{ lb/in}^2$$

Say 700 lb/in².

The ultimate fiber stresses for several woods and the resultant factors of safety for those woods are as follows.

Wood	Ultimate stress, lb/in²	Factor of safety
Northern white cedar	3,600	5+
Western red cedar	5,600	8
Long-leaf yellow pine	7,400	10+
Wallaba	11,000	15+

The total moment at the pole is 9591 ft·lb; say 9600; multiplied by a factor of safety of 2, it is 19,200 ft·lb; say 20,000.

From the ASA Standards Table of Wood Pole Classification, for a 40-ft class 5 pole, the resisting moments at 6 ft from the butt, nearest to this value but greater, are as shown in Table 5-2. Any of these poles will sustain the loading due to wind.

As a check, the pole circumference C, in inches, required to withstand a bending moment M, in foot-pounds, for a wood having a permissible fiber stress f, in pounds per square inch, is:

$$C = \sqrt[3]{\frac{3790M}{f}}$$

Assume long-leaf yellow pine, $f = 7400$ lb/in², is used:

$$C = \sqrt[3]{\frac{3790 \times 20,000}{7400}} = 21.71 \text{ in}$$

This is the minimum required, as opposed to the 31.5-in actual circumference. The difference provides a margin which insures the performance of the pole should it rot at the ground line and reduce the cross-sectional area there.

TABLE 5-2 RESISTING MOMENTS AND DATA FOR 40-ft POLES

| | | | Circumference, in | |
| | | Resisting | Minimum, | 6 ft |
Wood	Class	moment, ft·lb	at top	from butt
Northern white cedar	5	60,800	19	40.0
Western red cedar	5	60,700	19	34.5
Long-leaf yellow pine	5	60,900	19	31.5

From ASA Standards Table of Wood Pole Classifications.

Because of the wide discrepancies that occur in the structure of the wood, average values based on tests and experience are used in compiling standard values. Hence, the results from computations cannot be exact, but are sufficient for practical purposes of design.

EXAMPLE 5-3

While the ground line is almost always assumed to be the weakest cross section of the pole, theoretically this occurs at a point above the ground where the diameter is 1.5 times the diameter where the resultant load, i.e., the total moment divided by the total load, is applied. Refer to Fig. 5-2, where

P = total load applied where diameter is d_2

d = diameter at weak section at distance x from P

$$t = \text{taper of pole} = \frac{d_1 - d_2}{h} = \frac{d - d_2}{x}$$

At d, the bending stress is:

$$f_d = \frac{Px}{0.0982(d_2 + tx)^3}$$

For f to be at a minimum, the first derivative df/dx must equal 0. If this is so, then

$$d_2 + tx - 3tx = d_2 - 2tx = 0$$

$$d_2 = 2tx = 2(d - d_2)$$

$$d = \frac{3}{2}d_2$$

Substituting this value and expressing x in terms of h gives:

$$f_d = \frac{Ph}{0.662(d_1 - d_2)d_2^2}$$

in which the maximum unit stress at the weakest point is given in terms of the known dimensions h, d_1, and d_2.

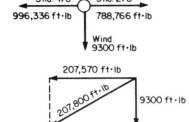

FIG. 5-3 Unequal loading on pole.

The conductors supported by a pole are in tension and also cause a loading to be applied on the pole at the first point of support. Where the conductors are in a straight line and the span lengths on both sides of the pole are equal, the loadings caused by the conductors are equal and opposite in direction and cancel each other. Where the conductors on each side of the pole are different, or where there is an offset or change in direction of the line, or where the conductors dead-end on the pole, the pole will be subjected to loadings for which provision should be made. The same principles are employed in obtaining moments acting on the pole, and this can best be illustrated by the following examples.

EXAMPLE 5-4

Assume that the 40-ft pole in Example 5-2 supports three no. 4/0 bare stranded copper conductors in one direction and three no. 2/0 bare stranded copper conductors in the other in a straight line; there are 150-ft spans in both directions. Refer to Fig. 5-3.

The moment on the pole caused by the wind for the three no. 2/0 conductors is as follows:

$$M = 3 \times \frac{1.418}{12} \times \frac{150}{2} \times 4 \times 34 = 3616 \text{ ft·lb}$$

For the three no. 4/0 conductors (from Example 5-2),

$$M = \tfrac{1}{2} \times 7792 = 3896 \text{ ft·lb}$$

For the 40-ft pole, from Example 5-2,

$$M = 1798 \text{ ft·lb}$$

Assume that the ultimate strength of the copper conductors is 37,000 lb/in^2 and that they are sagged to one-half their ultimate strength. The total moment on the pole caused by conductor tension, for the three no. 2/0 conductors, is:

$$M = 3 \times 0.418 \times \frac{37,000}{2} \times 34 = 788,766 \text{ ft.lb}$$

For the three no. 4/0 conductors,

$$M = 3 \times 0.528 \times \frac{37,000}{2} \times 34 = 996,336 \text{ ft·lb}$$

The difference between the two is 207,570 ft·lb (say 207,600 ft·lb). The total moment on the pole from conductor tension and wind is:

$$M = \sqrt{M_c^2 + M_w^2} = \sqrt{207.6^2 + 9.3^2} \times 10^3 = 207.8 \times 10^3$$

Multiplied by 2 for a factor of safety, $M = 415.6 \times 10^3$; say 420,000 ft·lb.

To find the pole circumference required to withstand this bending moment, assume $f = 7400 \text{ lb/in}^2$ for long-leaf yellow pine:

$$C = \sqrt[3]{\frac{3790 \times 420 \times 10^3}{7400}} = 59.91 \text{ in at ground line}$$

This is beyond the strength of a 40-ft pole of the maximum class, 00; a guy should be installed.

Check for the possibility of using a wallaba pole, $f = 11,000 \text{ lb/in}^2$:

$$C = \sqrt[3]{\frac{3790 \times 420 \times 10^3}{11,000}} = 52.50 \text{ in at ground line}$$

A 40-ft wallaba pole of maximum class 00 will not accommodate this loading; a guy is still required.

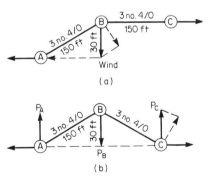

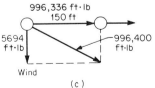

FIG. 5-4 Loadings due to (a) offset in line and (b) bend in line. (c) Loading due to dead end.

EXAMPLE 5-5

Assume the same 40-ft pole and three no. 4/0 stranded copper conductors, level with the top of the pole, with equal 150-ft spans on each side, as in Example 5-2, but with the line offset this time by 30 ft, as shown in Fig. 5-4a. The moments on pole B due to the conductors, from Example 5-4, from A toward B, and B toward C, are 996,336 ft·lb, and:

$$X = \sqrt{150^2 - 30^2} = 146.96 \text{ ft}$$

where X is the portion of the conductors exposed to wind. The moment on pole B, the portion in line with BC, is:

$$M_{BC} = \frac{146.96}{150} \times 996,336 = 976,144 \text{ ft·lb}$$

The moment on pole B due to wind for one 150-ft span, from Example 5-4, is 3896 ft·lb, and:

$$M_{Bw} = 3896 + \frac{146.96}{150} \times 3896 = 7713 \text{ ft·lb}$$

The total moment on pole B, in line, is 996,336 − 976,144, or 20,192 ft·lb. At right angles, it is 199,267 ft·lb, or (30 ft ÷ 146.96 ft) × 976,144. The wind moment is 7713 ft·lb, and the total right-angle moment is 206,980 ft·lb.

$$M_B = \sqrt{20,192^2 + 206,980^2}$$

say

$$\sqrt{20^2 + 207^2} \times 10^3 = 208,000 \text{ ft·lb}$$

Multiply by 2 for the factor of safety; the result is 416,000 ft·lb. For pine,

$$C = \sqrt[3]{\frac{3790 \times 416,000}{7400}} = 59.73 \text{ in at ground line}$$

This is beyond the strength of the 40-ft maximum pole class, 00; a guy is required.

Check for possible use of a wallaba pole, $f = 11,000 \text{ lb/in}^2$:

$$C = \sqrt[3]{\frac{3790 \times 416,000}{11,000}} = 52.33 \text{ in at ground line}$$

This is also beyond the strength of the 40-ft maximum pole class, 00; a guy is still required.

It must be noted that the right-angle and wind moment on pole B actually amount to less than what is indicated because the vector sum of the moments is that shown by the dashed line in Fig. 5-4a. Since this is less than was calculated above, the error introduced is on the safe side.

Similar procedures are used in determining the moments on, and the adequacy of, poles A, B, and C, as shown in Fig. 5-4*b*.

EXAMPLE 5-6

Assume that the 40-ft pole in Example 5-2 supports three no. 4/0 bare stranded copper conductors dead-ended on the pole, with a span of 150 ft, as shown in Fig. 5-4*c*.

The moment on the pole due to three no. 4/0 conductors with a 150-ft span, from Example 5-5, is 996,336 ft·lb. The moment on the pole due to wind on the conductors is 3896 ft·lb, and that due to wind on the pole is 1798 ft·lb. The total moment on the pole due to wind is 5694 ft·lb.

$$\text{Total moment on pole} = \sqrt{996.3^2 + 5.694^2} \times 10^3 = 996{,}316 \text{ ft·lb}$$

Multiply by 2 for a factor of safety:

$$M = 1{,}992{,}632 \text{ ft·lb}$$

For pine,

$$C = \sqrt[3]{\frac{3790 \times 1993 \times 10^3}{7400}} = \sqrt[3]{1{,}020{,}000} = 100.65 \text{ in}$$

This circumference is much beyond any 40-ft pole, pine or wallaba (from the conclusions in Example 5-5). Guying must be used or other means employed to accommodate the loading safely.

Equipment on Poles Poles supporting transformers, capacitors, regulators, switches, or other equipment support loads which are principally compressive in nature; the equipment also adds to the wind loading. Since the center of gravity projects out from the pole, a moment is created on the upper portion of the pole, pivoting about its lower point of support. Ordinarily, the pole class selected for supporting the conductors, with its factors of safety included, is capable of carrying this additional load safely. For larger-scale equipment installations, however, a pole one class greater than that adequate for the conductors is specified.

CROSS ARMS

Cross arms are now almost limited to carrying polyphase circuits in areas where appearance is not of paramount importance. They are also used where lines cross each other or make abrupt turns at large angles to each other. They are used as alley or side arms in which the greater part of their length extends on one side of the pole to provide adequate clearances where pole locations may be affected by limited-space rights-of-way. Cross arms are shown in Fig. 5-5.

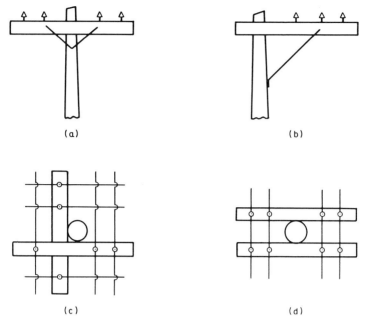

FIG. 5-5   Uses of cross arms: (*a*) line arm; (*b*) side arm; (*c*) buck arm; (*d*) double arms.

Loadings

The cross arm acts as a beam, supported at the point of attachment to the pole, and must be capable of being subjected to vertical loadings from the weight of the conductors (encased in ice) and a 225-lb worker (specified as an additional safety measure). It is also subjected to horizontal loadings stemming from winds and from tension in the conductors where the tensions on each side of the pole do not cancel each other; e.g., where spans or conductors are not the same on each side of the pole, at dead ends, bends, or offsets in the line, or where consideration is given to conductor breaking contingencies.

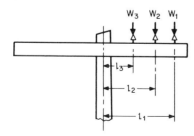

FIG. 5-6   Bending moment on cross arm.

Stresses

The same principles for determining stresses in beams as were applied in the case of poles may also be applied to cross arms; see Fig. 5-6.

Bending Moment The total bending moment M is equal to the sum of all the individual loads multiplied by their distances from the cross section under consideration. Ordinarily, the weakest section should be at the middle of the arm where it is attached to the pole. At the pin holes, however, the cross section of the cross arm is reduced and may, under unusual circumstances, be the weakest point in the cross arm. The determination can easily be made by computing unit fiber stress at the several points. Like the pole, the cross arm acts as a beam and the same formula for determining stresses may be employed:

$$f = \frac{M}{I/c}$$

where f = maximum unit fiber stress occurring at extreme
  edges of cross section, lb/in^2

M = total bending moment, in·lb

I = moment of inertia of cross section

  c = distance from neutral axis to extreme edge, in

The moment of inertia for a rectangular cross section is

$$I = \frac{1}{12}bd^3 \quad \text{and} \quad c = \frac{d}{2}$$

so that the section modulus

$$\frac{I}{c} = \frac{1}{6}bd^3$$

where the neutral axis is parallel to side d, as shown in Fig. 5-7.

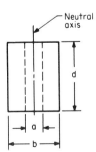

FIG. 5-7 Cross-arm cross section.

Where the cross section is reduced by a pinhole in the cross arm, the section modulus becomes:

$$\frac{I}{c} = \frac{1}{6}d\left(b^2 - \frac{a^3}{b}\right)$$

where a is the diameter of the hole.

EXAMPLE 5-7

Assume a standard 8-ft six-pin arm mounted at its center on a pole, supporting six conductors, each of which, with a half-inch coating of ice, has a maximum weight of 100 lb. The lengths, or moment arms, from the center of the arm to each of the pins are respectively 15, 29½, and 44 in. The moment *about* the first pin hole from the pole is:

$$M = 100 \text{ lb} \times 14\frac{1}{2} \text{ in} + 100 \text{ lb} \times 29 \text{ in} = 4350 \text{ in·lb}$$

With the 3½- × 4½-in cross section reduced by a 1-in pin hole, the section modulus at that point is:

$$\frac{I}{c} = \frac{1}{6}(3.5 - 1.0) \times 4.5^2 = 8.4375$$

and the fiber stress is:

$$f = \frac{M}{I/c} = \frac{4350}{8.4375} = 515.56 \text{ lb/in}^2$$

Double Arms

When fiber stresses approach the maximum safe values for a particular kind of wood (always keeping in mind a factor of safety of 2), two arms or double arms are used. Ordinarily, these are found at dead ends, at points where loads are greatly unbalanced (such as large offsets or bends in the line), and at intermediate points along a long line to limit damage in the event that conductor breaks create severe load unbalances on the supporting structures.

The two arms are placed one on each side of the pole and bolted together near the ends, and often at intermediate points. Properly constructed, with spacers of wood or steel between the arms, the structure created would act as a truss with strengths of 10 to 12 times that of a single arm, or 5 to 6 times that of the two arms considered individually. Since such quality trusses may not always be constructed in the field, prudence dictates that only the ultimate fiber strength equivalent to that of two cross arms be considered. Where the loadings on the arms may exceed their fiber strengths, the arms may be guyed, as shown in Fig. 5-8, or steel arms may be substituted for wooden ones.

Douglas fir and long leaf yellow pine are the most popular kinds of wood used for cross arms, though other kinds may also be found in use. Their ultimate bearing strengths are listed in Table 5-3.

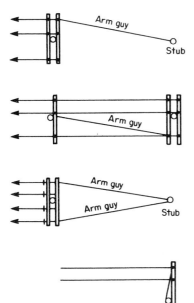

FIG. 5-8 Typical arm guys.

Bearing strength on an inclined surface at an angle to the direction of the grain is given as follows:

$$f_a = f \sin^2 a + n \cos^2 a$$

where a is the angle of inclination of the load to the direction of the grain.

Cross-Arm Braces

Cross arms fastened to poles are usually steadied in position by braces. Flat braces, usually flat strips of galvanized steel bolted to the cross arm and fastened

TABLE 5-3 ULTIMATE BEARING STRENGTH OF WOOD, lb/in²

Wood	End-grain bearing f	Cross-grain bearing n
Long-leaf yellow pine	5000	1000
Douglas fir	4500	800
Western red cedar	3500	700
Cypress	3500	700
Redwood	3500	700
Northern white cedar	3000	700

to the pole by a lag screw, are most commonly used. The support given the cross arm by the flat braces is questionable, and is usually neglected in determining the effects of loads on the cross arm. Where the cross arm is not symmetrically loaded on each side of the pole, the brace on the load side is in compression and is of little benefit because of its slenderness; the brace on the other side is in tension and aids in transmitting some of the load to the pole, but does not reduce the bending moment.

For heavier loads, a preformed brace made of angle iron, of larger cross section than the flat brace, aids in supporting the loads on the cross arm. Here, the moments acting on the cross arm are usually computed about the point of attachment of the angle brace to the cross arm.

For alley or side arms, an angle-iron brace is used; usually it is a straight length with the ends adapted to be fastened to the arm and the pole. The brace transmits a considerable part of the load on the arm to the pole; when all the load on the arm is beyond the brace, the vertical compressive stress on the brace may be even greater than the load. The points of attachment at the cross arm and at the pole are important; the smaller the angle between the pole and the brace, the greater the load transmitted to the pole. On the other hand, as the length of the brace increases, the slenderness ratio (the brace acting as a slender column) becomes larger and the effectiveness of the brace diminishes.

Bolts

The stability of a cross arm and its strength rely heavily on the strength of the bolts through which the stresses are transmitted. The distribution of stresses on the through bolt holding the cross arm to the pole are shown in Fig. 5-9a. The vertical load on the cross arm is transferred by the bolt to the pole.

The unit pressure on the bolt in the cross arm is:

$$P_a = \frac{W}{b_1 d}$$

where d is the diameter of the bolt; the maximum unit pressure in the pole is:

$$P_p = \frac{W}{b_2 d}$$

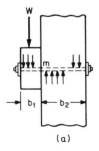

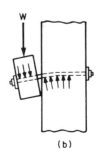

FIG. 5-9 Action on bolt holding cross arm to pole. m is point of maximum shear stress.

(a)   (b)

The maximum unit pressure must not exceed the bearing value of the wood, or distortion will take place. As the ultimate strength of the wood is approached, the bolt will tend to bend as the fibers of the wood begin to give way, as shown in Fig. 5-9*b*.

PINS

Loading

Pins are subject to both vertical and horizontal loadings. The vertical loading results from the weight of the conductor and its half-inch radial coating of ice. The horizontal loading stems from the wind, from differential tensions in adjacent conductor spans, from nontangent spans, or from broken-wire conditions in which the tensions in the conductor spans become unbalanced.

Under vertical load, the pin acts as a simple column, transmitting its load to the cross arm at the shoulders resting on the cross arm. The stress is equal to the load divided by the area under pressure, the area of the shoulder resting on the cross arm. This component is usually not large compared with the other components acting on the pin and is often neglected.

Under horizontal loadings, the pin acts as a cantilever beam, and the maximum stress occurs at the point where the pin rests on the arm; refer to Fig. 5-10. The bending moment M is equal to the load P multiplied by the distance of the conductor above that point (h):

$$M = Ph$$

The maximum fiber stress is usually considered to be at the point where the shoulder comes in contact with the cross arm. Its unit value, in pounds per square inch, may be calculated:

$$f = \frac{Ph}{0.0982h^3}$$

where d is the diameter of the shank of the pin.

Since the balancing of moments may occur at the edge of the shoulder of the pin (for either wood or metal) rather than at the edge of the shank of the pin, some crushing effect may take place on the wood of the cross arm, which may affect its strength substantially. In such instances, the weak point may occur at the cross section of the pin above the cross arm; the value may be found by

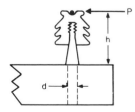

FIG. 5-10 Loading on pins.

substituting the diameter of the pin at about one-third of the distance down from the point of the conductor attachment, approximately the diameter at the weak point.

Double Pins   Double pins, one on each of the double arms, are used where the strength of one pin is inadequate.

Pins in Lieu of Cross Arms   The advent of wye primary systems employing a common neutral with the secondary (situated on the pole in the secondary position) allowed single-phase primary conductors to be supported on a steel ridge pin, as shown in Fig. 5-11. The vertical loading on the pin, the weight of the ice-coated conductor, is transmitted to the pole through the bolt by which the pin is attached to the pole. Horizontal loadings, from both wind and conductor tension, act on the pin as a cantilever beam, and the same analysis of stresses in the pin, bolt, and pole applies here as with cross arms.

In polyphase systems, the conductors may be supported on pins attached directly to the pole, eliminating the use of cross arms (this method of support is sometimes referred to as "armless construction"). This not only makes for a neater appearance, but, as indicated earlier, improves electrical performance by mitigating the voltage drop due to the reactance of the line. (The arrangement and spacing of the conductors accounts for the lessened reactance. Also, this construction employs bucket trucks or platforms for easier access to the conductors.)

In this instance, the vertical loading acts on the end of the pin, with the pin acting as a cantilever beam; the bending stress occurs at the pole, and the fiber stresses on the pin are calculated in the same manner described above for cross

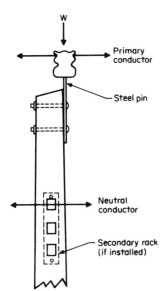

FIG. 5-11   Pin support for single-phase primary.

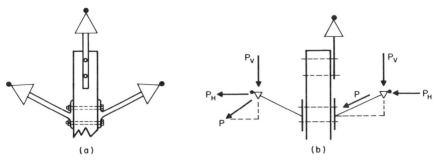

FIG. 5-12 Pins in lieu of cross-arm construction. (a) Use of long steel pins. (b) Stresses on long steel pins.

arms and poles. The horizontal loading acts to create compressive stresses in the pin, which acts as a slender column. Again, this component is small compared with the other components and is often neglected. On the other side of the pole, however, the horizontal forces act to pull the conductor away from the pole; while there is ample strength for the tensile stress imposed on the pin and the bolts with which it is attached to the pole, these forces do dictate that the conductors be so attached to the insulators as to ensure the conductor will not be separated from the insulator under the stresses imposed.

For higher voltages, additional space requirements may necessitate longer pins or the installation of insulated conductors (or both). The pins may be attached to the pole as indicated in Fig. 5-12. Here, the stresses on the pin from both the vertical and horizontal loads impose bending moments about the bottom of the pin at the pole. The moments and stresses on the pins and supporting bolts, indicated in Fig. 5-12, are computed in the same fashion as for the cases mentioned above.

For all applications of the pins, a minimum strength of 700 lb per pin is usually specified in withstanding unbalanced tension in a conductor supported by the pin.

SECONDARY RACKS

Secondary racks usually support three spool insulators mounted on a common shaft attached to a steel backing that is bolted to the pole by one or more bolts (though sometimes, depending on the stress to be accommodated, a lag screw may take the place of a bolt). The spool insulators are usually spaced 8 in apart, although sometimes less. If the rack is properly mounted, the insulators may fail before the rack.

Loadings

The loading on secondary racks has both vertical and horizontal components; see Fig. 5-13. The vertical loading consists of the weight of the conductors (with the coating of ice) of both the mains and services attached to them. With spool-

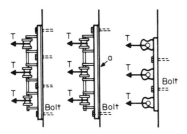

FIG. 5-13 Loading on secondary racks.

type racks, the load is carried by small arms acting as cantilever beams, supporting the insulators. Generally, the insulator is located between two such arms, but only the lower one carries the load. With the knob-type rack, the vertical load is transmitted to the rear plate by the U bolts attaching the insulators to the plate, and to the pole by the bolts attaching the rack to the pole.

The horizontal load is due to wind acting on ice-laden conductors and the tension in them, for both the secondary mains and the services. If the rack is attached to the pole by a bolt under each insulator, the strength of the back plate plays little part, as the stress is transmitted almost directly by the bolts. Where the spool-type rack is mounted with the bolts at the ends only (or by a bolt and a lag screw), the back of the rack acts as a cantilever beam supported at both ends, with the load applied at each of the points where the arms are attached to the back. The greatest stress is at the point in the back where either of the two arms supporting the middle insulator is attached, the bending moment being greatest at that point. In the knob-type rack, both the vertical and horizontal loadings are transmitted to the back of the rack and through the mounting and bolts to the pole.

Where the conductors dead-end on a rack mounted on the side of a pole (or where there is an unbalanced pull from broken conductors), the side pull on the rack is limited by the strength of the arms or the U bolts supporting the insulators. It is preferable to mount the rack on the face of the pole, with the arms and U bolts supporting the insulators in tension, and the stresses transmitted to the pole almost directly by the mounting bolts.

In general, like pins, conductor fastenings should be able to withstand a 700-lb stress.

Messenger Clamp As mentioned earlier, when the secondary conductors are cabled about the neutral conductor acting as the supporting messenger, a clamp supporting the neutral is bolted to the pole. Here, both the vertical and horizontal loadings are transmitted to the pole by the through bolt by which the clamp is attached to the pole.

INSULATORS

Insulators used on overhead distribution systems are made of porcelain, glass, and, more recently, synthetic materials. Glass insulators, though no longer widely installed, exist in abundant numbers and will probably remain in service for some time. Several types of insulators employed in distribution line designs are discussed below.

Loadings

In general, porcelain has relatively little tensile strength but excellent strength in sustaining compressive stresses, properties substantially true also of glass. Lines are therefore so designed that the insulator materials will be in compression when carrying the (mechanical) loads imposed on them.

Pin Type As the name implies, pin-type insulators are mounted on pins (of wood or metal) and the conductors are fastened to them. Their strengths (in compression) are usually greater than those of the pins upon which they are mounted. The dimensions of the insulating material necessary to meet the mechanical requirements are usually ample in meeting the electrical requirements, including surge voltages when wet.

Post Type The post-type insulator is essentially a pin-type insulator that incorporates its own steel pin. Vertical loads are provided for by the porcelain, while horizontal loads act to create a moment about the point of attachment to the pole. Stress is transmitted by its steel core (or pin) to the cross arm or pole.

The post-type insulator is meant to be installed in a vertical (or near-vertical) position but is also sometimes installed horizontally (or nearly so) in lieu of cross arms to carry the conductors of a polyphase primary circuit; see Fig. 5-14. Here, the horizontal loadings create compressive stresses on one side of the pole and tensile stresses on the other, both of which are transmitted through the steel core to the pole. Vertical loadings create a compressive stress in the porcelain between the conductors and the inner steel core or pin; this stress is transmitted to the steel pin, which in turn transmits it to the pole; see Fig. 5-12. In instances where the post insulators are mounted at an angle to the pole, the stresses created will be the horizontal and vertical vector components of the forces acting on the porcelain and the steel pin.

One advantage of the pin- or post-type insulator is that, if it is mounted on a cross arm, a shorter pole is required to achieve the same height of the conductor above ground than is needed for suspension- or strain-type insulators.

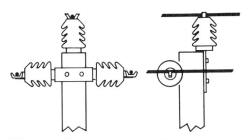

FIG. 5-14 Use of post-type insulators.

Suspension or Strain Type   The suspension- or strain-type insulators are also known as disk or string insulators. They are not generally used on distribution circuits except at turning points and at dead ends, where pins and pin-type insulators may not provide sufficient strength; they are particularly useful in providing for the unbalance of stresses caused by broken conductors where pins, and even double pins, may be inadequate and some form of insulated clamp or other means of attachment is required. Here the conductors are dead-ended on each side of the pole, and the disk insulator, designed for heavy loadings of 10,000 to 20,000 lb (and similar to those used for transmission lines, but smaller in diameter), serves this purpose. Often, two or more such disks are assembled as a unit to accommodate higher operating voltages.

Strain-Ball Type   The strain-ball type of insulator has been used to dead-end lower-voltage primary and secondary conductors, and as an insulator in guy wires in older installations; many still exist. Here, the porcelain is under compression, accommodating the stresses imposed on it. Standard ratings include strengths (in compression) of 10,000, 12,000, and 15,000 lb.

Spool Type   Insulators of the spool type are associated with secondary racks, described earlier, and are standardized in design. The compressive strength of the spool porcelain is usually greater than the strength of the other parts of the rack.

Other Types   Among the other types of insulators are knob types, sometimes used for services and on secondary racks, and other shapes for use as bushings, bus supports, and other purposes.

GUYS AND ANCHORS
Stresses
When the horizontal loads imposed on poles and cross arms may exceed the safe carrying strengths of the wood involved (or the holding power of the soil), guys of steel wire are usually installed to take up or counteract the excessive stress. The various types of guys are shown in Fig. 5-15. Note that the guys take up the horizontal stresses and distribute them to other cross arms, to other poles and into the ground, or directly into the ground. The structural and environmental conditions peculiar to each situation dictate the type of guy used.

Arm Guys   Several forms of arm guy have been previously discussed under the subject of cross arms. The guy is normally attached to the end of the cross arm having the heavier load and is carried down to an anchor or to another pole; sometimes the guy wire is attached to the lighter-loaded end of the cross arm and carried back to the adjacent pole in the direction in which the load is pulling. Where the cross arm, or the double arms, may be overstressed, both ends of the cross arm or arms may be guyed. Any deviation of these guys from the horizontal imposes a vertical compressive load on the cross arm (equal to the vertical com-

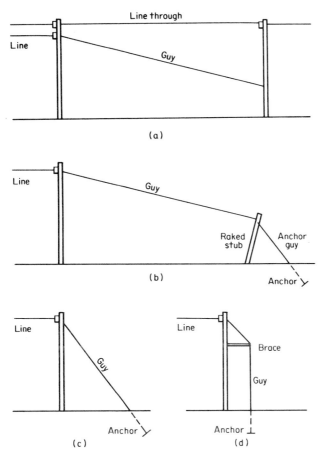

FIG. 5-15 Typical pole guys: (*a*) pole-to-pole guy; (*b*) pole-to-stub guy; (*c*) anchor guy; (*d*) sidewalk guy.

ponent of their tensions) and this loading must be included when considering the design of the poles.

Anchor Guy The most commonly used guy is the anchor guy, of which one end is fastened to the pole and the other end to a rod to which is attached an anchor that is buried in the ground. The anchor selected depends on the holding power of the soil: from poor in swamps to excellent in hard and dry earth.

Span Guy In the span guy, the guy wire extends from the head of the pole under load horizontally to an adjacent pole; since this merely transfers the load through the guy wire to the other pole, the receiving pole must be strong enough to take the additional load, or the receiving pole must be guyed.

Head Guy   In the case of the head guy, the guy wire extends from the head of the pole under load at an angle to a point on an adjacent pole at a height above ground sufficient to provide for clearance requirements. The load thus applied to the adjacent pole reduces the ground-line moment (the horizontal pull of the wire multiplied by its height on the pole) applied to this receiving pole.

Stub Guy   The stub guy is another form of head guy, the receiving pole being a stub or strut to maintain clearance; the stub usually has an anchor guy attached to it and may be raked so that the force acting on the stub will produce a compressive stress as much as possible.

Sidewalk Guy   Another form of anchor guy is the sidewalk guy, which uses a strut extending horizontally from the pole at the height for which clearance is specified, the anchor being buried as nearly vertical from the end of the strut as practical. This guy places a bending moment equal to the lateral component of support times the distance between the strut and guy attachments to the pole. The maximum bending moment at the pole base, however, is reduced by this type of guy.

Storm Guy   Long and straight pole lines having no side taps and unshielded from the elements are guyed four ways, both parallel to and at right angles to the line, at locations based on judgment and experience. This serves to minimize damage resulting from accidents or severe storms and caused by uneven stresses from broken poles and conductors.

Crib Bracing   Although not perfectly attainable, pole designs aim for a rigid structure, one that will not move in the ground when an unbalanced horizontal load is applied at its top. In large part, the depth setting of the pole and the kind of soil determine whether the full length of the pole can be utilized.

Where unbalanced loads cannot be served by guys because of space limitations, appearance, consumer requirements, or other reasons, resort is had to

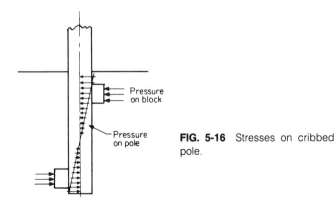

FIG. 5-16   Stresses on cribbed pole.

cribbing or *crib bracing*. Here, planks, logs, sections of poles, or steel anchors are installed at the base of the pole to provide additional holding power; see Fig. 5-16. The tension of the conductors on the pole is reduced by increasing their sag until the total moment at the ground line is within the strength limitations of the pole; the cribbing is then installed and the space about the pole refilled and tamped. The addition of cribbing does *not* add to the strength of the pole.

Loading

The loads imposed on guys are generally due to the tension in the conductors and the angle between adjacent conductor spans, if any; the magnitude of the tension depends on the size of the conductor, its loading (including wind and ice), and the sag in the span. The design limits are usually based on the elastic limit of the conductors (e.g., 50 to 60 percent of the ultimate strength of copper). The usual stress is generally less than that, as the design limits are based on the worst assumed loading conditions, which are approached only occasionally.

The stress on the pole at an angle in the line is also due to tension in the conductors, but only a component of that tension is handled by the guy; the amount depends on the size of the angle in the line. Refer to Fig. 5-17*a*.

If T is the total tension caused by all the conductors, and a is the angle of the line, the component of the total tension T in line with the guy is

$$T_a = T \sin \frac{a}{2}$$

and the total stress handled by the guy is twice that, or

$$T_{\text{guy}} = 2T \sin \frac{a}{2}$$

If the tensions in the two spans are not balanced, then the resultant stress will be the vector sum of the two, and will be the stress handled by the guy. Wind pressure on the pole itself must also be taken into account in determining the total load to be handled by the guy.

If the angle is large, usually more than 60°, the loading on the guy bisecting the angle will be greater than the dead-end loading of the line, and it is generally better to install two guys, considered as dead-end guys, if practical.

The guy should be attached as near as practicable to the center of loading of the loads it supports. Where the individual loads act at different elevations, they should be converted into an equivalent single load at the point of attachment of the guy. Referring to Fig. 5-17*b*, if T_p is the loading (say of the primary) at height h_p, T_s is the loading (say of the secondary) at height h_s, P_w is the wind pressure on the pole assumed to be concentrated at height h_w, and L_H is the equivalent horizontal loading, then

$$L_H = \frac{T_p h_p + T_s h_s + P_w h_w}{h}$$

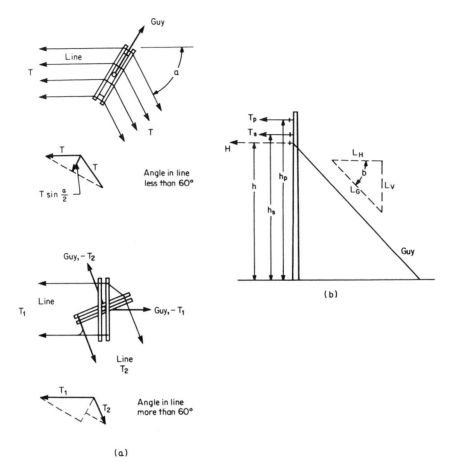

FIG. 5-17 Loading on guys. (*a*) Guys at angles. (*b*) Guy loading.

Since the guy is not usually horizontal, the actual tension in it will be greater than L_H. If b is the angle the guy makes with the horizontal, then the loading in the guy, L_G, is:

$$L_G = \frac{L_H}{\cos b}$$

The vertical component L_V is:

$$L_V = L_H \tan b \quad \text{or} \quad L_V + L_G \sin b$$

and acts as an additional vertical load on the pole.

EXAMPLE 5-8

Assume three no. 4 medium-hard-drawn copper primary conductors dead-ended at the top arm 33 ft above the ground, and 4 no. 2 soft-drawn copper cabled secondaries dead-ended 2 ft below, or 31 ft above the ground; a pole face area of 25 ft^2; and a wind pressure of 4 lb/ft^2 applied one-third the distance down from the top arm, or 22 ft above the ground. The guy is attached 30 ft above ground at an angle of 45°.

$$T_p = 3 \times 950 = 2850 \text{ lb}$$

$$T_s = 4 \times 990 = 3960 \text{ lb}$$

$$P_w = 25 \times 4 = 100 \text{ lb}$$

$$L_H = \frac{(2850 \times 33) + (3960 \times 31) + (100 \times 22)}{30} = 7300 \text{ lb}$$

$$L_G = \frac{7300}{\cos 45°} = \frac{7300}{0.707} = 10{,}325 \text{ lb}$$

If the guy is attached at a point too far from the center of the load, a significant stress may be imposed on the pole, as the section of the pole above that point acts as a beam, and the moment at that point will be approximately

$$M = T_p(h_p - h) + T_s(h_s - h) - P_w(h - h_w)$$

and the fiber stress in the pole at the point of attachment of the guy will be

$$f = \frac{M}{0.0982d^3}$$

where d is the diameter of the pole at this point.

Guys should be attached to the pole as near as possible to the center of the load; when installed, they should take the entire horizontal load, with the pole acting as a strut.

Guy Wires

Guy wire is made of stranded steel cable (usually 7 or 19 strands) so that the failure of one or two strands will not cause the immediate failure of the cable. The strands are usually galvanized or copper-clad to resist the effects of weather. The strands may be of mild steel or high-strength steel, but must be of sufficient strength to support the loads imposed on the guy. Such steel wires usually come in four grades of strength, and standards further specify that guy wires not be stressed beyond 75 percent of their ultimate strengths. Wires are manufactured in diameter differences or steps of $\frac{1}{32}$ in, but sizes less than $\frac{1}{4}$ in or greater than $\frac{1}{2}$ in are seldom, if ever, used, two or more guys being employed if stresses greater

TABLE 5-4 GUY WIRE CHARACTERISTICS

Wire Class	Ultimate strength, lb/in²	Elastic limit, lb/in²
Standard	47,000	24,000
Regular	75,000	38,000
High-strength	125,000	69,000
Extra high-strength	187,000	112,000

Weight of steel wire: 0.002671 lb/in³.

Modulus of elasticity: 29×10^6.

Coefficient of linear expansion: 11.8×10^{-6}/°C; 6.62×10^{-6}/°F

Within the ¼- to ½-in range, for the four classes of wire, ultimate strengths vary from a minimum of 1900 lb to a maximum of 27,000 lb.

than the maximum strength of the guy wire are required. In practice, however, only three sizes are usually stocked and specified: light, medium, and heavy; they are often referred to by their maximum permissible strengths, e.g., 6000-lb, 10,000-lb, and 20,000-lb (6M, 10M, and 20M).

The characteristics of steel wire used for guys are given in Table 5-4.

Guy wires are attached to poles and cross arms by eye bolts, thimbles, clamps, clips, hooks and plates, and special guy bolts which have the eye shaped and bent at an angle to accommodate the wire.

Push Braces

Where guys are impractical to install, push braces are sometimes installed. These are essentially compression-type "guys." The pole used for a brace must be of sufficient length for the purpose and its class must be capable of withstanding the compressive stress imposed on it. This stress is the vector sum of the horizontal and vertical loads on the pole being reinforced in the direction of the axis of the pole as a brace. Figure 5-18 illustrates such a brace and the stresses imposed on it.

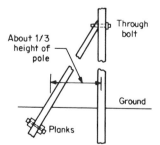

FIG. 5-18 Push brace.

Anchors

The holding power of the anchor should obviously match the strength of its associated guy wire. In general, the holding power will depend on the area the anchor offers the soil, the depth at which it is buried as a function of the weight or resisting force of the soil, and the kind and nature of the soil.

Types of Anchors   Anchors come in many shapes and types. They may be classified into four general types:

1. Buried logs, planks, or plates attached to the end of a rod.
2. Screw anchors, screwed into the soil at varying depths. A very large screw anchor, known as the swamp anchor, is used in swampy areas.
3. Expanding anchors, in which a plate in sections is folded into a small diameter, the unit set into a small-diameter hole (or at the bottom of a pole), and the anchor rod screwed or pounded into it so that the sections spread out, biting into the adjacent soil. If the expanding anchor plate is divided into eight sections, for example, the anchor is known as an eight-way expanding anchor.
4. Rock anchors, which are merely rods driven into the rock, hard shale, or hardpan, at approximately right angles to the guy wire. The depth at which they are installed will vary with the strength required and the character of the rock.

TABLE 5-5A   CLASSIFICATION OF SOILS

Class	Description
1	Hard rock: solid.
2	Shale, sandstone: solid or in adjacent layers.
3	Hard, dry: hardpan, usually found under class 4 strata.
4	Crumbly, damp: clay usually predominating. Insufficiently moist to pack into a ball when squeezed by hand.
5	Firm, moist: clay usually predominating with other soils commonly present. Sufficiently moist to pack into a firm ball when squeezed by hand (most soils in well-drained areas fall into this classification).
6	Plastic, wet: clay usually predominating as in class 5, but because of unfavorable moisture conditions, such as in areas subjected to seasonally heavy rainfall, sufficient water is present to penetrate the soil to appreciable depths and, though the area be fairly well drained, the soil becomes plastic during such seasons, and when squeezed will readily assume any shape (a soil not uncommon in fairly flat areas).
7a	Loose, dry: found in arid regions, sand or gravel usually predominating (filled-in or built-up areas in dry regions fall into this class, and as the name implies, there is very little bond to hold the particles together).
7b	Loose, wet: same as loose, dry for holding power; high in sand, gravel or loam content. Holding power in some seasons is good, but during rainy seasons soil absorbs excessive moisture readily with resultant loss of holding power, especially in poorly drained areas. This class also includes very soft wet clay.
8	Swamps and marshes.

Courtesy Long Island Lighting Co.

TABLE 5-5B SELECTION OF ANCHORS—APPROXIMATE HOLDING POWER, lb
Diameter of screw or expanded anchor plate; diameter and length of rod

			Type of anchor and rod size			
	*Screw**		*Expanding*		*Swamp***	
Soil class	*8-in 1 in × 5.5 ft*	*Eight-way 8-in ¾ in × 8 ft*	*Eight-way 10-in 1 in × 10 ft*	*Four-way 12-in 1¼ in × 10 ft*	*13-in 2-in pipe*	*15-in 2-in pipe*
1***	NR	NR	NR	NR	NR	NR
2***	NR	NR	NR	NR	NR	NR
3	NR	26,500	31,000	40,000	NR	NR
4	11,000	22,000	26,500	34,000	NR	NR
5	8,000	18,500	21,000	26,500	NR	NR
6	6,500	15,000	16,500	21,500	NR	NR
7	3,500	10,000	12,000	16,000	NR	NR
8	NR	NR	NR	NR	12,000	15,000

*Screw anchors are used especially in temporary installations because of easy removal.

**At least one 10-ft length of 2-in pipe should be installed; additional lengths should be installed until pipe can no longer be turned (say, by four workers operating the wrenches).

***Special rock anchors of varying holding power should be used.

NR—Not recommended.

Courtesy Long Island Lighting Co.

Soils

Soils may be classified very roughly as indicated in Table 5-5A; anchor requirements for the different classes of soil are shown in Table 5-5B. Moist soils will vary in their classification during the year because of changes in moisture content, and the worst condition should be considered for design purposes.

CONDUCTORS
Tensions and Sags

The stringing of conductors on an overhead system presents other problems than those created by their dead weight and the effects of wind and ice on them. If they are stretched too tightly between poles, the stresses imposed on the pole structure (including pins, insulators, cross arms, racks, and hardware) would be such as to render them impractical. The stresses on the conductors themselves increase rapidly as remaining sag is eliminated, causing them to exceed their elastic limits by any small movement of the pole or conductor. The result would be greater permanent elongation, a reduction in overall cross section of the conductor, and a greater possibility of conductor failure.

On the other hand, if they are stretched too loosely, the swaying or deflection of the conductors would necessitate exceedingly wide spacing, in both the horizontal and vertical planes, with the support system approaching impracticability. The final construction should, therefore, provide for sufficient sag so that the elastic limit of the conductors will not be exceeded by a sufficient margin, while maintaining clearances that may be required under the probable conditions that may be encountered.

The tension in a conductor may be controlled by maintaining a proper sag in it, the tension being approximately inversely proportional to the sag. The sag in a conductor must be determined not only by the loading conditions (i.e., light, medium, or heavy), but also by probable temperature variations; local physical conditions and restrictions of codes and regulations must also be taken into consideration.

Calculation

In calculating sags and tensions in a conductor, the loading is assumed to be applied uniformly over its length, with the conductor capable of assuming its final shape freely, that of a catenary. To simplify the calculation, the loading can be assumed to be uniformly applied over its span, as shown in Fig. 5-19, resulting in a parabola. For relatively short spans, such as are found in distribution systems, the error is well within the practical limits of construction practices in the field, and may be neglected. (For long transmission spans, however, the differences become significant and reference should be made to calculating catenary methods for such lines.)

For parabolas, the relation between sag or deflection d, tension T, and span length L is expressed:

$$d = \frac{wL^2}{8T}$$

where d and L are in feet, T in pounds, and w, the resultant load (conductor, ice, etc.), in pounds per foot.

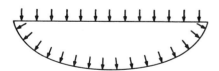

FIG. 5-19 Uniform loading on conductors.

For long spans, the horizontal component H of the tension in the conductor is given by the right-angle triangle relation:

$$H^2 = T^2 - (wx)^2$$

where x is one-half the length of the conductor in the span, but may be taken as approximately one-half the length of the span itself. The conductor, at the point of support where the tension T is maximum, is at an angle $\tan^{-1}$ *(wx/H)* to the horizontal; its length l compared to the span length L is given by

$$l = L + \frac{8d^2}{3L}$$

where d is the sag in feet.

The sags and tensions vary with temperatures; they are usually calculated at a "standard" temperature (say 60°F) and adjusted to other temperatures. As the temperature increases, the conductor expands and becomes longer; the sag increases and the tension in the conductor decreases. The elongation of a conductor from temperature *change* is equal to its length l multiplied by the change in temperature t and the coefficient of expansion e:

$$\text{Elongation} = lte$$

Likewise, if the tension is lessened, the elongation of the conductor decreases, as does the sag; if the loading on the conductor is increased, the tension and the accompanying elongation are increased, increasing in turn the sag and decreasing the tension. The elongation, or change in length, because of change in tension is given by:

$$\text{Change in length} = \frac{T_1 - T_2}{aE}$$

where $T_1 - T_2$ is the *change* in tension, a is the cross-sectional area, and E the modulus of elasticity.

Unequal Spans It is often impractical to have all spans of equal length, but each span in a level line should have approximately the same tension, with differing sags or severe stresses imposed on the supporting structure. Where the difference in adjacent span lengths is large, the conductors of the unequal adjacent spans should be dead-ended on the common pole.

Different Elevations Conductors suspended between supports at different elevations must be sagged so that the low point occurs between the center of the span and the lower support; if the low point is higher than the lower support,

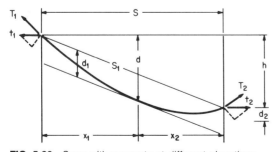

FIG. 5-20 Span with supports at different elevations.

an uplifting force will develop that may upset the balance of the lower support. As temperatures and loadings change, the low point will move horizontally along the span. For purposes of design, it may be assumed that this point remains fixed.

Referring to Fig. 5-20, the sags and tensions are based on the half-span between that point and the upper support, the sag being measured from a horizontal line through the upper support. The location of the low point may be determined approximately:

$$x_1 = \frac{S}{2} + \frac{ht}{Sw} \quad \text{or} \quad x_1 = S\frac{\sqrt{d}}{\sqrt{d-h} + \sqrt{d}}$$

$$x_2 = S - x_1 \quad \text{or} \quad x_2 = \frac{S}{2}\left(1 - \frac{h}{4d_1}\right)$$

and

$$d_2 = d_1\left(1 - \frac{h}{4d_1}\right)^2$$

The horizontal components of the conductor tensions t_1 and t_2 must be equal and the vertical component must be greater on the upper support; that is, t_1 must be greater than T_2 and

$$T_1 = T_2 + wh$$

where w is the weight per foot of conductor.

The sag in the inclined span is more nearly equal to the sag in a level span of a length equal to $S_1 + (S_1 - S)$. For practical purposes, however, the sag may be computed as if the supports were on the same elevation and the span length were S; the sag may be conveniently measured as the vertical distance d_1 from the line through the points of support.

Sag for ACSR

The processes described for determining sag pertain to conductors of a single, homogeneous material, e.g., copper, aluminum, or steel. For composite conductors, such as ACSR, the stresses on each of the different materials of the conductors are taken into consideration. At low levels of tension, the permanent set in the aluminum strands and the slight looseness of the steel cause all the load to be carried by the steel strands; the conductor stretches as if it consisted only of the steel strands. For greater loads, however, the aluminum and steel work together essentially as a single conductor.

Temperature changes change the length of the aluminum and steel strands in relation to each other, and also the point at which the steel and aluminum strands work together. The distribution of the tension between the steel and aluminum strands also changes, creating, in turn, a different expansion or contraction of the composite conductor. The coefficient of expansion for the com-

TABLE 5-6 TYPICAL CONDUCTOR SEPARATION

Voltage between conductors	*Minimum horizontal spacing*	*Minimum vertical spacing*
Up to 8700	12 in	16 in
8701 to 50,000	12 in, plus 0.4 in for each 1000 V above 8700 V*	40 in
Above 50,000	Same as for 8701 to 50,000 V	40 in, plus 0.4 in for each 1000 V above 50,000 V

*This is approximate. To determine spacing, the following formulas are specified. For conductors (copper or the equivalent) smaller than no. 2 AWG:

$$S = 0.3E + 7\sqrt{\frac{d}{3}} - 8$$

For conductors larger than no. 2 AWG:

$$S = 0.3E + 8\sqrt{\frac{d}{12}}$$

where the separation S is in inches, the line voltage E is in kilovolts, and the greatest sag d of the conductor at 60°F is in inches.

For current industry recommended values, refer to the latest revision of the National Electric Safety Code.

posite conductor is, therefore, not the same at all tensions as it is in conductors of one material only.

This coefficient of expansion for the composite conductor may be derived with sufficient accuracy for practical purposes from the coefficients of expansion a and the modulus of elasticity E and the percent area H for each of the metals involved:

$$E_{AS} = E_A H_A + E_S H_S$$

and

$$a_{AS} = a_A \frac{E_A H_A}{E_{AS}} + a_S \frac{E_S H_S}{E_{AS}}$$

EXAMPLE 5-9

For a no. 4/0 ACSR conductor,

$$E_A = 9 \times 10^6 \quad H_A = 0.857 \quad a_A = 12.8 \times 10^{-6} \text{ per °F}$$

$$E_S = 29 \times 10^6 \quad H_S = 0.143 \quad a_S = 6.6 \times 10^{-6} \text{ per °F}$$

$$E_{AS} = 9 \times 10^6 \times 0.857 + 29 \times 10^6 \times 0.143 = 11.860 \times 10^6$$

and

$$a_{AS} = 12.8 \times 10^{-6} \times \frac{7.713 \times 10^6}{11.860 \times 10^6} + 6.6 \times 10^{-6} \times \frac{4.147 \times 10^6}{11.860 \times 10^6}$$

$$= 10.632 \times 10^{-6} \text{ per } °F$$

Sag is a limiting factor in the allowable horizontal spacing between conductors because of the possibility of their whipping together in the wind. Typical minimum separation distances are shown in Table 5-6.

As a practical matter, for distribution systems, the construction, maintenance, and operating conditions, including climbing and working space, and experience—all are reflected in the standards for separation distances that exceed the minimum requirements (based on electrical considerations).

GRADES OF CONSTRUCTION

Good engineering practice recognizes different conditions and circumstances, in which prudence dictates that construction reflect the hazards involved and the potential consequences of failure. Construction is divided into three grades labeled B, C, and N, in declining order of importance.* These specify greater factors of safety in the utilization of materials, e.g., poles and conductors. See Table 5-7.

In general, grade B construction is specified for all supply circuits crossing over railroad tracks; for open-wire supply circuits of over 7500 V or constant-current circuits exceeding 7.5 A where crossing over communication circuits; and in urban and suburban districts.

Grade C construction is specified for open-wire supply circuits of over 7500 V in rural districts where crossing over or in conflict with supply circuits of 0 to 750 V, excepting services; and for open-wire supply circuits of 750 to 7500 V in urban districts under nearly all conditions except as noted for grade B construction, and also where crossing over or in conflict with communication circuits.

Grade N construction is specified for all other types of construction where grade B or C construction is not specified.

Conflict with communication lines includes not only those structures where the two are on the same pole, but also where they are so near each other that the overturning of one at the ground line would cause it to fall onto the other.

*The fourth edition (1940) of the National Electric Safety Code specified five grades of construction for power lines: A, B, C, D, and E. Later editions simplified these and grades A and E were eliminated; grade A was combined with B, grade E combined with D, and grade N established, generally defined as "safe construction." Grades B, C, and N apply to power lines, and are specified in later editions.

TABLE 5-7 TYPICAL SAFETY FACTORS FOR POLES AND CONDUCTORS

1. Safety factors for poles

Situation	When installed		At replacement: treated or untreated poles
	Treated poles	*Untreated poles*	

At crossings: pole lines of one grade of construction throughout

Grade B	4.0	4.0	2.67
Grade C	2.0	2.67	1.33
Grade N	1.33	1.33	0.67

Poles in isolated sections of higher-grade construction in pole lines of lower grade

Grade B	3.0	4.0	2.0
Grade C	2.0	3.0	1.33
Grade N	1.33	1.67	0.67

Elsewhere than at crossings

Grade B	4.0	4.0	2.67
Grade C	2.0	2.0	1.33
Grade N	1.0	1.33	0.67

2. Safety factors for conductors

Grade B 50 % of ultimate strength
Grade C 50 % of ultimate strength
Grade N 60 % of ultimate strength

For current industry recommended values, refer to the latest revision of the National Electric Safety Code.

CLEARANCES

Prudent design provides for the safety of both the public and the worker when in the vicinity of power lines; moreover, it is desirable that interruptions to electric service be held to a practical minimum. In addition to adequate construction, certain clearances should be maintained between energized conductors and the surrounding structures in both horizontal and vertical planes. The NESC recommends minimum clearances for most of the situations that may be normally encountered.

Clearances include those between conductors and the ground, along and across roads, walkways, driveways, and railroads, and those over buildings and structures; these are shown graphically in Fig. 5-21a. Clearances between conductors and other lines, guy wires, service conductors, and communication cir-

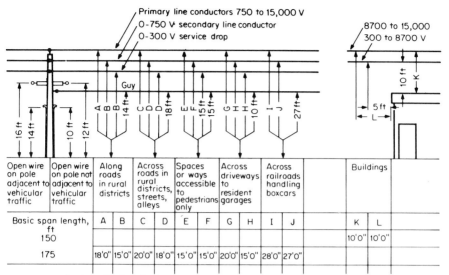

For heavy loading district except for 3-strand conductors each wire of which is 0.09 in or less in diameter.
Clearances in table are increased for spans above the basic span length.

(a)

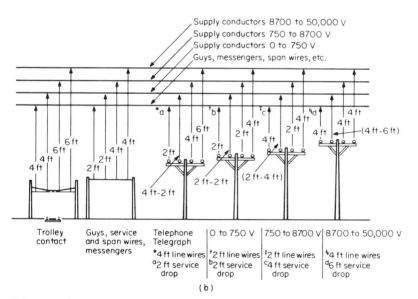

(b)

FIG. 5-21 (a) Typical minimum basic clearances, feet and inches. (b) Typical minimum wire crossing clearances. (From E. B. Kurtz, Lineman's Handbook, 3d ed., McGraw-Hill, New York, 1955.)

cuits are shown in Fig. 5-21*b*. These clearances are from conductors energized at both secondary and primary voltages at 60°F and no wind.

During high winds, conductors may sway into trees, or tree branches may fall and come in contact with the conductors. These may cause grounds and short circuits as well as physical breakages of lines that cause them to fall to the ground. It is, therefore, necessary to provide safeguards to prevent such occurrences by maintaining liberal clearances to trees.

JOINT CONSTRUCTION

Advantages

The joint use of poles for both electrical distribution and communication circuits, usually supplying the same consumers, has advantages that are well recognized. Not only are better appearance and cost savings achieved by a single line, but it is often easier to maintain proper clearances between the facilities of the several users when they are on the same pole and do not have to cross over or under each other. The largest employers of joint construction are the power and telephone utilities; others that may also be included are telegraph, traffic and special light controls, police and fire alarms, and (increasingly) cable television circuits. In rural areas, where appearance is not a paramount consideration, where long spans and lower clearances are permitted, and where services are few, joint construction may not prove practical.

In general, however, joint use appears desirable for urban and suburban areas, whether the facilities are located in streets and alleys, or on rear-lot lines and easements from which services to consumers are extended. Such construction, however, may require heavier loading of poles, frequent use of higher poles, often a greater grade of construction, additional guying, special maintenance procedures, and greater coordination between the joint users. Clearances between facilities and grades of construction are considered in the allotment of space on the pole and the distribution of costs and savings.

Stresses

In determining stresses imposed on poles, those imposed by all the users must be taken into consideration. Typical wind loadings for several sizes of telephone cable, with a wind of 4 lb/ft² on wires covered with a half-inch of ice, are given in Table 5-8. Tension acts on the pole from the messenger only. Typical maximum loads for several sizes of steel wire are given in the discussion of guys and anchors in this chapter.

Space Allocation

The space allotted on a pole for each purpose should be definitely specified. Usually the first point of attachment on a pole (from the ground up) is a communication conductor; its sag is taken into account to produce the minimum ground clearance at the center of its span. Other communication circuits follow

TABLE 5-8 LOADINGS ON TELEPHONE CABLES

Telephone cable sizes	Gauge (Cu)	Wind loading, lb/ft
50-pair, including messenger	24	1.02
	22	1.03
200-pair, including messenger	24	1.23
	22	1.27
600-pair, including messenger	24	1.53
	22	1.54

above, with the neutral zone separating them from the power circuit or circuits located at the top of the pole.

The division of space on a pole can be based on the assumption that the communication lines will take as much space as is necessary, with the power line taking the remainder above the neutral zone, making the construction conform to the space allowed. Refer to Fig. 5-22. Another method specifies a standard division of space on a "standard pole"; since minimum ground clearance is not the same for all locations, more space may be allotted to the communication circuits than may be required. If more space is required by the power conductors, a higher pole may be necessary, unless agreement can be reached decreasing the communication requirements.

Division of Costs The division of costs in joint use may be more complex than appears on the surface, and distribution engineers are usually assigned this chore. In general, there are two ways of accomplishing this: the pole can be jointly owned—i.e., each user owns a share of the pole; and it can be wholly owned by one user, who then rents space to the other users. While such divisions appear to be equitable and readily attainable, other factors enter to complicate the process.

Other Considerations In practice, poles are set and replaced by the power or telephone utilities. In many instances, the power utility has the equipment to set

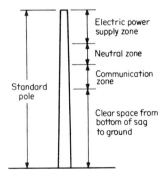

FIG. 5-22 Allotment of space on joint-use standard pole.

poles longer than 40 ft and is the sole utility stocking them. When emergencies occur, the power utility must respond as quickly as it can, while the telephone utility can defer its work until after the power crews complete their work. Where replacement of poles involving primary voltages above 5000 V is necessary, the power company is equipped to make the replacement, often cutting the old pole below the lowest power circuit, permitting the telephone crews to relocate their facilities to the new pole and remove the stub of the old pole without coming near the power lines. Also, tree trimming costs are almost always totally assumed by the power company although all users benefit.

Power lines on the top of the pole provide lightning protection for the facilities installed farther down. Obtaining rights-of-way, franchises, permits, etc., is usually done by one of the utilities. On the other hand, power lines may cause inductive interference in the operations of the other users, and can cause dangerous and widespread damage should they fall on the facilities of the others. Liability of each party in case of injury to workers or the public must be given consideration. These account for some—not all—the factors that the ingenuity of electrical distribution engineers and their counterparts in the communication utility must consider in order to come to an equitable and fair solution acceptable to all the parties concerned.

PRACTICAL DESIGN METHODS*
General

While a pole line must be designed pole by pole and the necessary strength must be built into each section of the line, much of the repetitious, time-consuming, and tedious calculation can be eliminated through the use of previously prepared tables and charts. Deviations from calculated theoretical values are compensated for by the tolerances incorporated in the standardization of the component parts, by the application of relatively liberal factors of safety, by provisions for future growth, and by the approximation of values inherent in the determination of the strength of materials, especially of wood products (and these carry an even greater margin of safety when treated by preservatives).

Design procedures containing necessary instructions and engineering data are prepared for selecting the proper pole, guy, and anchor for each segment of distribution pole line construction.

Design considerations should include conductors and other equipment to be added in the future, based on available information at the time of design. Also, the loading of communication (usually telephone) conductors must be taken into account in the selection of poles and anchors; sizes of guy wires are often specified to accommodate only the power conductors, while those for telephone conductors are specified and installed by the telephone utility.

---

*Adapted from Long Island Lighting Co. *Design & Application Standards.*

Use of the Tables

The tension at which a conductor is installed on a pole is a function of its material, its size, and the thickness of the covering, if any, over the conductor. This tension (in pounds) multiplied by the height of the conductor above ground (in feet) gives a value of foot-pounds of energy applied at the ground line of the pole. This is also called the ground-line moment, and poles are rated according to their ability to resist the moment.

In order to evaluate the total ground-line moment on a pole, multiply each conductor's tension by its respective height above ground and then add the results together.

To simplify calculations, the tables express values in *units*, making the job of analysis one of simple addition. All conductors—primary, secondary, telephone—are compared to *one* reference conductor at the primary level, having a design loading tension of 150 lb and a unit value of 1.0.

EXAMPLE 5-10

A no. 1/0 bare aluminum conductor, with a loading tension of 1200 lb, is given a unit value of 1200 divided by 150, or 8.0; 336,400-cmil aluminum with a loading tension of 2000 lb thus has a unit value of 2000 divided by 150, or 13.3 units; and so forth.

EXAMPLE 5-11

A three-conductor no. 1/0 aluminum secondary cable is converted first to its reference unit value, i.e., 2000 divided by 150, or 13.3 units. This is referenced up to the primary position by multiplying this unit value by the ratio of secondary height to primary height: for a 40-ft pole, this is 30.5 ft divided by 33 ft, or 0.92. The resulting unit value is 13.3 × 0.92, or 12.3 units. This means that a no. 1/0 triplex secondary conductor is equivalent to 12.3 primary conductors, each with a loading value of 150 lb of tension.

Selecting the Pole

The ability of a pole to be self-supporting depends on the class of the pole and the load it must carry (ignoring for the present the characteristics of the soil).

Classes of Poles All wood poles are divided into classes based on thickness and circumference. One system uses five classes: 5, 4, 2, 0, and 00, ranging from moderately thin (class 5) to extra heavy (class 00).

Knowing the load, it is possible then to select the proper class of pole for each location and degree of loading. If the number of units of loading is greater than the number of units the pole can support by itself, either the pole must be guyed or the conductors slackened to reduce the tension. Tables 5-9, 5-10, and 5-11 provide the necessary data for selecting the proper size of pole.

TABLE 5-9 GUYING TABLE: CONDUCTOR LOADING VALUES FOR 40-ft POLES (IN UNITS)*

Conductor size	Maximum loading, lbs	Turn angle, degrees						
		5	10	15	20	25	30	35
Primary								
Al 1/0 bare	1200	0.7	1.4	2.1	2.8	3.5	4.1	4.8
1/0 HDPE-PVC	1500	0.9	1.7	2.6	3.5	4.3	5.2	6.0
3/0 bare	1450	0.8	1.7	2.5	3.4	4.2	5.0	5.8
3/0 HDPE-PVC	1800	1.0	2.1	3.1	4.2	5.2	6.2	7.2
350,000-cmil bare	2000	1.2	2.3	3.5	4.6	5.8	6.9	8.0
350,000-cmil HDPE-PVC	2000	1.2	2.3	3.5	4.6	5.8	6.9	8.0
Cu 6 bare	640	0.4	0.7	1.1	1.5	1.8	2.2	2.6
3 bare	1070	0.6	1.2	1.9	2.5	3.1	3.7	4.3
2 PVC	1230	0.7	1.4	2.1	2.8	3.5	4.2	4.9
1/0 bare	1550	0.9	1.8	2.7	3.6	4.5	5.3	6.2
1/0 HDPE-PVC	2000	1.2	2.3	3.5	4.6	5.8	6.9	8.0
4/0 bare	1650	1.0	1.9	2.9	3.8	4.8	5.7	6.6
4/0 HDPE-PVC	2000	1.2	2.3	3.5	4.6	5.8	6.9	8.0
Secondary								
Al 3 c or 4c—all	2000	1.1	2.1	3.2	4.3	5.3	6.3	7.4
Cu 3c or 4c—6	640	0.3	0.7	1.0	1.4	1.7	2.0	2.4
3c or 4c—2	1500	0.8	1.6	2.4	3.2	4.0	4.8	5.5
4c—2	2000	1.1	2.1	3.2	4.3	5.3	6.3	7.4
Telephone								
6M	3600	1.8	3.4	5.1	6.8	8.5	10.2	11.8
10M	6000	2.9	5.7	8.6	11.4	14.2	17.0	19.8
16M	9600	4.6	9.2	13.7	18.2	22.7	27.2	31.6

*The conductor loading value, in units, appropriate to each combination of conductor size and turn angle is the resultant force on the pole due to conductors pulling in two directions.

**Conductor loading values for dead-end poles are in the 60° angle column.

Similar tables have been prepared for other pole sizes, from 25 to 55 ft.

Courtesy Long Island Lighting Co.

				Turn angle, degrees						
_40_	_45_	_50_	_55_	_60**_	_65_	_70_	_75_	_80_	_85_	_90_
5.5	6.1	6.8	7.4	8.0	8.6	9.2	9.7	10.3	10.8	11.3
6.8	7.6	8.5	9.2	10.2	10.7	11.5	12.2	12.9	13.5	14.1
6.6	7.4	8.2	9.0	9.7	10.4	11.1	11.8	12.5	13.1	13.7
8.2	9.2	10.1	11.1	12.0	12.9	13.8	14.6	15.4	16.2	17.0
9.1	10.2	11.2	12.3	13.3	14.3	15.3	16.2	17.1	18.0	18.9
9.1	10.2	11.2	12.3	13.3	14.3	15.3	16.2	17.1	18.0	18.9
2.9	3.3	3.6	3.9	4.3	4.6	4.9	5.2	5.5	5.8	6.0
4.9	5.5	6.0	6.6	7.1	7.7	8.2	8.7	9.2	9.6	10.1
5.6	6.3	6.9	7.6	8.2	8.8	9.4	10.0	10.5	11.1	11.6
7.1	7.9	8.7	9.5	10.3	11.1	11.8	12.6	13.3	13.9	14.6
9.1	10.2	11.3	12.3	13.3	14.3	15.3	16.2	17.1	18.0	18.9
7.5	8.4	9.3	10.2	11.0	11.9	12.6	13.4	14.1	14.9	15.6
9.1	10.2	11.3	12.3	13.3	14.3	15.3	16.2	17.1	18.0	18.9
8.4	9.4	10.4	11.3	12.3	13.2	14.1	15.0	15.8	16.6	17.3
2.7	3.0	3.3	3.6	3.9	4.2	4.5	4.8	5.1	5.3	5.6
6.3	7.0	7.8	8.5	9.2	9.9	10.0	11.2	11.8	12.4	13.0
8.4	9.4	10.4	11.3	12.3	13.2	14.1	15.0	15.8	16.6	17.3
13.5	15.1	16.6	18.2	19.7	21.1	22.6	24.0	25.3	26.6	27.8
22.4	25.1	27.7	30.3	32.8	35.2	37.6	39.9	42.2	44.3	46.4
35.9	40.8	44.6	48.5	52.5	56.4	60.2	63.9	67.5	70.9	74.2

TABLE 5-10 GUYING TABLE; TRANSVERSE LOADING VALUES FOR 40-ft POLES (IN UNITS)

Conductor size		Wind load, lb/ft	Average span length, ft						Long-span construction			
			75	100	125	150	175	200	250	300	350	400
Primary												
Al	1/0	0.566	0.3	0.4	0.5	0.6	0.7	0.8	0.9	1.1	1.3	1.5
	3/0	0.601	0.3	0.4	0.5	0.6	0.7	0.8	1.0	1.2	1.4	1.6
	336,000-cmil	0.655	0.3	0.4	0.5	0.7	0.8	0.9	1.1	1.3	1.5	1.7
Cu	3	0.524	0.3	0.3	0.4	0.5	0.6	0.7	—	—	—	—
	1/0	0.558	0.3	0.4	0.5	0.6	0.7	0.7	0.9	1.1	1.3	1.5
	4/0	0.609	0.3	0.4	0.5	0.6	0.7	0.8	1.0	1.2	1.4	1.6
Secondary												
Al 3c	4	0.572	0.3	0.4	0.4	0.5	0.6	0.7	—	—	—	—
	1/0	0.676	0.3	0.4	0.5	0.6	0.7	0.8	—	—	—	—
	3/0-1/0-1/0	0.725	0.3	0.4	0.6	0.7	0.8	0.9	—	—	—	—
	3/0	0.743	0.3	0.5	0.6	0.7	0.8	0.9	—	—	—	—
	336,000 cmil-3/0-3/0	0.749	0.3	0.5	0.6	0.7	0.8	0.9	—	—	—	—
Al 4c	1/0	0.708	0.3	0.4	0.5	0.7	0.8	0.9	—	—	—	—
	3/0-1/0-1/0	0.741	0.3	0.5	0.6	0.7	0.8	0.9	—	—	—	—
	3/0	0.780	0.4	0.5	0.6	0.7	0.8	1.0	—	—	—	—
	336,000 cmil-3/0-3/0-3/0	0.846	0.4	0.5	0.6	0.8	0.9	1.0	—	—	—	—
Cu 3c	6	0.519	0.2	0.3	0.4	0.5	0.6	0.6	—	—	—	—
	2	0.601	0.3	0.4	0.5	0.6	0.6	0.7	—	—	—	—
	1/0-2-2	0.636	0.3	0.4	0.5	0.6	0.7	0.8	—	—	—	—
	4/0-2-2	0.679	0.3	0.4	0.5	0.6	0.7	0.8	—	—	—	—

Cu 4c 6	0.556	0.2	0.3	0.4	0.5	0.6	0.7	—	—	—
2	0.622	0.3	0.4	0.5	0.6	0.7	0.8	—	—	—
1/0	0.657	0.3	0.4	0.5	0.6	0.7	0.8	—	—	—
4/0-1/0-1/0	0.707	0.3	0.4	0.5	0.7	0.8	0.9	—	—	—

Copperweld-Alumoweld

Three no. 6 to $\frac{7}{16}$ in	0.480	0.2	0.3	0.4	0.4	0.5	0.6	—	—	—

Unbalanced service

Concentric and twisted, 200 lb	—	1.2	1.2	1.2	1.2	1.2	1.2	—	—	—

Telephone

6M	1.20	0.5	0.7	0.8	1.0	1.1	1.3	—	—	—
10M	1.50	0.6	0.8	1.0	1.2	1.4	1.6	—	—	—
16M	1.80	0.7	1.0	1.2	1.5	1.7	2.0	—	—	—
25M	2.10	0.9	1.1	1.4	1.7	2.0	2.3	—	—	—
Twin	0.50	0.2	0.3	0.3	0.4	0.4	0.5	—	—	—

The transverse loading value, in units, is the unbalanced loading force on the pole due to the wind blowing against the pole and conductors.

Note: Where the computed average span length does not equal one of the numbers heading the column, use the column headed by the number *closest* to the computed span length.

Similar tables have been prepared for other pole sizes, from 25 to 55 ft.

Courtesy Long Island Lighting Co.

TABLE 5-11 GUYING TABLES: GROUND-LINE RESISTING MOMENTS FOR 40-ft POLES (IN UNITS)

| Pole class | Grade B construction | | | | | | | | Grade C construction | | | | | |
| | Longitudinal loading | | Dead-end loading | | Angle loading | | Vertical and transverse loading | | Dead-end loading | | Angle loading | | Vertical and transverse loading | |
	A	B	A	B	A	B	A	B	A	B	A	B	A	B
00	30.5	40.6	20.3	30.5	17.5	24.3	13.2	15.2	32.5	40.6	25.9	33.3	18.2	30.5
0	25.8	34.4	17.2	25.8	14.8	20.9	8.6	12.9	25.6	34.4	21.9	28.2	13.7	25.8
2	18.0	23.9	12.0	18.0	10.3	14.5	6.0	9.0	18.0	23.9	15.2	19.5	9.6	18.0
4	11.4	15.2	7.6	11.4	6.6	9.2	3.8	5.7	11.4	15.2	9.7	12.5	6.1	11.4
5	9.0	12.1	6.0	9.0	5.2	7.3	3.0	4.5	9.0	12.1	7.7	9.8	4.8	9.0

Ultimate strength values: A—when installed; B—at replacement.

Directions. Read down the appropriate column until a value is reached that is equal to or greater than the loading value computed from Tables 5-9 and 5-10. Looking left to the first column, read the class of pole to be used.

Note: If the pole comes out to be *greater* than a class 2, a smaller pole and guy should be specified. A class 5 pole is adequate for usual conditions; however, the pole selected should conform to other sections of this procedure. Use Table 5-12 to select the guy wire and Table 5-13 to select the anchor.

Similar tables have been prepared for other pole sizes, from 25 to 55 ft.

Courtesy Long Island Lighting Co.

Heavier-Class Poles   Poles one class heavier than the class specified by the tables should be used for each of the following purposes:

1. Junction poles

2. Poles supporting alley or side arms

3. Poles supporting line disconnects (except in-line types) or fuse cutouts

   In addition, a pole of at least class 4 should be specified for dead-end poles, angle poles, and transformer (or capacitor, regulator, or other equipment) poles.

Extra Heavy-Class Poles   Class 0 and class 00 poles are special oversize poles and are used primarily in the following situations:

1. In place of sidewalk guys, the most expensive guys to install

2. At an angle in the line in place of the combination of a span guy with a stub pole and sidewalk guy on the opposite side of the road

3. At T intersections, where normally a sidewalk guy or a span guy with a stub pole would be used

4. Where guying permission or rights cannot be obtained

5. To satisfy consumer complaints, by replacing an existing or proposed anchor guy

Selecting the Guy Wire

When the loading exceeds the strength of a pole, the pole should be guyed. The physical location of the pole in the field will determine what type of guy should be used: anchor, span, head, or sidewalk type. Tables 5-12A and 5-12B provide the necessary data for selecting the proper size of guy wire and lead. Guy protectors should be installed on all anchor guys accessible to the public.

   When a pole line changes direction and the turn angle is less than 60°, the corner, or "pivot," pole may be guyed with a single guy bisecting the angle. At angles greater than 60°, the pole should be guyed against the stress in each direction.

Selecting the Anchor

Anchors come in many types and sizes, each designed for certain soil and guying conditions. While each will do its specific job better than another design of anchor, most find use under more than one set of conditions. Table 5-13 provides the necessary data for selecting the type and size of anchor.

Sag Calculation

Occasionally, the stringing tension of the conductors must be lessened, either in conjunction with cribbing or because the wire tension is too great.

   Increasing the sag reduces the tension, and the procedure for calculating

TABLE 5-12A GUYING TABLE: GUY WIRE AND LEADS, ft, FOR 40-ft POLES—GRADE B CONSTRUCTION

Number of units	Alumoweld						Copperweld					
	Dead-end and angle loading			Transverse loading			Dead-end and angle loading			Transverse loading		
	6M	11M	17M	6M	11M	17M	6M	11M	17M	6M	11M	17M
10	11	5	3	—	10	7	13	6	4	—	12	8
11	12	6	4	—	11	8	14	7	5	—	13	9
12	13	7	5	—	13	8	16	8	5	—	15	9
13	14	8	5	—	14	9	18	9	5	—	17	10
14	15	8	5	—	15	10	19	9	6	—	18	11
15	17	9	6	—	17	11	21	10	6	—	21	12
16	18	9	6	—	18	11	—	11	7	—	—	13
17	21	10	6	—	20	12	—	11	8	—	—	14
18	—	11	7	—	—	13	—	13	8	—	—	15
19	—	11	7	—	—	14	—	13	8	—	—	16
20	—	12	8	—	—	15	—	14	9	—	—	18
21	—	13	8	—	—	16	—	15	9	—	—	18
22	—	13	9	—	—	17	—	16	10	—	—	20
23	—	14	9	—	—	18	—	17	10	—	—	—
24	—	15	9	—	—	19	—	18	11	—	—	—
25	—	15	10	—	—	20	—	18	11	—	—	—
26	—	16	10	—	—	—	—	20	12	—	—	—
27	—	17	11	—	—	—	—	—	13	—	—	—
28	—	18	11	—	—	—	—	—	13	—	—	—
29	—	19	12	—	—	—	—	—	13	—	—	—
30	—	20	12	—	—	—	—	—	14	—	—	—
31	—	—	13	—	—	—	—	—	15	—	—	—
32	—	—	13	—	—	—	—	—	15	—	—	—
33	—	—	13	—	—	—	—	—	16	—	—	—
⋮	⋮	⋮	⋮	⋮	⋮	⋮	⋮	⋮	⋮	⋮	⋮	⋮
38	—	—	16	—	—	—	—	—	19	—	—	—

Directions. (1) Read down the first column, number of units, to the total units of loading—power conductors only—computed from Tables 5-9 and 5-10. (2) Move to the right to the group of three columns that pertains to the type of loading on the pole; e.g., Alumoweld, transverse loading. (3) The number in each box is the ground-line distance ("lead" distance) in feet between the base of the pole and the anchor rod. (4) Looking up, the number at the head of each column (6M, 11M, 17M) corresponds to the guy wire.

Note: In certain cases of heavy loading, more than one (or one size of) guy wire may be required to do the job. Care should be taken to ensure that the lead distances are identical for the guy wires specified.

Similar tables have been prepared for other pole sizes, from 25 to 55 ft.

Courtesy Long Island Lighting Co.

TABLE 5-12B GUYING TABLE: GUY WIRE AND LEADS FOR 40-ft POLES—GRADE C
CONSTRUCTION

Number of units	Alumoweld						Copperweld					
	Dead-end and angle loading			Transverse loading			Dead-end and angle loading			Transverse loading		
	6M	*11M*	*17M*	*6M*	*11M*	*17M*	*6M*	*11M*	*17M*	*6M*	*11M*	*17M*
10	8	4	3	15	8	5	9	5	3	18	9	6
11	9	5	3	17	8	5	10	5	3	20	10	6
12	9	5	3	18	9	6	11	6	3	—	11	7
13	10	5	3	21	10	6	12	6	4	—	12	8
14	11	6	4	—	11	7	13	7	5	—	13	8
15	12	6	4	—	12	8	15	8	5	—	14	9
16	13	7	5	—	13	8	15	8	5	—	15	9
17	14	7	5	—	14	9	17	8	5	—	16	10
18	15	8	5	—	15	9	18	9	6	—	18	11
19	16	8	5	—	16	10	20	9	6	—	19	11
20	18	9	6	—	17	11	—	10	6	—	21	12
21	18	9	6	—	18	11	—	11	7	—	—	13
22	20	9	6	—	19	12	—	11	7	—	—	14
23	—	10	6	—	21	13	—	12	8	—	—	15
24	—	11	7	—	—	13	—	13	8	—	—	15
25	—	11	7	—	—	14	—	13	8	—	—	16
26	—	12	8	—	—	15	—	13	9	—	—	17
27	—	12	8	—	—	15	—	14	9	—	—	18
28	—	13	8	—	—	16	—	15	9	—	—	18
⋮	⋮	⋮	⋮	⋮	⋮	⋮	⋮	⋮	⋮	⋮	⋮	⋮
37	—	18	11	—	—	—	—	—	13	—	—	—
⋮	⋮	⋮	⋮	⋮	⋮	⋮	⋮	⋮	⋮	⋮	⋮	⋮
41	—	—	13	—	—	—	—	—	15	—	—	—
⋮	⋮	⋮	⋮	⋮	⋮	⋮	⋮	⋮	⋮	⋮	⋮	⋮
43	—	—	13	—	—	—	—	—	15	—	—	—
⋮	⋮	⋮	⋮	⋮	⋮	⋮	⋮	⋮	⋮	⋮	⋮	⋮
49	—	—	16	—	—	—	—	—	18	—	—	—

The procedure for using this table is the same as that described in the footnote to Table 5-12A.

Similar tables have been prepared for other pole sizes, from 25 to 55 ft.

Courtesy Long Island Lighting Co.

new stringing sags based on the tension desired in the conductor is given below.
Note that safe clearances must be maintained between conductors at all times.

Sag Calculation for Slack Sections   On dead-end poles where adequate guying
or cribbing is not practical, the stringing sags for slack sections, limited to 100-
ft spans, can be calculated by the formula:

$$S = \frac{wL^2}{8T}$$

TABLE 5-13 GUYING TABLES: ANCHOR HOLDING POWER FOR 40-ft POLES

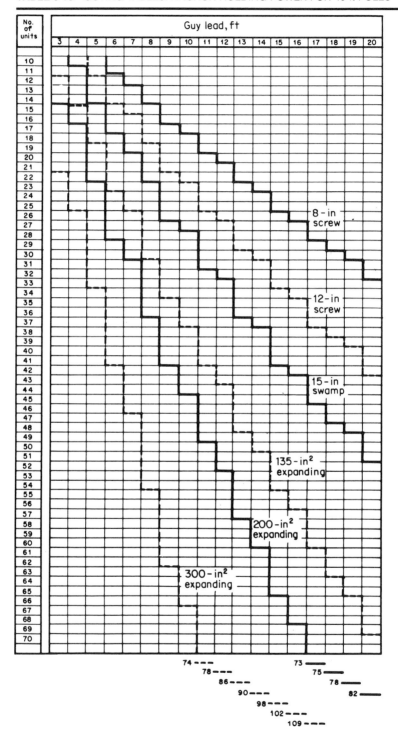

Directions. (1) Read down the first column to the loading value in units computed from Tables 5-9 and 5-10. This value should include telephone conductors. (2) Read right to the column for guy wire lead length, in feet, selected from Table 5-12A or 5-12B. (3) The first curve to the left of this point denotes

TABLE 5-14 CONDUCTOR HEAVY LOADING VALUES, lb/ft

Number of conductors	Type	Size	Heavy loading
Copper			
1	Bare, solid	6	0.916
		3	1.028
	Bare, stranded	1/0	1.269
		4/0	1.67
	Insulated (HDPE)	3	1.355
		1/0	1.60
		4/0	2.05
3	Insulated (XLPE)	6	1.38
		2	1.945
		1/0-2-2	2.15
		4/0-2-2	2.32
4	Insulated (XLPE)	6	1.545
		2	2.185
		1/0	2.81
		4/0-1/0-1/0-1/0	2.935
Aluminum			
1	Bare or stranded, alloy	1/0	1.129
		3/0	1.267
	Pure	336,400 cmil	1.490
	HDPE-insulated, alloy	1/0	1.39
		3/0	1.60
	Pure	336,400 cmil	1.85
3	XLPE-Insulated, pure, with alloy neutral	4	1.39
		1/0	1.822
		3/0-1/0-1/0	1.979
		3/0	2.15
	Pure	336,400 cmil-1/0-1/0	2.214
4	XLPE-insulated, pure, with alloy neutral	1/0	2.030
		3/0-1/0-1/0-1/0	2.295
		3/0	2.42
	Pure	336,400 cmil-3/0-3/0-3/0	2.77

HDPE = high-density polyethylene insulation.

XLPE = cross-linked polyethylene insulation.

Courtesy Long Island Lighting Co.

the minimum size of anchor to use. Anchor selection should be in accordance with other sections of the procedure.

Note: In cases where the loading value exceeds the strength of the desired anchor, an adjustment can be made by increasing the guy wire lead (see values above) or reducing the tension in the conductor. See text for calculation of slack sections.

Similar tables have been prepared for other pole sizes, from 25 to 55 ft.

Courtesy Long Island Lighting Co.

where S is the sag in feet, w is the heavy loading value in pounds per foot from Table 5-14, L is the length of the span in feet, and T is the maximum tension desired in the conductor in pounds.

EXAMPLE 5-12

Calculate the required sag for a 75-ft span of a three-phase four-wire section of no. 1/0 bare copper where the dead-end pole cannot be guyed. From previous calculations, it is necessary to limit the tension under loading conditions to 1000 lb per conductor.

$$w = 1.269 \text{ lb/ft} \qquad L = 75 \text{ ft}$$

$$S = \frac{1.269(75)^2}{8 \times 1000} = 0.9 \text{ ft or } 11 \text{ in}$$

Alternative Method

Some guying situations cannot be resolved using the method described above. Appendix 5A contains basic guying theory and a detailed method of determining pole, guy, and anchor sizes in terms of ground-line moments, pounds of tension in the guy wire, etc.

Examples As part of this procedure, examples are included illustrating problems in selecting poles, guys, and anchors for small-angle turns, for dead ends, and for right-angle turns in the line. These are contained in Appendix 5B.

APPENDIX 5A
PRACTICAL METHOD OF CALCULATING POLE AND GUY SIZES

INTRODUCTION

This Appendix describes a method of determining pole and guy sizes, including a step-by-step solution of loading problems. It covers the general run of design problems, but in some instances, more detailed calculations may be required.

POLE CLASS REQUIREMENTS
Factors Involved

The total loading on a pole is the accumulation of all the individual stress-producing forces that act on the pole. In this method of solution, eight factors, a through h, will determine the class of pole required:

a. Tension in wires or cables at dead ends: Table 5-17 (see the column for a 60° angle)

TABLE 5-15 TRANSVERSE LOADING PER CONDUCTOR

Type	Size	Loading, lb/ft
Primary wire		
PVC or HDPE: Cu	3	0.524
	1/0	0.558
	4/0	0.609
Al	1/0	0.566
	3/0	0.601
	336,400 cmil	0.655
Secondary mains		
XLPE: Cu, 3c	6	0.519
	2	0.601
	1/0-2-2	0.636
	4/0-2-2	0.679
Cu, 4c	6	0.556
	2	0.622
	1/0	0.657
	4/0-1/0-1/0-1/0	0.707
Al, 3c	4	0.572
	1/0	0.676
	3/0-1/0-1/0	0.725
	3/0	0.743
	336,400 cmil-1/0-1/0	0.749
Al, 4c	1/0	0.708
	3/0-1/0-1/0-1/0	0.741
	3/0	0.782
	336,400 cmil-3/0-3/0-3/0	0.846
Copperweld or Alumoweld		
	Three no. 8 to $\frac{7}{16}$ in	0.48 (max.)
Telephone and guy wire		
	6M	1.20
	10M	1.50
	16M	1.80
	25M	2.10
Twin		0.50

Courtesy Long Island Lighting Co.

b. Tension in wires or cables due to an angle in the line: Table 5-17

c. Transverse loading per wire or cable: Table 5-15

d. Transverse loading on poles: Table 5-19

e. Transverse loading on equipment: Table 5-18

f. Tension from unbalanced services: Table 5-16

g. Longitudinal loading at a change in grade: Table 5-17 (see the 60° angle column)

h. Factors of safety prescribed for poles: Table 5-19

Classification of Poles

To simplify the analysis, poles have been classified as:

1. Straightaway poles (in a straight line)

2. Angle poles (at angles or offsets in the line)

3. Dead-end poles

The procedures, each designed for a particular classification of pole, start on page 184.

TABLE 5-16 TENSION RESULTING FROM UNBALANCED SERVICES

Type	Number of conductors	Size, AWG	Tension, lb
Copper: open wire	1	8	75
	1	6	75
	1	3	100
	1	1/0	150
	1	4/0	300
Copper: concentric (cable)	2	8	100
	3	8	150
	3	6	180
	3	4	225
	3	2	300
Copper: twisted (cable)	3	6	450
	4	6	450
	3	2	450
	4	2	450
	4	1/0	450
Aluminum: twisted (cable)	3	4	450
	3	1/0	450
	4	1/0	450
	4	3/0	450
Telephone drop	2	—	75

Courtesy Long Island Lighting Co.

TABLE 5-17 TENSION IN CONDUCTORS—ANGLES AND DEAD ENDS—lb
For spans up to 200 ft

Type	Size	5°	10°	15°	20°	25°	30°	35°	40°	45°	50°	55°	60°*
All primary**													
	—	175	348	522	694	866	1036	1203	1367	1531	1640	1846	2000
Bare													
Cu	3	93	187	279	372	463	554	643	732	821	904	988	1070
	1/0	135	270	405	538	677	802	958	1060	1186	1310	1431	1550
	4/0	141	288	430	573	714	854	993	1129	1262	1394	1524	1650
Al	1/0	104	208	313	418	521	622	721	821	918	1015	1109	1200
	3/0	126	253	379	501	628	751	872	992	1110	1226	1339	1450
	336,400 cmil	175	348	522	694	934	1170	1380	1483	1609	1740	1840	2000
PVC and HDPE													
Cu	3	107	214	321	428	533	637	739	842	921	1041	1137	1230
	1/0	175	348	522	694	866	1036	1203	1367	1530	1690	1846	2000
Al	4/0	310	570	720	1086	1376	1684	1942	2209	—	—	—	—
	1/0	131	262	392	521	651	779	904	1030	1151	1270	1390	1500
	3/0	157	314	470	625	779	932	1083	1231	1390	1521	1662	1800
	336,400 cmil	175	348	522	694	866	1036	1203	1367	1530	1690	1846	2000
XLPE													
Cu	6	56	111	167	222	277	333	385	439	490	541	591	640
	3	131	262	392	521	651	779	904	1030	1151	1270	1390	1500
	1/0 1/0-2-2 4/0-2-2	187	373	569	743	928	1111	1289	1469	1589	1798	1931	2000
	4/0-1/0-1/0-1/0	175	348	522	694	866	1036	1203	1367	1530	1690	1846	2000
Al	1/0 3/0 1/0-1/0-1/0- 3/0-1/0-1/0-1/0 336,400 cmil-1/0-1/0 336,400 cmil-1/0-1/0-1/0	175	348	522	694	866	1036	1203	1367	1530	1690	1846	2000

179

TABLE 5-17 TENSION IN CONDUCTORS—ANGLES AND DEAD ENDS—lb (Continued)
For spans up to 200 ft

Type	Size	5°	10°	15°	20°	25°	30°	35°	40°	45°	50°	55°	60°*
Guy wire													
—	6M	319	628	940	1250	1559	1863	2165	2412	2755	3043	3324	3600
—	10M	524	1046	1541	2083	2548	3100	3609	4104	4585	5071	5540	6000
—	16M	837	1674	2507	3333	4156	4969	5774	6567	7346	8014	8865	9600
Angle (pull feet)*													
		4.35	8.7	13.1	17.4	21.7	25.9	30.0	34.2	37.3	42.3	46.2	50.0

*Use these values for dead ends.

**Spans longer than 200 ft.

***Determination of line angle in pull-feet (see diagram). Interpolate for other angles.

Table based on:

$P = \frac{1}{4}$ ultimate strength of conductor, or 2000 lb, whichever is less

R = Resultant tension due to angle

$R = 2P \sin \dfrac{a}{2}$

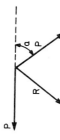

Adapted from Long Island Lighting Co. design standards.

TABLE 5-18 EQUIPMENT DATA AND MAXIMUM ALLOWABLE EQUIPMENT WEIGHTS ON POLES

Equipment		Weight lb		Allowable weight for transverse loading, lb	
Type	Size	Maximum	Minimum	Maximum	Minimum
Distribution transformers, 1ϕ	10 kVA	403	210	25	10
	15	649	265	30	15
	25	959	363	50	25
	37½	1139	540	55	35
	50	1305	640	65	40
	75	1869	825	95	50
	100	2240	1015	110	55
	167	2640	1430	130	70
	250	2200	1600	110	90
	333	2450	2100	110	105
	500	3100	2500	—	—
Regulators: induction, step	24 kVA	1215	—	65	—
	12	965 ⎫		50 ⎫	
	25	1200 ⎬ 1000		65 ⎬ 55	
	76.2	1700 ⎭		90 ⎭	
Reclosers	50–200 A	170	—	10	—
	560	755	—	30	15
Oil switch, 15-kV	200 A	60	—	5	—
	400	456	—	20	10
Capacitors, unit-type	25 kVAR	100	—	—	—
	50	90	65	5	—
	100	135	100	5	—
	150	110	80	5	—
Capacitors, switched	300 kVAR	1320	1090	65	60
	600	1290	875	65	50
Interrupter switch, 15-kV	600 A	450	150	55	—
Air-break switch, 34.5-kV	400 A	600	—	100	—
Street lighting transformer, types RO, ROC	10 kW	965	—	45	—
	15	1045	—	60	—
	20	1300	—	90	—

Maximum permissible weight of equipment on poles

Pole class	Weight, lb
5	900
4	1800
3	2500
2	3200
1	4000
0	5050
00	6750

Notes. (1) Pole classes specified are minimum to be used on new installations. Where poles already exist in the field, use of poles one class lighter than specified above is permitted. (2) Use 3 times the transverse loading for three transformers or other groups of equipment. (3) Allowance for transverse loading on equipment is equal to 4 lb/ft² on the projected area angles to the line.

Courtesy Long Island Lighting Co.

TABLE 5-19 GROUND-LINE RESISTING MOMENTS FOR CREOSOTED SOUTHERN PINE POL
In thousands of foot-pounds. Number above each column represents factor of safety.

		Pole			Grade C construction loading						
			Longitudinal		Dead-end		Vertical and transverse				
							Crossing		Other		
Height, ft	Class	Ultimate strength 1.0	A —	B —	A 1.33	B 1.0	A 2.66	B 1.33	A* 2.0	B 1.
30	00	154.2	No data		115.9	154.2	58.0	115.9	61.7	115
	0	124.0	required where		93.2	124.0	46.6	93.2	49.6	93
	1	103.7	grade C		78.0	103.7	39.0	78.0	41.5	78
	2	84.3	construction is		63.4	84.3	31.7	63.4	35.2	63
	3	67.5	used and no		50.9	67.5	25.4	50.6	27.0	50
	4	53.0	change in grade		39.8	53.0	19.9	39.8	21.2	39
	5	43.1	of construction		32.4	43.1	16.2	32.4	17.2	32
	6	34.4	exists.		25.9	34.4	12.9	25.9	13.8	25
35	00	176.5			132.7	176.5	66.4	132.7	70.6	132
	0	146.5			111.6	148.5	55.8	111.6	59.4	111
	1	123.6			93.0	123.6	46.5	93.0	49.4	93
	2	101.7			76.5	101.7	36.2	76.5	40.7	76
	3	82.5			62.0	82.5	31.0	62.0	33.0	62
	4	62.9			47.3	62.9	23.6	47.3	25.2	47
	5	51.7			38.9	61.7	19.4	38.9	20.7	38
	6	39.7			29.8	39.7	14.9	29.8	15.9	29
40	00	200.6			150.6	200.6	75.4	150.6	80.2	150
	0	170.1			127.9	170.1	63.9	127.9	66.0	127
	1	142.7			107.3	142.7	83.6	107.3	57.1	107
	2	110.6			89.2	110.6	44.6	89.2	47.4	89
	3	97.3			73.2	97.3	36.6	73.2	36.6	73
	4	75.1			56.5	75.1	26.2	56.5	30.0	56
	5	59.5			44.7	59.5	22.4	44.7	23.8	44
45	00	241.4			181.5	241.4	90.8	181.5	96.6	181
	0	200.1			150.5 ˙	200.1	75.2	150.5	80.0	150
	1	161.6			121.5	161.6	60.6	121.5	64.6	121
	2	135.2			101.7	135.2	50.6	101.7	54.1	101
	3	107.5			80.8	107.5	40.4	80.8	43.0	80
	4	87.6			65.9	87.6	32.9	65.9	35.0	65
50	1	181.9			136.8	181.9	68.4	136.8	72.6	136
	2	148.0			111.3	148.0	55.6	111.3	59.2	111
	3	118.5			89.1	118.5	44.5	89.1	47.4	89
	4	97.2			73.1	97.2	36.5	73.1	39.9	73
55	1	197.5			148.5	197.5	74.2	148.5	79.0	148
	2	161.5			121.4	161.5	60.7	121.4	64.6	121
	3	130.2			97.9	130.2	48.9	97.9	52.1	97
	.4	108.2			81.4	108.2	40.7	81.4	43.3	81

A = when installed; B = at replacement.

*These values include an additional factor of safety to allow for pole deflections and soil conditions peculiar t
territory.

Courtesy Long Island Lighting Co.

Grade B construction loading					
Longitudinal		Dead-end		Vertical and transverse	
A 1.33	B 1.0	A 2.0	B 1.33	A 4.0	B 2.66
115.9	154.2	77.1	115.9	38.6	59.0
93.2	124.0	62.0	93.2	31.0	46.6
78.0	103.7	51.9	78.0	25.9	39.0
63.4	84.3	42.2	63.4	21.1	31.7
50.8	67.5	33.8	50.8	16.9	25.4
39.6	53.0	26.5	39.6	13.3	19.9
32.4	43.1	21.6	32.4	10.8	16.2
25.9	34.4	17.2	25.9	8.6	12.9
132.7	176.5	88.3	132.7	44.1	66.4
111.6	149.5	74.3	111.6	37.1	55.6
93.0	123.6	61.8	93.0	30.9	46.5
76.5	101.7	50.9	76.5	25.4	36.2
62.0	82.5	41.3	62.0	20.6	31.0
47.3	62.9	31.5	47.3	15.7	23.6
36.9	51.7	25.9	36.9	12.9	19.4
29.8	39.7	19.9	29.8	9.9	14.9
150.8	200.6	100.3	150.8	50.2	75.4
127.9	170.1	85.1	127.9	42.5	63.9
107.3	142.7	71.4	107.3	36.7	53.6
89.2	116.6	59.3	89.2	29.7	44.6
73.2	97.3	48.7	73.2	24.3	36.6
56.5	75.1	37.6	56.5	18.6	26.2
44.7	59.5	29.6	44.7	14.9	22.4
181.5	241.4	120.7	181.5	60.4	90.8
150.5	200.1	100.1	150.5	50.0	76.2
121.5	161.6	60.8	121.5	40.4	60.8
101.7	135.2	67.6	101.7	33.8	50.8
80.8	107.5	53.6	80.8	26.9	40.4
65.9	87.6	43.8	65.9	21.9	32.9
136.8	181.9	91.0	136.8	45.5	68.4
111.3	148.0	74.0	111.3	37.0	55.6
89.1	118.5	59.3	69.1	29.6	44.5
73.1	97.2	48.6	73.1	24.3	36.5
148.5	197.5	98.6	148.5	49.4	74.2
121.4	161.5	80.8	121.4	40.4	60.7
97.9	130.2	65.1	97.9	32.6	48.9
81.4	108.2	54.1	81.4	27.1	40.7

Straightaway Poles

Step 1. Determine the grade of construction applying, B or C.

Step 2. From Table 5-19, determine the factor of safety required for transverse loading, and calculate the total load on the pole.
a. For grade B construction, the factor of safety is 4 and the total load is $4(c + d + e + f)$.
b. For grade C construction, the factor of safety is 2 and the total load is $2(c + d + e + f)$.
c. For grade C construction at crossings, the factor of safety is 2.66 and the total load is $2.66(c + d + e + f)$.

Step 3. Calculate the total ground-line moment on the pole by multiplying the total load (step 2) by the average height of the conductors or equipment.

Step 4. From Table 5-19, select the pole class having a ground-line resisting moment at least equal to that calculated in step 3.

Step 5. If the pole is to support equipment, from Table 5-18 determine the pole class required by the weight of the equipment. If it exceeds that selected in step 4, the larger class should be specified.

Step 6. If the transverse loading is too great for the unsupported pole, proceed to the section Guying Requirements in this appendix.

Angle Poles

Step 1. Determine the grade of construction applying.

Step 2. From Table 5-19 determine the factor of safety required for the type of loading and calculate the total load on the pole.
a. For grade B construction, the factors of safety are 4 and 2 and the total load is $4(c + d + e + f) + 2b$.
b. For grade C construction, the factors of safety are 2 and 1.33 and the total load is $2(c + d + e + f) + 1.33b$.
c. For grade C construction at crossings, the factors of safety are 2.66 and 1.33 and the total load is $2.66(c + d + e + f) + 1.33b$.

Step 3. Calculate the total ground-line moment on the pole by multiplying the total load (step 2) by the average height of the conductors.

Step 4. From Table 5-19 select the pole class having a ground-line resisting moment at least equal to that calculated in step 3.

Step 5. If the pole is to support equipment, from Table 5-18 determine the pole class required by the weight of the equipment. If it exceeds that calculated in step 4, the larger class should be specified.

Step 6. If the calculation in step 4 results in a class 5 pole, guying or cribbing will generally not be required unless soil conditions are poor. If the result is other than class 5, proceed to the section Guying Requirements.

Dead-End Poles

Step 1. Determine grade of construction applying, B or C.

Step 2. From Table 5-19 determine the factors of safety required for dead-end loading, and calculate the total load on the pole.
a. For grade B construction, the factor of safety is 2 and the total load is 2a.
b. For grade C construction, the factor of safety is 1.33 and the total load is 1.33a.

Step 3. Calculate the total ground-line moment on the pole by multiplying the total load (step 2) by the average height of the conductors.

Step 4. From Table 5-19 select the pole class having a ground-line resisting moment at least equal to that calculated in step 3.

Step 5. If the pole is to support equipment, from Table 5-18 determine the pole class required by the weight of the equipment. If it exceeds that calculated in step 4, the larger class should be specified.

Step 6. If the calculation in step 4 results in a class 5 pole, guying or cribbing will not generally be required unless soil conditions are poor. If the result is other than class 5, proceed to the section Guying Requirements.

GUYING REQUIREMENTS
General

Smaller poles may be specified on the assumption they will be used with guying; this is possible because the pole acts as a strut only, the horizontal load being supported by the guy wire. Poles supporting equipment, or placed at a turn angle greater than 10°, however, should not be smaller than class 4.

The steps enumerated above under Pole Class Requirements should cover all extreme cases of transverse or longitudinal loading. If the loading is too great for the unsupported pole, the following additional steps for each classification of pole, designed to determine guying requirements, should be considered.

Straightaway Poles

Step 7. From Table 5-19 determine the factor of safety required for transverse loading, and calculate the total load on the pole.
a. For grade B construction, the factor of safety is 2.66 and the total load is 2.66c.
b. For grade C construction, the factor of safety is 2 and the total load is 2c.

Step 8. From Table 5-20, calculate the guy tension.

Step 9. From Table 5-21, select the proper guy wire for the guy tension (step 8).

TABLE 5-20 MULTIPLIERS FOR DETERMINING TENSION IN GUY WIRE WHEN TENSION IN LINE WIRE IS KNOWN

Applicable to conditions A to D inclusive; see illustrations below.

H, ft	\multicolumn{13}{c}{D, ft}
	5	6	7	8	10	12	14	16	18	20	25	30	35
15	3.16	2.69	2.36	2.12	1.80	1.60	1.46	1.37	1.30	1.25	1.16	1.12	1.09
16	3.35	2.85	2.47	2.24	1.89	1.67	1.52	1.41	1.34	1.28	1.19	1.13	1.10
17	3.54	3.00	2.63	2.35	1.97	1.73	1.57	1.46	1.37	1.31	1.21	1.15	1.11
18	3.73	3.16	2.76	2.46	2.06	1.79	1.63	1.50	1.41	1.35	1.23	1.17	1.12
19	3.93	3.32	2.89	2.58	2.15	1.87	1.68	1.55	1.49	1.38	1.26	1.19	1.14
20	4.12	3.48	3.03	2.69	2.24	1.94	1.74	1.60	1.49	1.41	1.28	1.20	1.15
21	4.32	3.64	3.17	2.81	2.33	2.01	1.80	1.65	1.54	1.45	1.31	1.22	1.17
22	4.51	3.80	3.30	2.93	2.42	2.09	1.86	1.70	1.58	1.49	1.33	1.24	1.18
23	4.71	3.96	3.44	3.04	2.51	2.16	1.92	1.75	1.62	1.52	1.36	1.26	1.20
24	4.90	4.12	3.57	3.16	2.60	2.24	1.98	1.80	1.67	1.56	1.39	1.28	1.21
25	5.10	4.28	3.71	3.28	2.69	2.31	2.04	1.85	1.71	1.60	1.41	1.30	1.23
26	5.29	4.45	3.84	3.40	2.79	2.39	2.11	1.91	1.76	1.64	1.44	1.32	1.25
27	5.51	4.62	3.99	3.52	2.88	2.46	2.17	1.96	1.80	1.68	1.47	1.34	1.26
28	5.69	4.78	4.13	3.64	2.97	2.54	2.24	2.01	1.85	1.72	1.50	1.37	1.28
29	5.89	4.94	4.27	3.76	3.07	2.61	2.30	2.06	1.90	1.76	1.53	1.39	1.30
30	6.08	5.09	4.41	3.88	3.16	2.68	2.36	2.12	1.95	1.80	1.54	1.41	1.32
31	6.28	5.26	4.54	4.04	3.26	2.77	2.42	2.18	1.99	1.84	1.59	1.44	1.33
32	6.48	5.42	4.68	4.13	3.36	2.85	2.49	2.24	2.04	1.89	1.62	1.46	1.35
33	6.68	5.59	4.82	4.24	3.45	2.93	2.56	2.29	2.09	1.93	1.65	1.48	1.37
34	6.88	5.75	4.96	4.36	3.54	3.01	2.62	2.34	2.14	1.97	1.69	1.51	1.39
35	7.08	5.92	5.10	4.48	3.64	3.09	2.69	2.40	2.19	2.02	1.72	1.53	1.41
36	7.27	6.08	5.24	4.60	3.74	3.16	2.76	2.46	2.24	2.08	1.75	1.56	1.44
37	7.63	6.28	5.38	4.73	3.83	3.24	2.83	2.52	2.26	2.10	1.79	1.59	1.46
38	7.66	6.40	5.52	4.86	3.93	3.32	2.89	2.58	2.34	2.15	1.82	1.61	1.48
39	7.87	6.59	5.66	4.98	4.03	3.40	2.96	2.64	2.39	2.19	1.85	1.64	1.50
40	8.07	6.74	5.91	5.10	4.12	3.49	3.28	2.69	2.44	2.24	1.89	1.67	1.52

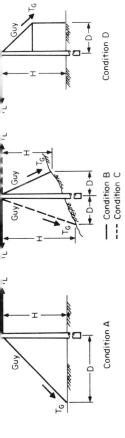

Condition A

Condition B
Condition C

— Condition B
--- Condition C

Condition D

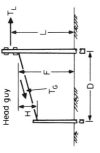

Condition E

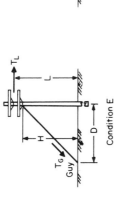

Head guy

Condition F

Example:

H = height of guy on pole = 25 ft
D = distance of guy from centerline of pole = 16 ft
T_L = tension in line wire = 2000 lb
T_S = tension in guy wire = T_L × multiplier (from table)
  = 2000 × 1.85 = 3700 lb

For condition E,

T_G = guy tension
T_L = resultant line tension
$$T_G = \frac{LT_LD^2 + H^2}{DH}$$

For condition F,

T_G = guy tension
T_L = resultant line tension
$$T_G = \frac{LT_LD^2 + H^2}{FD}$$

Courtesy Long Island Lighting Co.

TABLE 5-21 ALLOWABLE GUY TENSIONS

| Material | *Guy wire* | | | *Loading, lb* | | | |
| | *Reference* | *Strand* | *Ultimate strength, lb* | *Dead-end and angle* | | *Transverse* | |
				Grade C, SF = 1.14	*Grade B, SF = 1.50*	*Grade C, SF = 2.00*	*Grade B, SF = 2.66*
Copperweld	6M	3 #8	6,280	5,500	4,100	3141	2360
	11M	11/32″	11,280	9,860	7,520	5640	4240
	17M	7/16″	16,890	14,780	11,260	8445	6340
Alumoweld	6M	3 #8	7,210	6,330	4,810	3605	2710
	11M	11/32″	12,960	11,370	8,640	6480	4870
	17M	7/16″	19,060	16,720	12,700	9530	7160

Dead-end loadings or 60° angles: primary

Copper	*Loading, lb*	*Aluminum*	*Loading, lb*
No. 6 bare	640	No. 1/0 bare	1200
No. 3 bare	1070	No. 1/0 HDPE or PVC	1500
No. 3 PVC	1230	No. 3/0 bare	1450
No. 1/0 bare	1550	No. 3/0 HDPE or PVC	1800
No. 1/0 HDPE or PVC	2000	336,400 cmil bare	2000
No. 4/0 bare	1650	336,400 cmil HDPE or PVC	2000
No. 4/0 HDPE or PVC	2000		

Secondary

3 c or 4c no. 6	640	All triplex or quadruplex
3c or 4c no. 2	1500	secondaries are 2000 lb
Over 4c no. 2	2000	

For open wire secondary, use primary tensions.

Note: All 1/0, 3/0 and 336,400-cmil Al or 1/0 and 4/0 Cu, where the span exceeds 200 ft, are to be considered at 2000 lb of tension.

All triplex or quadruplex unbalanced services to be considered at 250 lb of tension.

Courtesy Long Island Lighting Co.

GUYING REQUIREMENTS (cont. from page 185)

Angle Pole

Step 7. If the pole class calculations result in a pole within the range from class 4 to class 2, the pole should be cribbed.

Step 8. If the result is a pole larger than class 2, a smaller pole and guying should be specified; a class 5 pole may be specified.

Step 9. From Table 5-19, select the factors of safety and calculate the total load on the pole.
a. For grade B construction, the total load is $(1.78c + b)1.50$.
b. For grade C construction, the total load is $(1.78c + b)1.14$.

Step 10. From Table 5-20 calculate the guy tension.

Step 11. From Table 5-21 select the proper guy wire for the guy tension (step 10).

Step 12. If the pole class calculations result in a pole larger than class 2 and the pole cannot be guyed, specify a class 0 or class 00 pole and determine if it has sufficient strength to handle the ground-line resisting moment, with cribbing specified if necessary.

Step 13. If even a class 00 double-cribbed pole is inadequate for the ground-line resisting moment and the pole cannot be guyed, a lighter pole and cribbing should be specified, with guying of the adjacent pole or poles and sagging of the line sufficiently to relieve the strain.

Dead-End Pole

Step 7. If the pole class calculations result in a pole within the range from class 4 to class 2, the pole should be cribbed.

Step 8. If the result is a pole larger than class 2, a smaller pole and guying should be specified. A class 5 pole will be adequate where either primary or secondary conductors terminate; where both terminate, a class 4 pole should be specified.

Step 9. From Table 5-19 select the factors of safety and calculate the total load on the pole.
a. For grade B construction, the total load is $1.50a$.
b. For grade C construction, the total load is $1.14a$.

Step 10. From Table 5-20 calculate the guy tension.

Step 11. From Table 5-21 select the proper guy wire for the guy tension (step 10).

Step 12. If the calculations above result in a pole larger than class 2 and the pole cannot be guyed, specify a class 0 or class 00 pole and determine if it has sufficient strength to handle the ground-line resisting moment, with cribbing specified if necessary.

Step 13. If even a class 00 double-cribbed pole is inadequate for the ground-line resisting moment and the pole cannot be guyed, a lighter pole and cribbing should be specified, with guying of the adjacent pole or poles and sagging of the line sufficiently to relieve the strain.

EXAMPLE 5-13   Straightaway pole, grade C construction, transverse loading, safety factor of 2

Assume a 40-ft pole line, 100-ft spans, a three-phase no. 3/0 aluminum-HDPE primary at 33 ft; a four-conductor no. 3/0 aluminum-XLPE secondary at 25 ft; three 15-kVA transformers at 30 ft; and a 10,000-lb telephone cable at 20 ft.

Solution

Use the values in Tables 5-15 and 5-18. (Weight of cross arms, insulators, etc., is ignored, but is taken into account with factors of safety, rounded values, and other considerations.)

The transverse loading on wires and cable is taken from Table 5-15. For the no. 3/0 aluminum-HDPE, $0.601 \times 100 \times 3 = 180$ lb. For the four-conductor no. 3/0 secondary, $0.782 \times 100 \times 1 = 78$ lb. The loading of one 10M telephone cable is $1.50 \times 100 \times 1 = 150$ lb.

For the moment at the ground line:

Primary	$180 \times 33 =$	5,940 ft·lb	
Secondary	$78 \times 25 =$	1,950	
Telephone	$150 \times 20 =$	3,000	
Total		10,890 ft·lb	

From Table 5-18, for the maximum transverse loading of three 15-kVA transformers, $30 \times 3 \times 30 = 2700$ ft · lb. Thus the total moment is 10,890 + 2700 = 13,590 ft · lb.

From Table 5-19, the ground-line resisting moment for a 40-ft class 5 pole is 23,800 ft · lb. Therefore a 40-ft class 5 pole is adequate for transverse loading. Since it is, check for vertical loading from Table 5-18.

$$649 \text{ lb} \times 3 = 1947 \text{ lb}$$

Since the class 5 limit is 900 lb, a class 3 or class 2 pole must be used.

EXAMPLE 5-14   Dead-end pole, grade C construction, safety factor 1.33

Assume a 40-ft pole line; a three phase no. 3/0 aluminum-HDPE primary at 33 ft; a four-conductor no. 3/0 aluminum-XLPE secondary at 30 ft; one 10M telephone cable at 20 ft; and a 14-ft lead available. The soil is class 5.

Solution

First check that a normal pole class cannot hold the conductors.

Table 5-21 gives the dead-end loadings. For the primary,

$$3 \times 1800 \text{ lb} \times 33 \text{ ft} = 178,000 \text{ ft·lb}$$

For the secondary,

$$1 \times 2000 \text{ lb} \times 30 \text{ ft} = 60,000 \text{ ft·lb}$$

This gives a total loading of 238,000, multiplied by a safety factor of 1.33. This exceeds any pole class from Table 5-19. Guying will have to be used, as follows.

From Table 5-21, the tension at the guying points is found. For the three-phase primary,

$$3 \times 1800 \text{ lb} = 5400 \text{ lb}$$

For the three-phase secondary,

$$1 \times 2000 \text{ lb} = 2000 \text{ lb}$$

The total tension is 7400 lb.

From Table 5-20, for a 14-ft lead and a 32-ft height and a factor of safety of 1.14, the multiplier is 2.49, so guy tension is $1.14 \times 2.49 \times 7400 = 18,400 \times 1.14$. Since the tension is 18,400 lb, use two 11M guy wires attached to the same anchor plate. From Table 5-5B, for class 5 soil, use a 10-in expanding anchor plate. If a telephone cable is attached, use a 12-in expanding anchor plate.

Set the next line pole and install a head guy at a 10-ft height on it. The tension at 10 ft will be 7400 lb; the total moment will be $7400 \times 10 = 74,000$ ft·lb. From Table 5-19, with a safety factor of 1.33, a class 2 pole will be required.

EXAMPLE 5-15   20° angle pole, safety factor of 1.33 for angle loading and 2.00 for transverse loading

Assume a 35-ft pole with 100-ft spans; a two-phase no. 1/0 aluminum-HDPE primary at 28 ft; a one-phase three-conductor no. 1/0 aluminum-XLPE secondary at 25 ft; and a 6M telephone cable at 20 ft.

Solution

The problem requires combining transverse and angle loadings. For transverse loadings, the values are given by Table 5-15. For the no. 1/0 aluminum-HDPE,

$$100 \text{ ft} = 0.566 \text{ lb/ft} \times 2 \times 28 \text{ ft} = 3170 \text{ ft·lb}$$

For the three-conductor no. 1/0 aluminum-XLPE,

$$100 \text{ ft} \times 0.676 \text{ lb/ft} \times 25 \text{ ft} = 1690 \text{ ft·lb}$$

For the 6M telephone cable,

$$100 \text{ ft} \times 1.2 \text{ lb/ft} \times 20 \text{ ft} = 2400 \text{ ft·lb}$$

The total is 7260 ft · lb; the factor of safety of 2 gives 14,520 ft·lb.

For angle loading (Table 5-17), for the no. 1/0 aluminum-HDPE,

$$2 \times 521 \text{ lb} \times 28 \text{ ft} = 29,176 \text{ ft·lb}$$

For the three-conductor no. 1/0 aluminum-XLPE,

$$1 \times 694 \text{ lb} \times 25 \text{ ft} = 17,350 \text{ ft·lb}$$

For the 6M telephone cable,

$$1 \times 1250 \text{ lb} \times 20 \text{ ft} = 25,000 \text{ ft·lb}$$

The total angle loading is 71,526 ft·lb; when multiplied by the factor of safety 1.33, it is 95,130 ft · lb; this plus the transverse loading of 14,520 ft · lb gives a total loading of 109,650 ft·lb.

From Table 5-19, under the ultimate strength of a 35-ft pole (since the safety factors have already been multiplied into the loadings), a load of 109,650 ft · lb requires a class 0 pole. If guying is possible, then use a class 5 pole and calculate the guy wire size for an 8-ft lead attached at 26 ft.

From Table 5-15, the transverse load at a factor of safety of 2.00 is found as follows. For the no. 1/0 aluminum-HDPE,

$$2 \times 100 \text{ ft} \times 0.566 \text{ lb/ft} = 113 \text{ lb}$$

For the three-conductor no. 1/0 aluminum-XLPE,

$$100 \text{ ft} \times 0.676 \text{ lb/ft} = 68 \text{ lb}$$

(The transverse loading of the 6M telephone cable need not be considered here; it is standard practice for each pole user to provide the guying only for its own facilities.) The load is thus 181 lb; multiplying by the safety factor of 2.00 gives 362 lb.

The angle load is determined from Table 5-17:

No. 1/0 aluminum-HDPE	2×521 lb $=$	1042 lb
Three conductor no. 1/0 aluminum-XLPE	1×694 lb $=$	694 lb
		1736 lb

(Again, the loading of the telephone cable is ignored.)

The sum of the two loadings is $1736 + 362 = 2098$ lb. From Table 5-20, the multiplier for an 8-ft lead at 26 ft is 3.40. Since the guy wire safety factor is 1.14,

$$3.40 \times 2098 \text{ lb} \times 1.14 = 7133 \text{ lb} \times 1.14$$

From Table 5-21, an 11M guy is required.

APPENDIX 5B
EXAMPLES

EXAMPLE 5-16

A 40-ft pole suporting a no. 1/0 aluminum-HDPE primary conductor and three-conductor no. 1/0 aluminum triplex secondary is situated where the pole line makes a 30° turn; all poles are 100 ft apart. Select the proper class of pole for this installation.

Solution

From Table 5-9, conductor loading values for a 40-ft pole, loading at 30°,

No 1/0 aluminum-HDPE conductor	5.2 units
Three-conductor no. 1/0 triplex aluminum	6.3
Total	11.5 units

From Table 5-10, transverse loading values for a 40-ft pole, for poles 100 ft apart (average span length in feet),

No. 1/0 aluminum-HDPE conductor	0.4 units
Three-conductor no. 1/0 triplex aluminum	0.4
Total	0.8 units

Total loading force = 11.5 + 0.8 = 12.3 units.

From Table 5-11, ground-line resisting moments, for a 40-ft pole, assuming grade C construction, the smallest pole capable of supporting 12.1 units is a class 2 pole, which should be specified.

EXAMPLE 5-17

A 40-ft dead-end pole must support three no. 4/0 copper-HDPE primary conductors, a three-conductor no. 2 copper triplex cable, and a 10M telephone messenger cable; the poles are spaced 100 ft apart. Select the proper class of pole and guy, if necessary, for this installation.

Solution

From Table 5-9, conductor loading values for a 40-ft pole for dead-end loading (see the 60° column),

No. 4/0 copper-HDPE conductor	13.3 × 3 =	39.9 units
Three-conductor no. 2 copper triplex		9.2
Subtotal		49.1 units
10M telephone cable		32.8
Total		81.9 units

From Table 5-10, of transverse loading values for a 40-ft pole, for a 100-ft average span length,

No. 4/0 copper-HDPE conductor	3 × 0.4 =	1.2 units
Three-conductor no. 2 copper triplex		0.4
Subtotal		1.6 units
10M telephone cable		0.7
Total		2.3 units

Since the transverse loading is very small in comparison to the conductor loading and acts in a direction perpendicular to the larger force, it is insignificant and can be neglected.

From Table 5-11, of ground-line resisting moments, for a 40-ft pole, assuming grade C construction, the heaviest class of pole, 00, cannot support 81.9 units. A smaller pole, class 4, should be specified, with guying. The guy wire selected supports only power conductors, which (as computed earlier in this example) apply a load of 49.1 units to the pole. From Table 5-12B, of guy wires and leads for a 40-ft pole and grade C construction, for 49.1 units, in the column Copperweld: Dead-End and Angle Loading, note that neither the 6M nor the 11M wire can handle the load, and the 17M wire requires a 13-ft lead.

If a shorter lead is desired, move up the 17M column to a shorter lead length, say, 11 ft. Moving left to the first column, note that a 17M wire with an 11-ft lead can support only 43 units of loading. Another guy wire is needed. Moving over to the 6M column, note that with an 11-ft lead, this wire can handle 16 units of loading. The two wires, 17M (43 units) + 6M (16 units), can handle 59 units. Both a 17M and a 6M Copperweld wire should be specified. The anchor, however, must support both power and telephone guy wires.

From Table 5-13, of anchor holding power for 40-ft poles, the total loading on the pole is 81.9 units and the lead length is 11 ft; move down the 11-ft lead column to 81.9 loading units; the anchor capable of meeting these conditions is the 300-in² expandable anchor, and this should be specified. Summing, a class 4 pole with two Copperweld guy wires, 6M and 17M, with a lead length of 11 ft and a 300-in² expandable anchor should be specified.

EXAMPLE 5-18

A single-phase pole line with a no. 1/0 aluminum-HDPE primary conductor, a three-conductor no. 1/0 triplex aluminum secondary cable, and a 6M telephone messenger makes a 90° turn; the poles are 40 ft and spaced 125 ft apart. Select the corner pole.

Solution

From Table 5-9, of conductor loading values for a 40-ft pole,

No. 1/0 aluminum-HDPE conductor	14.1 units
Three-conductor no. 1/0 aluminum triplex	17.3
6M telephone cable	27.8
Total	59.2 units

From Table 5-10, of transverse loading values for a 40-ft pole, for a 125-ft average span length,

No. 1/0 aluminum-HDPE conductor	0.5 units
Three-conductor no. 1/0 aluminum triplex	0.5
6M telephone cable	0.8
Total	1.8 units

The total loading force is 59.2 + 1.8 = 61.0 units.

From Table 5-11, of ground-line resisting moments, the heaviest pole, class 00, can only support 30.5 units. A class 4 pole with guying should be specified.

From Table 5-12B, of guy wires and leads for 40-ft poles and grade C construction, since this is a corner pole, it should be guyed in each direction. The conductor loading in each direction will act on the pole as a dead-end condition. From Table 5-9, the 60° column,

No. 1/0 aluminum-HDPE conductor	10.2 units
Three-conductor no. 1/0 aluminum triplex	12.3
6M telephone cable	19.7
Total	42.2 units

Reading down in the first column of Table 5-12B, to 23 units (the total for the power conductors only), moving right to Alumoweld: Dead-End and Angle Loading, note that the 6M wire cannot handle the load, the 11M wire needs a 10-ft lead, and the 17M wire needs a 6-ft lead. Unless field conditions conflict, specify the smaller wire, 11M with a 10-ft lead. (Where the pole cannot be guyed in each direction, one guy may be used; follow the procedure set forth in Table 5-10.)

From Table 5-13, of anchor holding power for a 40-ft pole, using a lead length of 10 ft, and using the first column (number of units) as a guide, move down the 10-ft column to the level of 42 units, and left to the first curve intersected, the 200-in^2 expandable anchor, which should be specified.

Summing a class 4 pole with one Alumoweld 11M wire in each direction, each with a 10-ft lead, and a 200-in^2 expandable anchor on each, should be specified.

APPENDIX 5C
CONCRETE AND METAL POLES

INTRODUCTION

While the overwhelming number of poles employed in electric distribution systems in the United States are made of wood (and are likely to remain so for the indefinite future), poles made of reinforced concrete and of steel and aluminum are also to be found where, for one reason or another, wooden poles are considered unsuitable.

In general, concrete poles are used in those areas, swampy and persistently wet, where the soils greatly shorten the life expectancy of wood poles. Moreover, in such instances, the rate of decay of wood may be so erratic and uncertain as to permit unsafe conditions to arise that may not be discovered before accidents result. Concrete poles are sometimes also specified in areas of chemical contamination and pollution that may cause rapid deterioration of wood.

Concrete poles and poles made of steel or aluminum are specified in special situations where poles of unusually high strength are required, beyond the range of wood poles, and where guying may be difficult or unobtainable.

Again, concrete and metal poles may be specified where their appearance is preferred to poles made of wood. They find particular application for street and highway lighting, for parking lot and security lighting, and for recreational and industrial lighting, applications often included in the planning and design functions of the distribution engineer.

In general, the first cost of such poles is greater than that of wooden poles of equal strength. Improved methods of manufacture, however, have already made these concrete and metal poles competitive for the larger transmission pole requirements, and prospects for distribution application appear promising. In addition, they are much heavier than corresponding wood poles and consequently much harder to handle, especially when they are to be replaced in the field.

CONSTRUCTION
Concrete Poles

Concrete poles are manufactured in several cross-sectional shapes: round, square, and polygons (usually six or eight sides). Moreover, they may be solid or hollow. The method of pouring the hollow poles involves spinning the form while the concrete is being poured, and forcing it to the outside while leaving the center hollow; the result is a highly uniform, compact, prestressed concrete of high strength and texture.

As concrete has a higher strength under compressive loads than under tensile loads, the overall strength of concrete poles depends a great deal on the steel reinforcement. Its strength also depends considerably on the mixture of cement and how it is cured.

Where cross arms or other distribution devices and equipment are to be mounted directly on the pole, provision is made for the necessary bolt holes when the concrete is being poured.

Besides their appearance and greater strength, hollow poles provide a means for electrical risers to be installed inside the pole out of sight and not readily accessible to the public. Other advantages claimed include resistance to fire, birds (especially woodpeckers), rot, and vandalism; also the elimination of harm and clothing damage to the public from contact with some wood preservatives. A variety of colors and finishes also is possible with concrete poles; these may contribute to their acceptance.

Metal Poles

Usually made of steel or aluminum, metal poles are made in the same cross-sectional shapes as the concrete poles: round, square, polygonal. In addition, they may also be fabricated in the form of angles, channels, or tees, or sometimes two of these forms tied together with lacing of some kind; expanded metal is also sometimes used. They may be bolted, welded, or riveted together, the method selected depending on a number of factors—the required strength, desired appearance, cost, and maintenance requirements.

INSTALLATION
Poles

Concrete poles, whether solid or hollow, are installed in the same manner as wood poles, enabling the same methods and equipment to be used for setting this type of pole.

Steel and aluminum poles may also be installed in this same fashion. In many instances, however, they are bolted to a base plate that is anchored in a concrete footing buried in the ground.

Attachments

Concrete cross arms, frequently used in transmission structures, are seldom used in distribution systems. The regular wood cross arm may be bolted to the concrete pole in the same manner as to wood poles; it may also be attached to the pole by metal U-shaped brackets which are clamped around the pole and to which the cross arm is bolted at the proper height.

Transformers, capacitors, cutouts, insulator pins, and other devices are attached to the concrete poles in a similar manner. Where bolt holes are provided, they may be bolted directly to the pole; otherwise brackets, as described above, may be used.

In many instances, steps are incorporated in the concrete poles, or provision is made for their temporary installation to permit workers to reach the upper parts of the pole.

Attachment to metal poles may be made in the same manner. Because drilling of holes in the metal pole is more readily accomplished than in concrete poles, bolted attachments to such poles are usually preferred.

DESIGN
Stresses

Stresses imposed on concrete and metal poles are calculated in the same manner as for wood poles: as if they were a cantilever beam fixed at one end. These values, with proper factors of safety applied, are used in the selection of the pole.

TABLE 5-22 HOLLOW ROUND CONCRETE POLE DESIGN DATA

Wood, ASA class	Reinforced concrete class	Ultimate strength, lb	Allowable strength, SF = 2,* lb
—	A	3200	1600
—	B	2700	1350
1	C	2250	1125
2	D	1850	925
3	E	1500	750
4	F	1200	600
5	G	950	475

Note: Assume load is applied 2 ft from the top of the pole and break occurs at the ground line.

*Safety factor of 2; these working loads are equivalent to ASA standard classes for wood poles.

Courtesy Centrecon, Inc.

Selection of Poles

In selecting a concrete pole, in addition to the ultimate ground-line moment and overall pole length requirements, similar to those required in selecting a wood pole, attention is also given to the size and location and number of holes, if they are included in the pole standards, and to the length of "embedment," or depth of burial in the ground. As a rule, this embedment is taken as 10 percent of the pole length plus 2 ft.

Concrete poles have not yet been standardized into classes as have wood poles, but there are indications pointing to that goal. Standards of one manufacturer are shown in Table 5-22, which includes a general comparison of ASA class standards for wood poles with round, hollow, steel-reinforced concrete poles. Additional data on these and some square-cross-section, hollow, reinforced concrete poles are contained in Chap. 10.

Metal poles also find widespread use as supports for a variety of lighting purposes as well as for supports for signs and platforms, as flag and pennant staffs, and in many other fields. Such poles are usually designed specially for the use to which they will be put, though some relatively minor standardization of sizes, diameters, thicknesses, etc., exists.

CHAPTER 6

MECHANICAL DESIGN: UNDERGROUND

The placing of cables and equipment underground presents problems completely different from those of installing similar facilities overhead. For downtown areas of high load density served by low-voltage secondary networks, underground service can be justified economically, satisfying the rather rigid reliability and environmental demands associated with this type of area. Such reasoning, however, can hardly justify the installation of the same kind of underground service in less dense urban areas and in suburban areas, which call for different construction, maintenance, and operating practices. The discussion, therefore, will be divided into two parts: that for city areas, particularly the downtown areas; and that for urban and suburban areas of lighter load densities. In general, cables and transformers in the former are placed in ducts, manholes, and vaults; while in the latter they are buried directly in the ground, with transformers installed in metal enclosures situated at ground level, or totally or partially buried in the ground.

HIGH-DENSITY LOADS: CITY AND DOWNTOWN AREAS

In congested areas, it is essential that conductors, transformers, and other equipment can be installed, maintained, repaired, and replaced, that faults can be located without causing undue inconvenience to the public. Cables are placed in conduits or ducts and spliced in manholes and service boxes, and transformers and equipment are installed in manholes and vaults. Ducts and manholes must be designed not only with regard to the facilities they do and will contain, but also with regard to the type of soil and terrain they are installed in and the existence of other utilities and obstructions (e.g., water, gas, sewer, telephone,

199

and railway). The design engineer's work is made more difficult by the economic desirability of installing prefabricated (precast) ducts, service boxes, and manholes, as well as the need to comply with codes and regulations at all governmental levels: national, state, county, and local city or village.

Ducts

The number of ducts installed should provide for whatever growth may be reasonably expected at the time of installation, but it may be limited by the space available. Ducts may vary in size, but those of a 4-in diameter are most commonly used. Ducts may be of wood, tile, fiber, plastic or other compounds, steel and iron pipe, or concrete; precast reinforced concrete appears to be greatly preferred. Some ducts are encased in concrete envelopes, and others are formed from concrete poured over cores (usually made of rubber or plastic) which are removed after the concrete has set; precast reinforced concrete sections are laid in the ground and connected together, formed into banks. The arrangement of ducts in a bank, though often dictated by the space available, is important from the viewpoint of dissipation of heat and its effect on the current-carrying ability of the cables installed within; see Fig. 6-1.

Ducts should be installed at a depth below the surface to avoid possible damage from accidental driving of so-called bullpoints and, if possible, should be below

	Cost of duct construction	Ability to radiate heat	Cable support and racking conditions in manhole
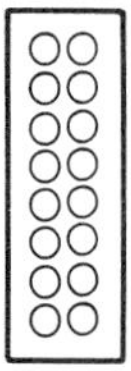	Expensive	Best	Best
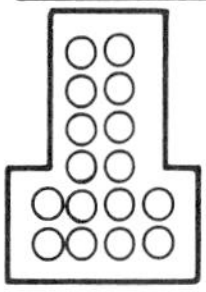	Moderate	Very good	Very good
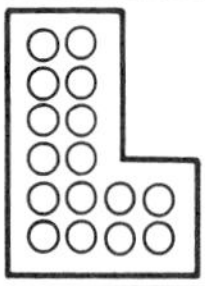	Moderate	Very good	Good
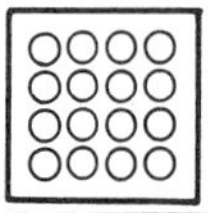	Cheapest	Very poor	Very poor

FIG. 6-1 Comparative duct characteristics. *(From EEI Underground Systems Reference Book.)*

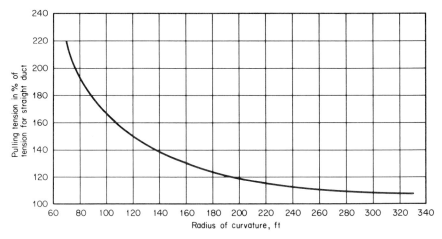

FIG. 6-2   Effect of radius of curvature of conduit on pulling tension (applicable to a conduit consisting of one continuous curved section). *(From EEI Underground Systems Reference Book.)*

the frost line to prevent dislocation from motion caused by severe changes in temperature. They should also be graded between manholes so that water will not accumulate in them. Moreover, where there are bends or offsets in the line, ducts should be gradually bent to accommodate the minimum bending radii of the largest cables to be installed; this bending radius should be from 7 to 20 times the diameter of the cable, depending on its size, voltage classification, type of insulation, type of sheath, and other characteristics. The bends in the duct line should preferably be located near the manholes, as each bend will increase the pulling tension on the cable and thus reduce the maximum length or distance between manholes; see Fig. 6-2.

Service Boxes

Secondary mains are installed in ducts buried at shallower depths than those carrying primary conductors. These terminate in distribution or service boxes from which service cables to consumers emanate. Constructed of precast reinforced concrete, they are usually standardized at dimensions approximately 4 ft square and 2 to 4 ft deep, and are thus suitable for secondary distribution in four directions. They are usually located so as to accommodate the secondary mains and the largest number of services without overcrowding the box and without introducing too many bends in the service conduits. The entrance (or "throat" or "chimney") to these boxes is usually quite large, generally about 3 ft square, providing ample room for the worker to work from ground level or standing erect with head and shoulders above ground. Covers are usually made of steel, with an inner cover installed to prevent dirt from entering, and may be locked to keep out unauthorized persons.

Cable Manholes

Cable manholes are usually larger in size, standardized at different dimensions and shapes, with headroom of about 6 ft or better. They may be rectangular in shape for straight-line conduit construction, square for accommodating cables from four directions, or L-shaped where there is a turn in the duct line. They may accommodate ducts containing secondary mains at upper levels and ducts carrying primary cables at lower levels. The latter proceed from manhole to manhole, bypassing the service boxes, and the conduits are sometimes referred to as trunk ducts.

The manholes are made of reinforced concrete with facilities included for installing hangers to support the cables and splices along their walls. The size of the manhole shaft through the roof (or throat or chimney) is generally about 3 ft square or 3 ft in diameter.

The manholes are spaced as far apart as is practical to hold down the number of splices required. Proper selection of locations for the manholes will also reduce the number of bends in the conduit or duct system to a minimum.

The various shapes of manholes also take into account the need for training, splicing, and racking of cables; typical "standard" shapes are shown in Fig. 6-3. The essential difference among them is the number of conduit lines entering

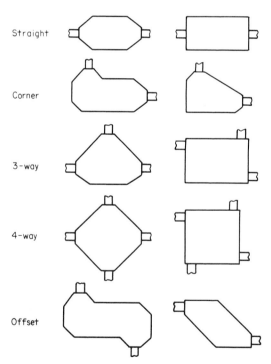

FIG. 6-3 Typical shapes of cable manholes. *(From EEI Underground Systems Reference Book.)*

the manhole and the angle at which they enter in the walls in relation to each other.

Where the standard precast manholes cannot be installed for any reason, manholes of various shapes and sizes may be constructed in the field.

It is generally desirable to keep the earth fill over the manhole at a minimum, not only for economic reasons, but also for making the installation of local services and streetlight connections more practical. Local ordinances and existing subsurface structures, however, may dictate the depth at which the manhole roof is located.

Transformer Manholes

Transformer manholes are designed to contain transformers and other equipment required for radial or network systems. Their dimensions depend on the location and the equipment they are to contain. Standard transformer manholes

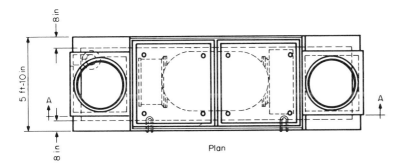

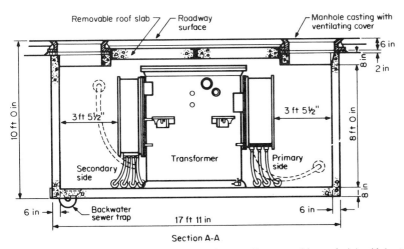

FIG. 6-4 Transformer manhole under roadway, with removable roof slab. *(Adapted from EEI Underground Systems Reference Book.)*

of reinforced prefabricated concrete are generally approximately 6 ft by 17 ft, by 8 ft of headroom, essentially to accommodate a 500-kVA three-phase network unit, as shown in Fig. 6-4; the manhole may also house other types of transformers and switches as part of a radial distribution system. The dimensions provide space for workers to operate and maintain switches on the primary side and network protectors on the secondary side.

Design Loading

The loading on the several parts of the manhole depends on the maximum load imposed on the street surface. The live load on the surface affects both the design of the roof slab and the walls. Wheel loads of 21,000 lb and impacts of 50 percent are typical values for heavily traveled streets over which truck traffic may be concentrated. For conservative values, wheel areas as little as 6 by 12 in, or a surface area of 0.5 ft², are also considered. The concentrated load may then be as much as 63,000 lb/ft²:

$$\text{Concentrated load} = \frac{\text{wheel load} \times (1 + \% \text{ impact} \div 100)}{\text{wheel area}}$$

$$= \frac{21,000(1 + 0.5)}{0.5} = 63,000 \text{ lb/ft}^2$$

The type of pavement, the nature of the soil beneath the pavement, and the depth (thickness) of the soil above the roof of the manhole serve to mitigate the actual effect of the concentrated load. The reduction in effective pressures at different depths below the surface is shown in the diagrams of Fig. 6-5a and b and the associated Tables 6-1A and 6-1B.

Roofs

Manhole roofs are designed as a series of structural steel beams or rails, or reinforced concrete with extra-heavy steel reinforcement or structural steel to support the manhole frames. Where installed in sidewalks or other areas not subjected to heavy vehicular traffic, roof designs may take into account the lighter loading. If there is any possibility of its being subjected to loads approximating street loadings, the design should be based on the heavier loadings.

Walls

Manhole wall designs are based on the horizontal component of the effect of both live and dead loads acting on the walls. The horizontal forces will depend on the surface, the angle of repose of the soil, and the effect of the water table.

At depths below about 5 ft, as shown in Table 6-1B, for the spread of wheel loads, the weight of the earth above the manhole predominates. Here, the average of the live-load effects approximates 450 lb/ft²; it appears to be constant at lower depths.

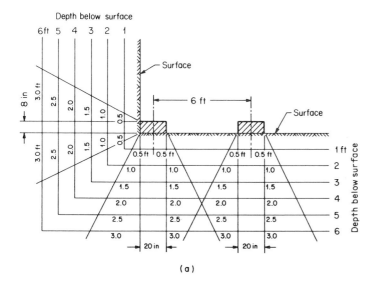

(a)

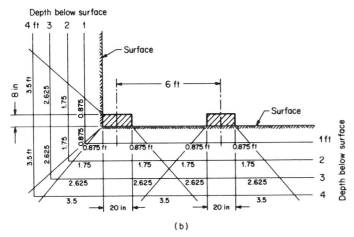

(b)

FIG. 6-5 Diagram showing area of spread of wheel loads (a) based on 1:1 spread and (b) based on 1¾:1 spread. *(From EEI Underground Systems Reference Book.)*

The dead loads at various depths and various horizontal pressures as a percentage of the vertical pressure are shown in Table 6-2, which extends the tabulation associated with Fig. 6-5a and b. Table 6-2 serves as a guide in determining the horizontal pressures with various headrooms and depths for the several corresponding angles of repose of the soil and pressures from the hydrostatic head of the water table.

TABLE 6-1A PRESSURE CALCULATIONS BASED ON A 21,000-lb WHEEL LOAD—
1:1 WHEEL SPREAD

Depth from surface, ft	Area of spread			Live-load pressure on cover*	Pressure on roof		
	Length L, ft	Width W, ft	Area, ft²		Live load**	Surcharge	Total
0	0.67	1.67	1.12	18,500	5200	0	5200
1	1.67	2.67	4.46	4,710	2100	150	2250
2	2.67	3.67	9.80	2,140	1310	250	1560
3	3.67	4.67	17.10	1,230	960	350	1310
4	4.67	5.67	26.50	790	750	450	1200
5	5.67	6.67	37.80	560	620	550	1170
6	6.67	7.67	51.10	410	520	650	1170
7	7.67	8.67	66.50	320	450	750	1200

*Average pressure P_{av} that might be imposed on cover by maximum concentrated load, or (21,000 lb)/area.

**The surface concentrated load uniformly distributed over the width of the manhole, or $P_{av}W/6$.

From *EEI Underground Systems Reference Book*.

Floors

In the design of manhole floors, the load-bearing power of the soil and the height of the water table play an important part. The soil must support the weight of the manhole structure, its contents, and any imposed surface live loads. In firm soils, the earth is capable of supporting the structure and any additional weight. The floor, therefore, is often poured after the walls are in place, adding to the strength of the walls. The floor wells are 4 to 6 in thick.

Where the earth is not capable of supporting the loading of the walls, the floor is used as a means of spreading the load. Here, the floor is poured before

TABLE 6-1B PRESSURE CALCULATIONS BASED ON A 21,000-lb WHEEL LOAD—
1¾:1 WHEEL SPREAD

Depth from surface, ft	Area of spread			Live-load pressure on cover*	Pressure on roof, lb/ft²		
	Length L, ft	Width W, ft	Area ft²		Live-load**	Surcharge	Total
0	0.67	1.67	1.12	18,500	5200	0	5200
1	2.42	3.42	8.27	2,500	1430	150	1580
2	4.17	5.17	21.5	970	840	250	1090
3	5.92	6.92	41.0	510	590	350	940
4	7.67	8.67	66.5	315	450	450	900

*Average pressure P_{av} that might be imposed on cover by maximum concentrated load, or (21,000 lb)/area.

**The surface concentrated load uniformly distributed over the width of the manhole, or $P_{av}W/6$.

From *EEI Underground Systems Reference Book*.

TABLE 6-2 HORIZONTAL EARTH PRESSURES AT VARIOUS DEPTHS

		No live load			Live and dead loads				
		Horizontal pressure, lb/ft²				Total	Horizontal pressure, lb/ft²		
Depth	Dead load	25%	30%	35%	Live load 1¾:1	live and dead load	25%	30%	35%
0	0	—	—	—	5200	5200	—	—	—
1	150	38	45	53	1430	1580	395	474	553
2	250	63	75	88	840	1090	273	327	382
3	350	88	105	123	590	940	235	282	329
4	450	113	135	158	450	900	225	270	315
5	550	137	165	193	450	1000	250	300	350
6	650	162	195	228	450	1100	275	330	385
7	750	187	225	262	450	1200	300	360	420
8	850	212	255	298	450	1300	325	390	455
9	950	237	285	333	450	1400	350	420	490
10	1050	263	315	367	450	1500	375	450	525
11	1150	288	345	402	450	1600	400	480	560
12	1250	312	375	438	450	1700	425	510	595
13	1350	338	405	472	450	1800	450	540	630
14	1450	352	435	507	450	1900	475	570	665
15	1550	387	465	542	450	2000	500	600	700

From *EEI Underground Systems Reference Book.*

the walls are installed. Similar measures are employed in areas of high water table. Such floors are usually made of reinforced concrete, a minimum thickness of 6 in, and are constructed with a keyway for the walls. Where the hydrostatic pressure is high, an additional pour of 2 to 4 in of concrete is added on top of the floor.

Prefabricated manholes may be completely precast in one piece, or in a caisson type in which the roof and floor are separate. The caisson walls are sunk in place, the precast floor is placed within it (or a floor is poured), and a precast roof is installed in keys in the walls provided for that purpose (such roofs are also installed in other types of manhole construction). Small manholes or service boxes may also be completely precast or formed from precast individual pieces.

Frames and Covers

The frames and covers are made of cast iron, malleable iron, or steel, and are designed to withstand the loadings mentioned earlier; covers infrequently used may be made of reinforced concrete. Depending on the area over which the load is applied, frames and covers may have to withstand wheel loads from 50,000 to 200,000 lb, though sidewalk covers may be designed for lowered loadings. Although covers may be either square or round, the latter shape is preferred to insure against their falling into the manhole when being replaced. Frames and covers for transformer manholes may be of the completely fabricated grating type, or of the combination type—part solid and part grating—that is specified for roadway use, but they will be of lower loading rating.

Transformer manholes are usually built with a removable roof slab covering an opening capable of admitting large distribution and network units. The manholes are of reinforced concrete with the slabs sealed and made watertight, the pavement being replaced after the transformers are installed. The pavement is cut and the slab removed when transformer replacements are required. When the transformer manhole is located under the sidewalk, the roof slabs are flush and made part of the sidewalk surface, and are readily removed when necessary. Prefabricated manholes may be completely precast or formed from precast individual floors, walls, and roofs.

Ventilation

The principal source of heat within a transformer manhole is that caused by losses in the core and windings of the transformer, losses which can be obtained from the manufacturer's data or can be calculated approximately. The dissipation of heat is to some extent based on the area of the enclosing walls and the nature of the adjacent soil conditions. For proper operation of transformers, the manhole should have sufficient cubic content supplemented with natural ventilation to keep the temperature within prescribed limits. Air temperatures in the manhole should not exceed 40°C, mainly occurring at periods of maximum load. The approximate number of cubic feet of air per minute to dissipate the heat may be found conveniently from the curve in Fig. 6-6. When such limits cannot be attained by normal circulation of air between the two ventilating gratings of the transformer manhole, it may be necessary to provide some means of forced ventilation, such as blowers, water coolers, etc.

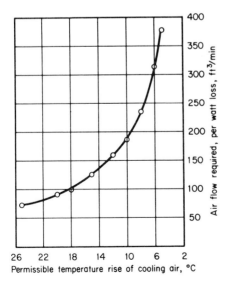

FIG. 6-6 Airflow requirements for limiting temperature rise in transformer vaults. *(From EEI Underground Systems Reference Book.)*

The ventilation of transformer manholes must not only provide sufficient cooling to keep the transformer within proper limits even under extreme operating conditions, but must provide a sufficient vent area outlet to prevent the explosion of gases or oil vapors that may develop from transformer failure. This vent area must bear a relationship to the volume of the transformer manhole: the smaller the ratio of the volume of air in the manhole to the total ventilating area (or area of openings), the less will be the pressure developed in the manhole and the lighter the construction necessary to withstand the force of the possible explosion.

Gratings situated at either end of the manhole, generally over the cleared areas reserved for workers, not only provide access but help to ventilate the transformer and manhole.

Transformer Vaults

Where the transformers are to be installed inside a building, or sometimes outside under a sidewalk, vaults are constructed. They are usually of reinforced concrete, the dimensions of which are dependent on the transformers and accessories to be installed, the space available and its location, the adjacent structures and substructures, and applicable code requirements and local ordinances. The same minimum ventilation and vent-volume requirements apply as for transformer manholes. Access for both equipment and personnel from the outside are usually sought, but may be substituted by adequate internal means of access. In very tall buildings, such vaults may be located on upper floors, in addition to the usual vault at basement level.

Vault ceilings or roofs should be designed to take into account possible fire and explosion from transformer failure or other causes.

Cables

Multiple-conductor cables are preferred for installation in duct systems because of the advantages of handling one cable in the field. Where the conductors are rather large, each single-conductor cable may be installed in a separate duct. Primary mains are almost always three-conductor cables. Single-conductor cables are generally used for secondary mains; they are also used for the secondary feeds from transformers, for streetlight circuits, and in special instances for primary mains.

Cable Insulation For both primary and secondary cables, plastics such as cross-linked polyethylene are now almost exclusively used; some oil-impregnated paper is still installed in special cases where much of that kind of insulation exists. Varnished cambric and rubber, for primary, and rubber, for secondary cables, were widely used, and many cables using these types of insulation are still in use. Some polyethylene (PE) and polyvinyl chloride (PVC) insulation is also being used, principally for secondary cables. The trend toward plastic insulations, it appears, will continue, with new and better compounds certain to make their appearance.

Cable Sheaths   Lead sheaths, again, though still used in decreasing amounts, appear to be giving way to plastics. Many lead-sheathed cables exist and will exist for a long time, but new installations are almost exclusively plastic, principally cross-linked polyethylene, used as both insulation and sheath.

Splices

Splices or joints in cables with nonplastic insulation and with metallic sheaths are much more complex and require greater time and skill to make than those in plastic-type cables. Such a splice is shown in Fig. 6-7. Cables with plastic insulation and plastic sheaths are simpler and quicker to splice as, generally, the splice consists of a wrapping of tape of the same material over the connectors, crimped on the two conductors being spliced. The plastic tape eventually solidifies into a homogeneous mass integral with the insulation or sheath.

Conductor connectors are usually copper or aluminum tubes crimped onto the ends of the conductors to be joined. Some of the older connectors were tubes split horizontally and squeezed onto the ends of the conductors, with solder

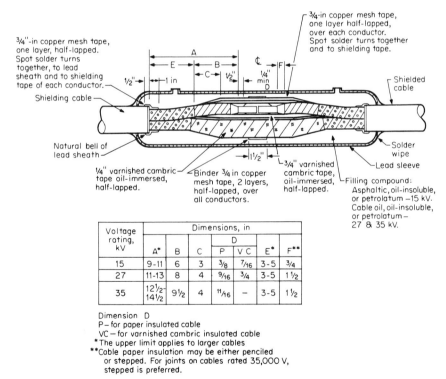

Voltage rating, kV	Dimensions, in						
	A*	B	C	D		E*	F**
				P	V C		
15	9-11	6	3	3/8	7/16	3-5	3/4
27	11-13	8	4	9/16	3/4	3-5	1 1/2
35	12 1/2-14 1/2	9 1/2	4	11/16	–	3-5	1 1/2

Dimension D
P – for paper insulated cable
VC – for varnished cambric insulated cable
*The upper limit applies to larger cables
**Cable paper insulation may be either penciled or stepped. For joints on cables rated 35,000 V, stepped is preferred.

FIG. 6-7   Straight joint for three-conductor shielded paper- or varnished-cambric-insulated lead-covered cable, 15 to 35 kV. *(From EEI Underground Systems Reference Book.)*

poured into the slit and over the connector. Soldered connectors are now almost entirely obsolete.

Underground Equipment

Transformers, oil-filled cutouts, and oil switches for use underground are hermetically sealed so as to be waterproof. Such submersible equipment is usually of welded construction. Wiping sleeves are welded or brazed directly to the tank or terminal chamber, to which cable sheaths are attached. Barriers in the conductors prevent the equipment oil from being siphoned into the cables.

In low-voltage network areas, network transformer units are installed, comprising switching facilities on the primary side and a network protector on the secondary, low-voltage side of the transformer.

PRACTICAL MANHOLE DESIGN PROCEDURE*

Practical procedures containing instructions and technical data for designing underground facilities have been prepared, reducing the work necessary in preparing plans and orders to the field. A procedure outlining the basis of the structural design for field-poured or precast reinforced concrete manholes and vaults used in electric distribution systems follows.

DESIGN LOADING
Live Load

Live-load requirements are based on those of the American Association of State Highway and Transportation Officials (AASHTO) for an HS-20 or H20 16,000-lb wheel load plus a 30 percent impact factor, or a maximum wheel load of 20,800 pounds. They must also meet prevailing local building codes and ordinances that usually specify 600 lb/ft². The basis for design is the greater of these two requirements.

Live-Load Criteria

1. *Sidewalk and driveway:* A 600 lb/ft² uniform live load or a 16,000-lb wheel load without impact, whichever produces the greatest stresses
2. *Roadway:* A 600 lb/ft² uniform load or a 16,000-lb wheel load with a 30 percent impact factor, or 20,800 lb, whichever produces the greatest stress

Wheel Load   The 16,000-lb (HS-20 or H20) wheel load shall be taken as acting on a 12- by 20-in area, and no more than two such loads shall be acting at a spacing of 5 ft; see Fig. 6-8.

*The material from this point through the end of the section Construction Practices on p. 232 is adapted from *Application & Design Manual,* courtesy Consolidated Edison Co. of New York.

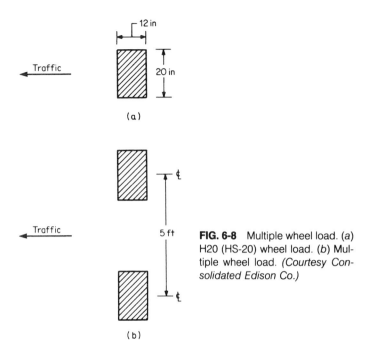

FIG. 6-8 Multiple wheel load. (*a*) H20 (HS-20) wheel load. (*b*) Multiple wheel load. *(Courtesy Consolidated Edison Co.)*

Wheel Load Distribution   The effective area over which a wheel load acts at any depth below the pavement surface shall be determined by spreading the wheel load through pavement (concrete or asphalt) and soil in the following manner:

1. *Pavement:* The wheel load is to be spread at a 45° angle with the vertical in all directions.

2. *Soil:* The wheel load is to be spread at a 30° angle with the vertical in all directions.

The wheel load area as a function of depth for a 9-in pavement is shown in Fig. 6-9.

Surcharge Load   The uniform 600 lb/ft^2 live load shall be taken, when applicable, as acting on the surface as a surcharge load.

Earth Pressure Coefficient   The active earth pressure coefficient shall be taken as $K_a = 0.33$, which assumes an angle of repose or internal friction of 30°.

Dead Load

The unit weights used for computing dead load shall be as follows:

1. Soil—100 lb/ft^3 wet soil weight

2. Soil—65 lb/ft^3 submerged soil weight

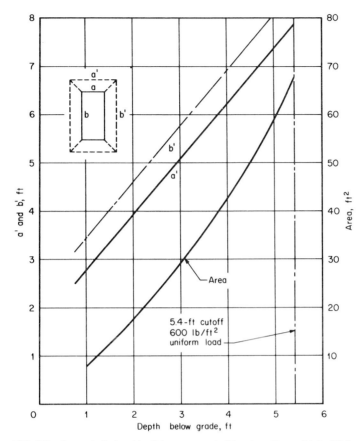

FIG. 6-9 Spread of wheel for 9-in pavement. *(Courtesy Consolidated Edison Co.)*

3. Pavement (concrete or asphalt)—144 lb/ft^3
4. Concrete (plain)—144 lb/ft^3
5. Concrete (reinforced)—150 lb/ft^3

DESIGN STRESS BASES

Allowable stresses are given in terms of the following quantities:

f_c allowable compressive strength in concrete, lb/in^2

f_c' 28-day compressive strength of concrete, lb/in^2

f_s allowable strength in steel, lb/in^2

n ratio of modules of elasticity of steel to that of concrete

V_c allowable shear stress in concrete

Allowable Stresses—Concrete

TABLE 6-3 ALLOWABLE STRESSES—CONCRETE

Type of concrete	n	f'_c	f_c	V_c
Precast plant concrete	8.5	3500	1575	118
	7.0	5000	2250	141
Field-placed concrete	8.5	3500	1100	83

Allowable Tensile Stresses—Reinforcing Steel

For ASTM A615 grade 40 deformed-billet steel bars, f_s = 20,000 lb/in². For ASTM A615 grade 60 deformed-billet steel bars, f_s = 24,000 lb/in².

Allowable Steel Stresses

Structural steel elements are to be designed in accordance with the latest revision of the AISC *Manual of Steel Construction*. All solid steel covers and steel gratings subjected to repeated traffic loading are to be designed in accordance with the AASHTO *Requirements for Design of Repeated Loads* for 500,000 cycles of load.

Allowable Bearing Pressures for Soil

Unless organic clays or silts are encountered, a value of 1.5 tons/ft² may be used as a conservative bearing value. If a manhole or vault is to be installed on clay, clayey soils, or organic material, careful evaluation should be made of the potential for settlement. The use of a crushed-stone base or piles may be required and soil borings may be necessary.

WALL DESIGN
Rigid Horizontally Reinforced Frame

The wall loading for lateral earth pressure due to live and dead loads shall be taken from the design chart in Fig. 6-10. The frame shall be analyzed using conventional indeterminate structural techniques. Midspan moments and corner moments may be calculated by using formulas given in the sample design problem in this chapter or by using the moment coefficients for each moment given in Figs. 6-11, 6-12, and 6-13.

Simply Supported Vertically Reinforced Structure

The wall loading for lateral earth pressure due to live and dead loads shall be taken from the design chart in Fig. 6-10. The wall should be analyzed as a simply supported strip with a height equal to the headroom of the manhole plus one-

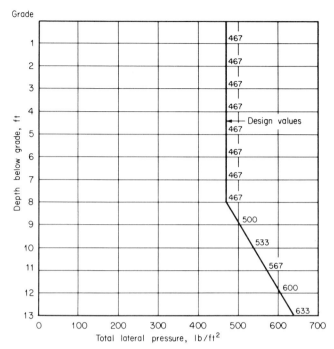

FIG. 6-10 Design chart—lateral pressures on walls. *(Courtesy Consolidated Edison Co.)*

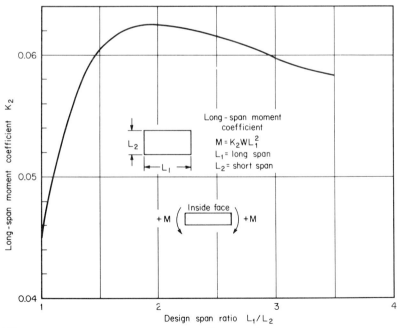

FIG. 6-11 Long-span moment coefficient. *(Courtesy Consolidated Edison Co.)*

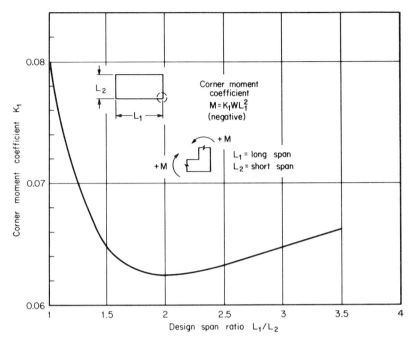

FIG. 6-12 Corner moment coefficient. *(Courtesy Consolidated Edison Co.)*

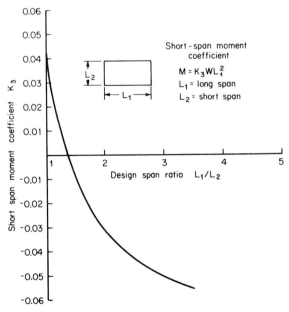

FIG. 6-13 Short-span moment coefficient. *(Courtesy Consolidated Edison Co.)*

half the sum of the floor and roof thicknesses. This method is to be used with field-poured manholes *only* and requires a field-poured roof connected to the walls and able to carry the wall reaction.

Combination of Horizontally Rigid Frame and Vertically Reinforced Designs

A method that combines the two methods described above may be used when conditions indicate that there are areas where the reinforcing for both the vertical and horizontal methods is severely interrupted by openings.

Partition Walls in Vaults

Partition walls in field-poured vaults which house oil-type transformers inside the consumer's property shall be designed for an internal blast load of 600 lb/ft².

Other Requirements

All field-poured vertically reinforced vaults and manholes shall have a minimum thickness of 6 in.

All field-poured horizontally reinforced rigid-frame-type vaults and manholes shall have a minimum wall thickness of 8 in.

Where watertight construction is required, the walls shall be monolithically poured with the floor for a minimum distance of 12 in above the top of the floor level. A water stop shall be inserted at this location, as shown in Fig. 6-14. A minimum wall of 10 in is required for all vault construction under these conditions, and 5000 lb/in² concrete may be used.

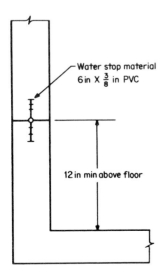

Water stop material
6 in X ⅜ in PVC

12 in min above floor

FIG. 6-14 Water stop detail; water stop to be lap-spliced 4 in on each side of vertical joints and continuous around all corners. *(Courtesy Consolidated Edison Co.)*

ROOF DESIGN

Live Loads

All roof structures for manholes or vaults shall be designed to carry the live loads specified above in the section Design Loading.

Wheel-Load Area

The design wheel load (HS-20 or H20) shall be taken acting on an area which is to be determined using the method of spreading a concentrated load defined under Wheel Load Distribution in the section Design Loading, above.

Field-Poured Manholes

All field-poured manhole roofs shall be designed using structural steel sections around the roof opening to support the manhole frame.

Precast Manholes

Precast manhole roofs may be designed using a simply supported reinforced concrete beam around the opening to support the manhole frame.

Roof Slabs

Roof slabs shall be designed as one- or two-way reinforced concrete slabs; where the ratio of short span to long span exceeds 0.5, a one-way slab design shall be used.

One-Way Slab   The design moment shall be determined using a simply supported beam strip loaded with the effective live-load intensity and uniform dead load. Table 6-4 gives the design moments as a function of depth.

TABLE 6-4   SIMPLE-SUPPORT ROOF SLAB MOMENT, ft·lb

	Design span				
Design depth, ft	*4 ft*	*5 ft*	*6 ft*	*7 ft*	*8 ft*
0.75 to 1.75	5,500	7,860	10,270	12,740	15,260
1.75 to 5.0	2,860	4,690	6,330	8,130	10,460

Two-Way Slab   The design moment shall be determined using Table 6-4 for a one-way slab and then proportioning the one-way slab design moments for the short-direction and long-direction moments using the conversion constants in Table 6-5.

TABLE 6-5 CONVERSION FACTORS FOR TWO-WAY SLAB MOMENTS

Ratio of clear spans	Long-span moment factor K_l	Short-span moment factor K_s
1.0	0.500	0.500
0.9	0.396	0.604
0.8	0.295	0.709
0.7	0.194	0.806
0.6	0.114	0.886
0.5	0.059	0.940

Short-span moment $M_s = K_s \times$ simple-span moment
Long-span moment $M_l = K_l \times$ simple-span moment

Above-Grade Vault Roofs

All above-grade vault roofs shall be designed for a uniform dead load plus a live load of 30 lb/ft^2 of projected area plus internal blast load.

Other Requirements

The minimum thickness of a precast roof shall be 6 in. The minimum thickness of a field-poured roof shall be 8 in.

FLOOR DESIGN
Loading

The loading on the floor of a manhole or vault depends on the size of the transformers or equipment to be installed. The weights of some transformers are listed in Table 6-6, together with some approximate areas upon which the loads are imposed.

TABLE 6-6 TRANSFORMER WEIGHTS AND LOAD AREAS

Transformer size, kVA	Weight or load, lb*	Area for direct load
300	9,000	2 ft $\times$ 3½ ft
500	12,000	2 ft $\times$ 4 ft
1000	16,000	2 ft $\times$ 4 ft
2000	24,000	3½ ft $\times$ 4½ ft
2500	30,000	3½ ft $\times$ 4½ ft

*Add 3000 lb for network protectors.

Manholes

Soils In all soil conditions *other* than soft clay, medium clay, organic material, and a water condition where the groundwater level is higher than the top of the floor level, the floor shall be designed as a reinforced concrete slab capable of resisting a bending moment produced by a partial uniform load which results from dividing the entire dead and live load by the area bounded by the outside perimeter of the structure and of the strip extending 24 in inward. The resulting partial uniform load shall be applied to the floor design span and a design moment shall be calculated.

In areas of soft clay, medium clay, organic material, and a water condition where the groundwater is higher than the floor level, the floor shall be designed as a reinforced slab resisting a uniform load which is the result of the total dead and live load divided by the total area of the base of the manhole. The floor slab may be designed as either a one- or a two-way slab. In severe water areas (tidal zones, high water table) provision shall be made for moment continuity at the junction of the floor and wall. Negative-moment steel shall be provided by using one-sixth the clear floor span plus 12 in as the reinforcing rod length to be carried into the wall and the floor. The rod size and spacing shall be taken as the same as those used for the floor reinforcing.

Precast Floors   Precast floors shall be designed for soil conditions as described above. If poor soil conditions exist, discretion should be used in specifying a precast manhole. A layer of crushed stone or concrete may be required to improve the foundation characteristic of the soil.

Precast floors shall be designed for lifting, using a dynamic load factor of 1.1.

Other Requirements   Minimum floor thickness, regardless of loading, shall be as follows:

Precast structures—6 in

Field-poured manholes—8 in

Field-poured vaults on soil—10 in

Field-poured vaults on rock—8 in

Field-poured vaults on steel frames—6 in

REINFORCING SPECIFICATIONS
Minimum Protection of Beams and Reinforcement Bars

The minimum concrete protection or cover of beams and reinforcement bars (in inches) is shown in Table 6-7. At no time, however, shall the required reinforcing in the roof, walls, or floor slab be less than that required for temperature reinforcing as defined in the American Concrete Institute Code, latest revision.

GRATINGS
Loadings

Loadings for gratings are shown in Table 6-8.

TABLE 6-7 MINIMUM CONCRETE PROTECTION, in

	Field-poured			*Precast*		
Structure	*Bottom*	*Sides*	*Top*	*Bottom*	*Sides*	*Top*
Roof beams—Manhole	1.0	1.0	1.0	1.0	1.0	1.0
Vault*	2.0	2.0	2.0	2.0	2.0	2.0
Roof slabs—Manhole	1.0	2.0	2.0	1.0	2.0	2.0
Vault*	1.0	2.0	2.0	1.0	2.0	2.0
Floors—Manhole	3.0	2.0	1.0	1.0	2.0	1.0
Vault:* Soil	3.0	2.0	1.0	2.0	2.0	1.0
Rock	3.0	2.0	1.0	—	—	—
Steel	0.75	—	—	—	—	—
Walls—Clear of inside face	—	1.0	—	—	1.0	—
Clear of outside face	—	2.0	—	—	1.0	—

*Vault includes transformer manhole.

Sidewalk Areas

Gratings over vaults and transformer manholes in sidewalk areas shall be formed of bearing bars riveted to reticuline bars at 7-in centers and, for grating doors, at $3\frac{1}{2}$-in centers.

TABLE 6-8 LOADINGS FOR GRATINGS

Type	*Location*	*Minimum live-load requirement*
Roadway	Vehicular way	20,800-lb wheel on 12- × 20-in arc
Heavy-duty	Driveway	Same as for vehicular way
Medium-duty	Sidewalk, nonindustrial	600 lb/ft²

CONSTRUCTION PRACTICES
Precast Manholes and Vaults

Precast manholes and vaults shall be constructed as follows:

1. Three separate precast parts consisting of a four-wall unit, a floor unit, and a roof unit.

2. Two parts, a lower and an upper part horizontally joined at the centerline of the walls, to be used for manholes only.

3. Precast structures shall be designed for a 4-ft depth of soil cover.

Field-Poured Manholes and Vaults

Field-poured manholes and vaults shall be designed and constructed as follows:

1. Field-poured roof with vertically reinforced walls
2. Precast roof with horizontally reinforced walls where size and conditions permit

Other Requirements

1. Precast and field-constructed transformer manholes located in roadways shall have grating roofs. Transformer manholes located in sidewalks or driveways shall have removable precast roofs incorporating grating sections.
2. Precast or field-poured transformer vaults are to be designed and constructed with partially or entirely removable roofs with an open area sufficient to clear transformer passage.
3. Precast vaults with dimensions exceeding transportation height limitations may be designed with top and bottom wall units for field assembly.

REINFORCED CONCRETE DESIGN*

General

Beam design must provide against failure by tension, compression, and diagonal tension. Longitudinal tension reinforcement is always required. Compression reinforcement is sometimes necessary where the dimensions of the member are limited. Reinforcement against diagonal tension, where necessary, is provided by stirrups, bent-up bars, or both. Beam formulas employed here are based on the following assumptions:

1. The concrete provides no tensile resistance in flexure.
2. The concrete is perfectly bonded to the reinforcing steel.
3. There are no initial stresses.

Standard Notation—Rectangular Beams

a area of stirrup steel in one plane
A_s area of longitudinal reinforcement
b width of beam
C total compressive stress in concrete

*This section is adapted from L. S. Marks, ed., *Mechanical Engineers' Handbook*, McGraw-Hill, New York.

C' total compressive stress in steel
d effective depth of beam, from compression face to centroid of steel
d' distance from compressive face to center of compressive steel
f_c compressive unit stress in concrete
f_s tensile unit stress in steel
f_s' compressive unit stress in steel
jd arm of resisting couple
kd distance from compression face to neutral axis
M resisting moment
n ratio of modulus of elasticity of steel E_s to that of concrete E_c
p steel ratio, A_s/bd
p' steel ratio for compression steel
R coefficient in formula $M = Rbd^2$
s horizontal spacing of stirrups
T total tensile stress in steel bars
u bond stress per unit area of bar
v shearing unit stress
V total vertical shear
V' vertical shear, in excess of that allowed on unreinforced web
z distance from compression face to resultant of compressive stresses; the resultant of C and C'
Σ_0 sum of perimeters of all longitudinal bars at a section

Resisting Moments

Referring to Fig. 6-15, the resisting moment for tension, in terms of steel stress, is given by:

$$M_s = A_s f_s jd$$

The resisting moment for compression, in terms of concrete stress is given by:

$$M_c = \frac{f_c kjbd^2}{2}$$

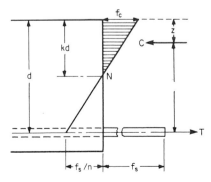

FIG. 6-15 Distribution of stress in reinforced concrete beam. *(From L. S. Marks, ed., Mechanical Engineers' Handbook, McGraw-Hill, New York.)*

For a balanced design:

$$M_s = M_c = Rbd^2$$

where $R = f_s pj = \dfrac{f_c kj}{2}$

$$k = \sqrt{2pn + (pn)^2} - pn = \frac{1}{1 + f_s/nf_c}$$

$$j = 1 - \frac{k}{3}$$

For fiber stresses:

$$f_s = \frac{M}{A_s jd} = \frac{M}{pjbd^2}$$

$$f_c = \frac{2M}{jkbd^2}$$

For balanced reinforcement:

$$p = \frac{1}{(2f_s/f_c)(f_s/nf_c + 1)}$$

Table 6-9 gives values of the several factors for rectangular beams. For shearing unit stress in a rectangular beam or slab:

$$v = \frac{V}{bjd}$$

For bond stress in a rectangular beam or slab:

$$u = \frac{V}{\Sigma_0 jd}$$

TABLE 6-9 DESIGN FACTORS FOR RECTANGULAR BEAMS

f_c	f_s	k	2000-lb concrete, $n = 15$		2500-lb concrete, $n = 12$		3000-lb concrete, $n = 10$		4000-lb concrete, $n = 7.5$	
			p	R	p	R	p	R	p	R
18,000	$0.4f_c'$	0.400	0.0088	138	0.0110	173	0.0134	208	0.0178	277
18,000	$0.45f_c'$	0.428	0.0107	165	0.0134	207	0.0161	248	0.0214	330
20,000	$0.4f_c'$	0.375	0.0075	131	0.0094	164	0.0112	197	0.0150	262
20,000	$0.45f_c'$	0.403	0.0090	157	0.0112	196	0.0135	236	0.0180	314

From L. S. Marks, ed., *Mechanical Engineers' Handbook*, McGraw-Hill, New York.

Stirrups—Web Reinforcement Where the shearing stress exceeds $0.02f_c'$, or $0.03f_c'$ where the reinforcement bars are anchored at the ends, the spacing is given by:

$$s = \frac{af_s j d}{V'}$$

This may be multiplied by $\sqrt{2}$, or 1.41, where the stirrups are inclined 45°.

Design Procedure for Rectangular Beams

1. Determine the appropriate ratio b/d to be used. This often varies from $\frac{2}{3}$ to $\frac{1}{2}$, and may be less for very large beams.

2. Determine the area bd required by the allowable shearing stress v.

3. From steps 1 and 2, select b and d.

4. From the equation for steel fiber stresses f_s, determine A_s and p. From the equation for steel ratio, for balanced reinforcement, determine the value of p. If the first value of p is equal to less than the second value of p, the tensile resisting moment governs and the depth d is satisfactory. If the first value of p exceeds the second value, increase d or provide compression reinforcement.

5. Determine bar sizes to provide the area of longitudinal reinforcement A_s, and check to see that the width b will permit proper bar spacing. Bar spacing should be at least $2\frac{1}{2}$ diameters for round bars and 3 times the width for square bars; clear spacing between bars should be at least $1\frac{1}{2}$ times the maximum width of the aggregate particles used.

6. Determine the shear and bond stresses to check that they are within the allowable values for the rectangular beam of the size selected.

SAMPLE DESIGN PROBLEM

This problem is meant to demonstrate the methods outlined in the procedure described above; some simplifications have been made for the sake of brevity.

The manhole to be designed is shown in Fig. 6-16a, together with the properties of the concrete and steel to be used.

Roof Design

For the steel beam design, refer to Fig. 6-16b. The design span is 4 in + 5 ft + 4 in or 5.67 ft.

Load Spread For the load spread, refer to Fig. 6-16c.

$$w_1 = 7 \text{ in} + 12 \text{ in} + 7 \text{ in} = 26 \text{ in}$$

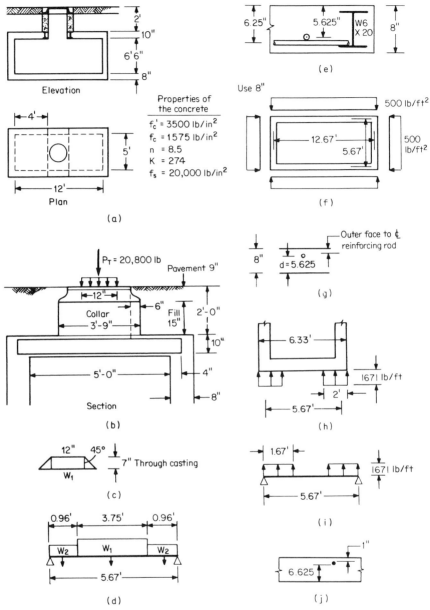

FIG. 6-16 Sample design problem. (a) Assumed manhole design and properties of the concrete. (b) Assumed manhole roof design and loading. (c) Diagram showing assumed load spread through casting at top of manhole collar. (d) Design showing dead loads imposed on manhole roof. (e) Assumed roof slab design showing steel beam and reinforcing rods. (f) Approximate wall dimensions and assumed lateral pressures on walls. (g) Assumed reinforcing rod cover for 8-in concrete beam. (h) Assumed manhole floor design and reaction areas. (i) Diagram showing moments acting on manhole floor. (j) Assumed reinforcing requirements for 8-in floor slab. *(Courtesy Consolidated Edison Co.)*

Continue the spread at 45° through the concrete portion of the collar (see Fig. 6-16*b* and the section Design Loading in this chapter):

$$24 \text{ in} - 7 \text{ in} = 17 \text{ in}$$

$$w_2 = 26 \text{ in} + 34 \text{ in} = 60 \text{ in}$$

This dimension is greater than the width of the collar, 45 in. The wheel is therefore reacted over 45 in times the wall thickness of the collar, i.e., 6 in (Fig. 6-16*b*).

Dead Load on Roof, Fill, and Pavement Refer to Fig. 6-16*b* and the section Design Loading. The load of the pavement is 9 in (i.e., 0.75 ft) × 144 lb/ft³ or 108 lb/ft². The loading of the fill is 15 in times 100 lb/ft³, or 125 lb/ft²; and that of the roof is 10 in times 150 lb/ft³, or 125 lb/ft². The total dead load of roof, fill, and pavement is 358 lb/ft².

Weight Distribution The weight distribution of the collar, in pounds per foot, is 17 in × 6 in × 150 lb/ft³, or 106 lb/ft. Assume four frames and casting weigh 660 lb; then

$$\frac{660 \text{ lb}}{4} = 165 \text{ lb}$$

and

$$\frac{165}{3.75} = 44 \text{ lb/ft}$$

For the pavement,

$$9 \text{ in} \times 6 \text{ in} \times 144 \text{ lb/ft}^3 = 54 \text{ lb/ft}$$

The weight distribution is as shown in Fig. 6-16*d*.

Assume w_1 acts on one-half of the 4-ft-wide slab (one of three slabs). Then the dead load on roof, fill, and pavement is 2 ft × 358 lb/ft², or 716 lb/ft. To this, add collar, frame, and pavement load of 106 + 44 + 54 = 204 lb/ft, and the wheel load of 20,800 lb/3.75 ft, or 5547 lb/ft. The total is 6467 lb/ft.

Assume that w_2 acts on half the 4-ft slab and the other part of the slab on either side of the steel beams. Then the dead load on roof, fill, and pavement is 4 ft × 358 lb/ft², or 1432 lb/ft. For the maximum moment on the roof slab (from one end), see Fig. 6-16*d*:

$$1432 \text{ lb} \times (0.96 + 3.75 + \tfrac{1}{2} \times 0.96) \text{ ft} = 5.19 \times 1432 = 7432 \text{ ft·lb}$$
$$6467 \text{ lb} \times (0.96 + \tfrac{1}{2} \times 3.75) \text{ ft} = 2.835 \times 6467 = 18,334 \text{ ft·lb}$$
$$1432 \text{ lb} \times (\tfrac{1}{2} \times 0.96) \text{ ft} = 0.48 \times 1432 = 687 \text{ ft·lb}$$

The total moment is 26,453 ft·lb.

From AISC Standard 5-17, $F_b = 24,000$ in·lb and $S_{xx} = (26,453 \times 12)/24,000 = 13.2$, where S_{xx} (a ratio) is the section modulus about the vertical axis. From the U.S. Steel *Structural Steel Handbook*, use a W6 × 20-in I beam whose $S_{xx} = 13.4$.

Roof Slab   The roof design span (see Fig. 6-16b) is 4 in + 5 ft + 4 in = 5.67 ft. From Table 6-4 and the design depth of 2.0 ft, and interpolating, the maximum moment is 5789 ft·lb.

In the two-way slab design (for a 4- × 5-ft slab; see Fig. 6-16a), the short-long ratio is 4 ft/5 ft, or 0.80 (use clear span). From Table 6-5, $K_l = 0.295$ and $K_s = 0.709$. To determine the design moments, for the short span,

$$M_s = 0.709 \times 5789 = 4104 \text{ ft·lb}$$

For the long span,

$$M_l = 0.295 \times 5789 = 1708 \text{ ft·lb}$$

To determine reinforcing, see Fig. 6-16e, and check for adequacy. For the short span,

$$d = 6.25 \text{ in}$$

Try a $\frac{5}{8}$-in bar with 8-in spacing.

From *Structural Steel Handbook*, $A_s = 0.47$ in^2, from which are derived: $p = 0.0063$, $k = 0.283$, $j = 0.905$, $jd = 5.66$ in, and

$$C = T = \frac{4104 \text{ ft·lb} \times 12}{5.66 \text{ in}} = 8700 \text{ lb}$$

$$f_s = \frac{M}{A_sjd} = \frac{Cjd}{A_sjd} = \frac{C}{A_s}$$

$$f_s = \frac{8700 \text{ lb}}{0.47 \text{ in}^2} = 18,511 \text{ lb/in}^2$$

as opposed to the 20,000 lb/in^2 specified. For the 12-in strip,

$$b = 12$$

$$f_c = \frac{2M}{jkbd^2} = \frac{2Cjd}{(jk)(12d^2)} = \frac{C}{6kd}$$

$$f_c = \frac{8700 \text{ lb}}{6 \text{ in} \times 0.283 \times 6.25 \text{ in}} = 820 \text{ lb/in}^2$$

as opposed to the 1575 lb/in^2 specified.

For the long span:

$$d = 6.25 \text{ in} - 0.625 \text{ in} = 5.625 \text{ in}$$

Try a $\frac{5}{8}$-in bar with 12 in spacing. From *Structural Steel Handbook*, $A_s = 0.31$ in^2, from which are derived: $p = 0.00459$, $k = 0.245$, $j = 0.918$, $jd = 5.16$ in, and

$$C = T = \frac{1708 \text{ ft·lb} \times 12}{5.16 \text{ in}} = 3972 \text{ lb}$$

$$f_s = \frac{3972 \text{ lb}}{0.31 \text{ in}^2} = 12,812 \text{ lb/in}^2$$

as opposed to the 20,000 lb/in^2 specified.

$$f_c = \frac{3972 \text{ lb}}{6 \times 0.245 \times 6.25 \text{ in}} = 432 \text{ lb/in}^2$$

as opposed to the 1575 lb/in² specified.

Wall Design

Use the rigid-frame method. The procedure is as follows:
1. Obtain the total uniform design load from Fig. 6-10.
2. Compute corner moments M_c by using the following formula, or Fig. 6-12:

$$M_c = \frac{(L_1 \times \text{FEM}_1) + (L_2 \times \text{FEM}_2)}{L_1 + L_2}$$

where FEM = fixed end movement for span = $\frac{1}{12}wL^2$

L_1 = long span

L_2 = short span

M_p = positive moment for span = $\frac{1}{8}wL^2 - M_c$

3. From the corner moment compute the design moments or use Figs. 6-11 and 13.

For the wall depth from street level, see Fig. 6-16a:

$$24 \text{ in} + 10 \text{ in} + 78 \text{ in} = 112 \text{ in or } 9.33 \text{ ft}$$

The design depth = 9.33 ft; use 9 ft. From Fig. 6-10, w = 500 lb/ft².

$$M = \frac{1}{12}wL_1^2 = \frac{1}{12} \times 500 \times 12^2 = 6000 \text{ ft·lb}$$

$$d = \sqrt{\frac{M}{K}} = \sqrt{\frac{6000}{274}} = 4.67 \text{ in} \qquad (\text{use 8 in})$$

$$\text{FEM}_1 = \frac{1}{12}(500 \text{ lb/ft})(12.67 \text{ ft})^2 = 6688.7 \text{ ft·lb}$$

$$\text{FEM}_2 = \frac{1}{12}(500 \text{ lb/ft})(5.67 \text{ ft})^2 = 1339.5 \text{ ft·lb}$$

Corner moment M_c:

$$M_c = \frac{(12.67 \text{ ft} \times 6688.7 \text{ ft·lb}) + (5.67 \text{ ft} \times 1339.5 \text{ ft·lb})}{12.67 \text{ ft} + 5.67 \text{ ft}}$$

$$= 5034.9 \text{ ft·lb}$$

Say 5035 ft·lb. From Fig. 6-12,

$$\text{Span ratio} = \frac{12.67 \text{ ft}}{5.67 \text{ ft}} = 2.23$$

$$K_1 = 0.0627$$

$$M_c = 0.0627 \times 500 \text{ lb} \times (12.67)^2 = 5032.6$$

or approximately 5035. The long-span positive moment is:

$$M_p = \tfrac{1}{8}wL_1^2 - 5035 = \tfrac{1}{8}(500)(12.67)^2 - 5035 = 4998 \text{ ft·lb}$$

The short-span midspan moment is:

$$M_s = \tfrac{1}{8}wL_2^2 - 5035 = \tfrac{1}{8}(500)(5.67)^2 - 5035 = -3025.7 \text{ ft·lb}$$

Reinforcing is required for the corner moment; refer to Fig. 6-16g.

A 2-in cover is required on the corner rods. Try a $\tfrac{5}{8}$-in bar with 6-in spacing (because of the 8-in wall); $d = 5.625$ in.

From *Structural Steel Handbook*, $A_s = 0.62$ in^2, from which are derived: $p = 0.62/(5.625 \times 12) = 0.0092$, $k = 0.40$, $j = 0.87$, $jd = 4.90$ in, and

$$C = T = \frac{5035 \text{ ft·lb} \times 12}{4.9 \text{ in}} = 12{,}330 \text{ lb}$$

$$f_s = \frac{12{,}330 \text{ lb}}{0.62 \text{ in}^2} = 19{,}887 \text{ lb/in}^2$$

as against the 20,000 lb/in^2 specified.

$$f_c = \frac{12{,}330 \text{ lb}}{6 \text{ in} \times 0.40 \times 5.625 \text{ in}} = 913 \text{ lb/in}^2$$

as opposed to the 1575 lb/in^2 specified.

Shear

$$V = \tfrac{1}{2}wL_1 = \tfrac{1}{2}(500 \text{ lb/ft})(12.67 \text{ ft}) = 3167.5 \text{ lb}$$

$$v = \frac{V}{bjd} = \frac{3167.5}{12 \times 4.90 \text{ in}} = 53.9 \text{ lb/in}^2$$

Allowable $v = 0.03f_c' = 0.03 \times 3500$ lb/in^2 (from the concrete specifications in Fig. 6-16a), or 105 lb/in^2.

Bond   For a $\tfrac{5}{8}$-in bar with 6-in spacing, $\Sigma_0 = 3.9$ (the circumference of the rod).

$$u = \frac{V}{\Sigma_0 jd} = \frac{3167.5}{3.9 \times 0.87 \times 5.625} = 165.9 \text{ lb/in}^2$$

Allowable u is up to but not in excess of 200 lb/in^2.

Floor Design

Calculate the total dead load and live load on the manhole floor:

Dead Load   The approximate dead load for the pavement is

$$(5.00 + 1.33) \text{ ft} \times (12.00 + 1.33) \text{ ft} = 84.37 \text{ ft}^2$$

At 108 lb/ft^2, that comes to 9112 lb. The weight of the fill, 84.37 ft^2 at 125 lb/ft^2, is 10,547 lb.

The dead load of the manhole is calculated as follows:
For the 10-in-thick roof,

$$84.37 \text{ ft}^2 \times 125 \text{ lb/ft}^2 = 10{,}547 \text{ lb}$$

For the (8-in-wide) long sides,

$$6.5 \text{ ft} \times (12 + 1.33) \text{ ft} \times 2 \times 125 \times 0.8 \text{ lb/ft}^2 = 17{,}329 \text{ lb}$$

For the short sides,

$$6.5 \text{ ft} \times (5 \text{ ft}) \times 2 \times 125 \times 0.8 \text{ lb/ft}^2 = 6500 \text{ lb}$$

The total manhole dead load without the floor is 34,375 lb. The total dead load is 54,035 lbs.

Live Load The weight of two wheels at 20,800 lb each is 41,600 lb; or 600 lb/ft^2 $\times$ 13.33 ft $\times$ 6.33 ft = 50,627 lb. Use 50,627 lb

The *total* of live load and dead load is 50,627 + 54,035 = 104,662 lb. The reaction area is assumed to be at 2 ft from the outer edges of the floor slab. See Fig. 6-16*h*, and for weight distribution, Fig. 6-16*i*.

$$\text{Area} = (2 \times 13.33 \text{ ft} \times 2 \text{ ft}) + (2 \times 2.33 \text{ ft} \times 2 \text{ ft})$$

$$= 62.64 \text{ ft}^2$$

$$\text{Uniform load} = \frac{104{,}662 \text{ lb}}{62.64 \text{ ft}^2} = 1671 \text{ lb/ft}^2$$

$$\text{Design span} = 6 \text{ ft } 4 \text{ in} - 8 \text{ in} = 5 \text{ ft } 8 \text{ in or } 5.67 \text{ ft}$$

The maximum moment (from one end) is the weight times the moment arm:

$$M = 1671 \text{ lb/ft}^2 \times 1.67 \text{ ft} \times \frac{1.67 \text{ ft}}{2} =$$

$$= 2790 \times 0.833 = 2324 \text{ ft·lb}$$

Reinforcing is required; the floor is assumed to be 8 in thick; refer to Fig. 6-16*j*.

Try a $\frac{5}{8}$-in bar with 15-in spacing; $d = 6.625$ in. From *Structural Steel Handbook*, $A_s = 0.25$ in^2, from which are derived: $p = \dfrac{0.25 \text{ in}^2}{6.625 \text{ in} \times 12 \text{ in}} = 0.003$, $k = 0.227$, $j = 0.89$, $jd = 5.90$ in, and

$$C = T = \frac{2324 \text{ ft·lb} \times 12}{5.90 \text{ in}} = 4727 \text{ lb}$$

$$f_s = \frac{4727 \text{ lb}}{0.25 \text{ in}^2} = 18{,}909 \text{ lb/in}^2$$

compared with the 20,000 lb/in^2 specified;

$$f_c = \frac{4727 \text{ lb}}{6 \text{ in} \times 0.227 \times 6.625 \text{ in}} = 524 \text{ lb/in}^2$$

compared with the 1575 lb/in^2 specified.

As a check:

$$A_s \text{ (required)} = \frac{4727 \text{ lb}}{20,000 \text{ lb/in}^2} = 0.236 \text{ in}^2$$

Use a $\frac{5}{8}$-in bar with 15-in spacing.

For further details, reference should be made to mechanical and civil engineering texts on reinforced concrete design.

UNDERGROUND RESIDENTIAL DISTRIBUTION (URD)

The development of plastics having high dielectrical and mechanical strengths, and the more widespread use of aluminum conductors, have contributed to economically practical underground systems for application in areas of light load density, including residential areas. The use of such plastics as the protective sheath for cables has also made practical their direct burial in the ground without the need for ducts and manholes. Splicing procedures for such cables are also more simple and less costly. Moreover, the installation of long lengths of cable capable of being plowed directly into the ground or placed in narrow and shallow trenches naturally reduces installation and maintenance costs.

Such designs make the consideration of this type of underground system economically more competitive with overhead systems, often replacing them where new construction is contemplated. Further economies may be realized if other facilities, such as telephone, CATV, gas, and water facilities, are installed simultaneously. With greater emphasis placed on appearance and reliability, studies comparing costs in particular areas and conditions should be made to determine the feasibility of this type of distribution.

Cables

Primary and secondary cables are designed specifically for this type of system. For a single-phase primary supply, the cable is designed with a stranded conductor completely embedded in plastic insulation, which also serves as the protective sheath, and a neutral of bare wire or ribbon wrapped concentrically over a layer of semiconducting Mylar. This one cable is easier to handle than two, and the neutral acts as protection during installation. Electrically, the reduced reactance from this arrangement results in a reduced voltage drop, while the neutral, acting as an electrostatic shield around the cable, tends to distribute the electrostatic stresses in the insulation more uniformly, eliminating sharp peaks that may cause insulation failure.

The neutral conductors are often exposed to electrolytic action and corrosion from stray currents and chemicals in the soil, resulting in voltage fluctuations from corroded neutrals and interference with nearby communication circuits. To prevent this, the entire cable is covered with an insulating jacket of plastic that is impervious to moisture and chemicals; see Fig. 6-17a.

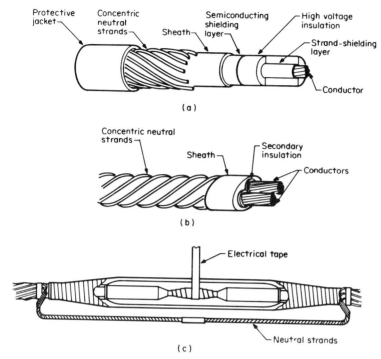

FIG. 6-17 (a) Primary concentric neutral underground cable. (b) Secondary concentric twin underground cable. (c) Typical single-conductor-cable splice. *(Courtesy Long Island Lighting Co.)*

Secondary cables likewise may consist of one or more insulated conductors and a neutral wrapped concentrically about them. Like the primary, the entire cable including the neutral may be jacketed for protection from corrosion, or it may have the neutral exposed. Refer to Fig. 6-17b.

Both primary and secondary cables have identification marks—colored plastic and metal tags—attached to them.

Cable Insulation

Cross-linked polyethylene (XLPE) is the material predominantly used for insulation purposes. Other insulating plastics include high-molecular-weight polyethylene (HMWP), also sometimes called high-density track-resistant polyethylene (HDPE), and ethylene propylene rubber (EPR). The earlier neoprene, polyvinyl chloride (PVC), and polyethylene (PE) cable sheaths have been replaced with the improved plastic materials mentioned. Plastic ducts made of polyvinyl chloride are sometimes installed to carry the cables or as spare ducts to make replacement of such cables easier.

Splices

The procedure for splicing such cables is the same for both primary and secondary cables. The conductors are connected together usually by a copper or an aluminum crimped sleeve. The insulation around the connector is built up by tape made of plastic similar to the cable insulation, wound around the connector to a specified thickness; a preformed insulation placed around the connector may also be used in place of tape. A semiconducting tape, usually of Mylar, is wound over the insulating tape, and another tape, made of the same material as the sheath, is wound over the entire assembly for protection. The concentric neutrals are bundled together to one side and mechanically connected together and bundled with the other conductor splices; they are then covered by a plastic tape to protect them from electrolytic action and corrosion. See Fig. 6-17c.

Terminal Devices

The cables may also be connected together through a terminal block containing a number of stud connectors, the entire assembly covered by molded insulation. The conductor is connected to the stud and the insulation and sheath of the cable are taped to the insulation and molded protection of the terminal block. The terminal may also consist of molded load-break elbows and bushings, enabling the conductor to be disconnected safely by means of an insulated hook stick which pulls the stud from the receptacle. In some instances, the molded terminal block may be combined with similar types of terminals of a transformer or switch. These disconnecting devices may be used to rearrange, deenergize, or energize primary circuits and equipment; see Fig. 6-18.

Risers

Risers, connecting underground lines to overhead lines, have been described earlier in the discussion of overhead systems.

Transformers

Transformers for this type of underground residential system are hermetically sealed against moisture, including their bushings and terminals. The terminals may contain one or more insulated disconnecting elbows, enabling the disconnection of the transformer and the sectionalizing of the primary circuit.

The transformers may be completely buried or installed on ground-level pads or semiburied pads, as shown in Fig. 6-19. Some have connections made behind an insulating panel with only the insulated disconnecting elbows protruding so that no energized parts are exposed when the enclosure is opened; these are known as *dead-front* units, in contrast to those in which the energized parts may be exposed when the enclosure is opened.

Transformers may be of the conventional type with associated oil-fuse cutouts

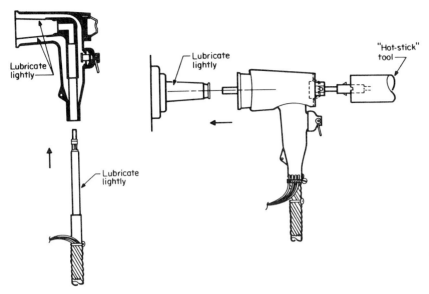

FIG. 6-18 Load-break elbow-type cable tap assembly. *(Courtesy Long Island Lighting Co.)*

or switches, or they may be of the completely self-protected (CSP) type with an internal weak link and secondary circuit breakers; they may also have taps on the primary coil for voltage ratio adjustment.

Corrosion

Corrosion of metallic parts may take place when they are exposed to chemicals and stray currents in the soil. This is especially critical in the case of neutral conductors and the tanks of transformers and other equipment.

Chemicals may exist naturally in the soil or may find their way there from sources of pollution. These may attack the metallic parts, resulting in the gradual destruction of the metal. Such action may be prevented or slowed by coating the metallic parts with preservative paints or, in the case of neutrals, covering them with an insulating plastic covering.

Stray currents may result from railroad return dc circuits, from galvanic action that takes place between dissimilar metals, especially in wet or damp soils, or from bacteria and other causes. Two or more metallic objects, reasonably close together and immersed in the chemical solutions present in the soil, will have an electric current flow between them. This flow of current causes a flow of ions (molecules carrying an electric charge) from one to the other, a phenomenon known as galvanic action or electrolysis, resulting in metal traveling or flowing from one object and depositing itself on the other object or structure. The direction of this flow will depend on the relative voltages between the two objects,

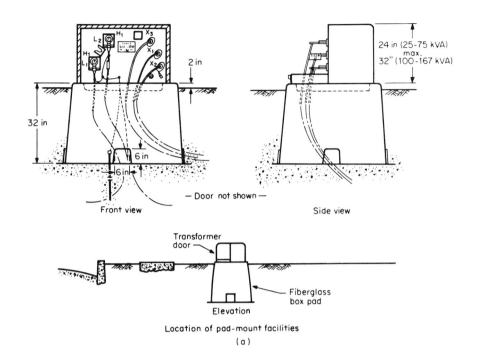

— Door not shown —

Front view   Side view

Location of pad-mount facilities

(a)

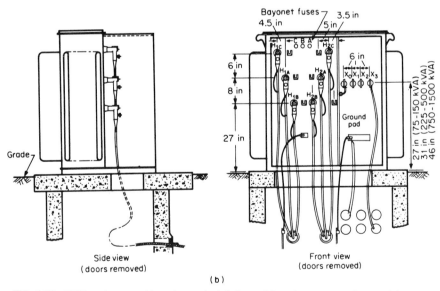

Side view
(doors removed)

Front view
(doors removed)

(b)

FIG. 6-19 URD underground transformer installations: (a) pad-mount transformer; (b) three-phase, 4-kV or 13-kV dead-front metal-clad transformer. *(Courtesy Long Island Lighting Co.)*

236

the flow being from the one of higher voltage to that of lower voltage; typical voltages for different materials are shown in Table 6-10.

The amount of corrosion that will occur due to electrolytic action will depend on:

1. The direction of the flow of current

2. The duration of the flow

3. The magnitude of the current flowing

4. The current density in the area over which the current flow takes place

5. The moisture content of the earth with which the objects are in contact

6. The chemical properties of the solutions through which the current flows

The material of a higher-voltage metallic object, such as the steel tank of an underground transformer or other equipment, will be drawn away from it, and in time the corrosion will cause failure. Although anticorrosive and insulating paints are of some usefulness in slowing down the process, they are uncertain in their results. A more positive means is to connect to the steel tank a metal,

TABLE 6-10 GALVANIC SERIES

Material	Approximate potential with respect to a saturated $Cu\text{-}CuSO_4$ electrode, V*
Commercially pure magnesium	-1.75
Magnesium alloy (6% Al, 3% Zn, 0.1% Mn)	-1.6
Zinc	-1.1
Aluminum alloy (5% Zn)	-1.0
Commercially pure aluminum	-0.8
Cadmium	-0.8
Mild steel (clean and shiny)	-0.5 to -0.8
Mild steel (rusted)	-0.2 to -0.5
Cast iron (not graphitized)	-0.5
Lead	-0.5
Tin	-0.5
Stainless steel, type 304 (active state)	-0.5
Copper, brass, bronze	-0.2
Mild steel (in concrete)	-0.2
Titanium	-0.2
High-silicon cast iron	-0.2
Nickel	$+0.1$ to -0.25
Monel	-0.15
Silver solder (40% Ag)	-0.1
Stainless steel, type 304 (passive state)	$+0.1$
Carbon, graphite, coke	$+0.3$

*These values are representative of the potentials normally observed in soils and waters which are neither markedly acid nor alkaline.

From *EEI Underground Systems Reference Book.*

TABLE 6-11 TYPICAL RATES OF METALLIC
CORROSION

Anode metal	Density, lb/in³	Penetration (in/yr) caused by discharge of 1 mA/in²
Magnesium	0.063	0.139
Zinc	0.258	0.091
Aluminum	0.098	0.065
Steel	0.284	0.071
Lead	0.409	0.182
Copper	0.323	0.142

From *EEI Underground Systems Reference Book.*

such as magnesium, that will produce a voltage higher than that produced by the steel tank. This will cause a current, and a flow of material, to flow from it instead of the steel tank. The attached metal, called an anode, will be sacrificed to protect the steel tank. The anodes should be inspected regularly and replaced when necessary.

The rate of penetration of a metal (in inches per year) is roughly inversely proportional to the area from which the current discharge takes place. This assumes the current is constant, i.e., the rate of corrosion varies directly with the intensity of current discharge. Theoretical rates of corrosion for some metals are given in Table 6-11.

If dissimilar metals are used, good design favors coupling a large anode with a small cathode (the receiving metallic object), rather than the reverse. When using coatings to prevent galvanic action caused by dissimilar metals, it is usually preferable to coat the cathode rather than the anode.

DESIGN OF DIRECT-BURIED ELECTRICAL DISTRIBUTION SYSTEMS*
General

The procedure described here covers the design of direct-buried electrical distribution systems to supply service to new residential development areas consisting of five or more new single-family homes. Similar procedures have been prepared for multiple-dwelling development areas and for commercial and industrial development areas.

Assumptions

1. The homes to be supplied are situated on individual building lots fronting on residential streets.
2. No electrical distribution system exists within the development.

*Adapted from Long Island Lighting Co. design standards. Other voltages are in use in other areas of the country, but this procedure may serve as a guide.

3. Only single-phase 120/240 V service is required by the consumers.

4. A 4-kV or 13-kV primary system exists near the development area.

Basic Design Considerations

Radial Design The radial design is usually more economical than the loop design and is permitted (but not mandatory) in any case where loop design is not indicated for any of the reasons specified below.

Loop Design Loop design shall be used in all cases where any of the following conditions apply:

1. The existing (external) primary distribution system, to which the development area system will be interconnected, is underground.

2. The area requires more than two transformers, or a total transformer capacity rating exceeding 150 kVA.

3. Two transformers are, or will be, needed that will be spaced not more than 1200 ft apart.

Radial and Loop Designs Both designs apply where any of the following conditions are to be met:

1. The character of the land area adjoining the development area, or any other factor, indicates a potential future need for further extension of the new underground system.

2. For both designs, if the existing primary distribution line is an overhead system remote from the perimeter of the development area, an overhead line extension shall be constructed to approach the perimeter.

3. The selection of existing primary lines to be used in supplying the new system shall be made with the usual consideration of physical availability, adequacy of load capability, the desired allocation of new loads among existing circuits and substations, primary circuit voltage drop, etc., consistent with long-range development plans for the area primary system.

4. When the existing distribution is 4 kV, consideration should be given to the possibility of conversion to 13 kV prior to initial operation of the new system, which would avert the need for transformers to operate initially at 4 kV.

5. All primary system conductors shall be insulated with cross-linked polyethylene (XLPE), the insulation rated at 15 or 24 kV.

6. Riser pole locations shall be selected with due consideration to accessibility to personnel and vehicles for operating purposes. Locations prone to vehicular accidents or tree damage shall be avoided.

7. The trench routes for installation of the primary cable shall be run parallel to the curb line or roadway edge and may be located either in the roadway or in the area behind the curb. Installation in the roadway is usually preferred

because of least construction cost and probable least cost for future operation and maintenance, but extenuating circumstances may sometimes justify installations behind the curb.

8. In choosing the transformer locations, care shall be taken to avoid conflicts with driveways, trees, fire hydrants, storm drains, etc.

9. Transformer load estimates and kVA ratings shall be selected in accordance with Tables 6-12 and 6-13 and Fig. 6-20; see Appendix 6A.

Radial System Design

Riser Interconnection   The new radial primary system shall originate at a riser interconnection to the existing primary system at a single-phase riser. This interconnection shall include a fused cutout to provide an overhead sectionalizing point to protect the overhead system from underground system faults.

Primary Conductors

1. The standard size for all radial system primary conductors shall be no. 2 aluminum.

2. The primary cable run shall be extended to the transformer nearest the riser pole. It shall be interconnected to that transformer by means of a load-break elbow terminator. The transformer shall be either a pad-mount (PM) type or a below-grade, semiburied (SB) type.

3. The location of the riser cable termination facility—the transformer or the load-break elbow-type cable tap assembly (LETA)—shall be selected so as to avoid existing or foreseeable obstacles to inspection, operation, and maintenance of the terminal equipment. Locations subject to soil erosion, flooding, collection of debris, obstruction by vehicles, etc., shall be avoided.

4. If the distance from the riser to transformer exceeds 1200 ft, a load-break cable tap assembly (LBCTA) shall be installed to limit the riser cable run and the distance between the LBCTA itself and the nearest transformer each to 1200 ft or less. (The 1200-ft limit is based on the length of cable available for on-the-ground installation between two switch points in emergencies.) Primary cable from this point is then extended to the first transformer, which shall be of the PM or SB type.

Loop System Design

Switch Points

1. Loop design provides two underground radial primary lines, each originating at a different interconnection to the external primary system, and routed to meet at a common normally open switchable tie point within the development area. This switch point and other (normally closed) switch points on the loop configuration formed by the pair of radial lines are provided by means of LBCTA-type terminations installed at the primary terminals of all SB- or

PM-type transformer installations. These switch points, and switching devices at the interconnections to the external system, allow switching operations on the loop system whereby any one transformer, or a section of primary cable between switch points, can be isolated while all other parts of the loop are able to be maintained energized from one or the other of the two loop sources.

2. In all loop systems, a sufficient number of switch points must be provided to limit the distance between any pair of switch points associated with the same cable section to 1200 ft or less. Where the required locations for loop interconnections (e.g., risers) and transformers of the PM or SB type are spaced too far apart to comply with this requirement, the LETA or LBCTA installations are used at intermediate locations to reduce the spacing of switch points to within the 1200 ft limit.

3. The switch point selected to be maintained normally open shall be located so as to divide the total loop load into two approximately equal parts.

Conductor Sizes Standard conductor sizes for primary loop system cables shall be no. 2 and no. 1/0 aluminum. The choice between these two sizes will depend upon the area dimensions and load density. This choice shall be determined by the use of a primary conductor selection chart and procedure; refer to Fig. 6-21*a* and *b* and Appendix 6A. The selection chart is designed to indicate the choice of conductor required to limit voltage drop on either segment of the loop to 1.5 V or less (on a 120-V base) under peak load conditions. The chart allows for a moderate imbalance in the sharing of load between the two sections and in the lengths of sections. It limits the no. 2 aluminum to cases in which the total loop load current will not exceed 150 A when the loop is operated in a contingency mode, i.e., with all load supplied from one end of the loop (tie point closed). No. 1/0 aluminum is permitted by the chart to carry load up to 200 A in the contingency operation mode. Use of the chart also indicates when neither conductor size is adequate for supplying the area load with a single-loop system, in which case alternative layouts, employing two or more loops, must be tried.

Multiple-Loop Systems

1. Load density and load distribution in some areas may require the installation of two-, three-, or more loop systems in the same trench route; loads (transformers) shall be allocated as evenly as practical among the loop systems to maintain load balance.

2. Only pad-mount (PM) or semiburied (SB) transformers shall be installed along the trench routes occupied by multiple-loop systems; direct-buried units shall not be installed. (This is to facilitate connection of the 1200-ft emergency cable.)

3. The 1200-ft limit between switch points is satisfied when each switch point is located not more than 1200 ft from another switch point on the same loop or on another loop in the same trench route.

Interconnections to External Systems   Whenever the external primary system is underground, consideration should be given regarding the facilities to be used (existing or required) for interconnections of the loop system.

1. Equipment to be used will generally be pad-mounted switch gear having in-and-out gang-operated three-phase switches rated for 600-A interruption duty. These are to provide main line sectionalizing points. They also provide for one, or two, three-phase sets of single-phase, fusible tap connections, rated for 200-A duty, which can be used as interconnection points for one or more loop systems.

2. A loop system interconnected to an underground external system shall not have more than one of its two terminals located in the same pad-mount facility.

3. Fuses used at these interconnections shall be selected in accordance with a predetermined schedule.

4. Rules governing the pairing of loop terminations with external source circuits and phases, where pad-mount-type interconnections are employed, are similar to those that apply where riser installations are used, as described in items 5, 7, 9, and 10.

5. When the external primary system is overhead, loop system interconnections shall be made via riser installations. The pair of risers required for each loop system shall be such that the two risers are installed on different poles.

6. Both interconnections must be of the same operating voltage class, i.e., 4 kV or 13 kV, and both must be interconnected to the same phase of the same primary circuit, or the same phase of two primary circuits supplied from the same substation, or the same phase of two primary circuits supplied from different substations supplied at the same transmission voltage, attention being paid to the phase relationships.

7. In areas requiring more than one loop system, terminals of more than one loop may be located on the same riser pole, either in separate riser conduits, or in the same conduit (e.g., a three-phase riser may be used for the riser cable runs of three different loops). However, as previously indicated, no one loop may have both terminals installed on the same pole.

8. Each loop interconnection shall be individually fused using a cutout fused in accordance with a predetermined schedule.

9. The riser cable run from each terminal of each loop system shall be extended to the first loop switch point as described for radial system design. No tee splices shall be installed in a riser cable run.

10. To facilitate future operation and maintenance, riser cable runs from the same or different loops should be routed to minimize confusion in locating each line. Crossovers of these cables should be avoided if possible.

Radial Lateral Extensions from Loop Systems   Where the main route of a loop system intersects a side street (or a cul-de-sac, etc.) in which transformers are required, it may be economical to avoid routing the loop into and out of the side street by installing a tee splice in the main run and extending a radial lateral line from the splice to the required location, subject to the following conditions:

1. Such extensions shall not be made to serve more than two transformers, nor a total transformer capacity rating exceeding 150 kVA.

2. The first transformer location on the extension must be within 1200 ft or less, measured along the most direct street route, from at least one of the two switch points that bound the loop section.

3. The radial line shall not be extended to a second transformer location if that location is more than 1200 ft from the first transformer (i.e., if the second transformer cannot be otherwise supplied, the loop systems must be extended into the side street).

4. The first transformer on a radial lateral line shall be of the switchable type appropriate to the area (PM or SB).

5. If a second transformer is required and permitted, the second unit shall be either a pad-mount (PM) unit in areas where PMs are used as switch points, or a direct-buried unit where below-grade units are used.

Transformer Installations

Transformers used in residential developments may be of two general types: pad-mount (PM), or a below-grade type, either semiburied (SB) or direct-buried (DB).

1. The PM type, designed for installation at grade, is contained within a metal enclosure that provides access to primary and secondary terminal compartments. The unit is mounted on a below-grade box-pad, and primary cable connections are of the load-break elbow terminator type.

2. In SB installations, primary cable connections are of the load-break elbow terminator type. Two primary terminal bushings are available, permitting the unit to be connected to either one or both of the primary loop sections adjoining the unit. To enable use of these connections as loop system switch points, SB installations employ a vertical cylindrical steel enclosure, with a removable cover placed level with grade, which provides acess to the elbow terminators.

3. In DB installations, primary cable connections are made using dead make–dead break type terminators, and these are completely buried (no enclosure provided). Direct-buried units must be deenergized, and reexcavation is required, whenever it is necessary to disconnect these units from the primary cable system after initial installation.

4. Transformers capable of operation at both 4 kV and 13 kV should be installed where such conversion appears probable. Since operation of the voltage selection mechanism to change from one voltage level to the other requires physical access to the transformer, no DB units are to incorporate this feature.

5. All transformer primary terminals not connected to primary cables shall be provided with insulating caps.

6. Below-grade transformers shall be installed between the curb line and the front property line, near the intersection of the front property line with a common side property line of the two adjoining lots.

7. Pad-mounted transformers shall be installed within a 10- by 10-ft easement behind the front property line and adjacent to the common side property line between two adjoining lots.

Secondary System Design

Splice Boxes   A secondary splice box shall be used at each below-grade-type transformer to house the connections between the transformer secondary leads and the secondary main conductors or service conductors. A removable cover allows for future access to the connections.

Main and Service Conductor System

1. Secondary main cables are those which supply two or more consumers. Cables supplying only one consumer are classified as service cables. Cables shall be either of the following:

   a. Three-single-conductor no. 1/0 aluminum triplex 600-V (with no. 2 neutral)

   b. Three-single-conductor no. 3/0 aluminum triplex 600-V (with no. 1/0 neutral)

2. Allowable limits for voltage drop and voltage dips on motor starts are the same as for overhead systems.

3. Subject to the voltage limits indicated, the preferred arrangements for secondary systems are those that result in the least cost for trenching, cable installation, and splicing, and with due regard for avoidance of conditions that would impede future operation and maintenance of the conductor systems.

4. For economy reasons, cable runs should use a common trench for road crossings and in connecting to meters located on private property.

5. Consideration should be given to secondary main and service runs on private property to consumers not ready to receive service at the time of initial construction. In all cases, the entrance onto private property shall be made at a point on the front property line not more than 5 ft away from the intersection with the side property line. Front-to-rear trenching on private property shall be extended in a line parallel to the side property line for the distance required to reach the point(s) opposite the meter locations.

6. At those locations that are not ready to receive service, trenching and cables shall be terminated just short of the front property line. This trenching will be extended onto private property when the premises become ready to receive

service. The cables are left disconnected at their splice box terminations or at the secondary terminals of PM units.

7. When meter location(s) are ready to receive service, trenching and installation of cables shall be completed in their entirety, including those portions between the front-to-rear trench run and the meter location(s). These cables are energized by making appropriate connections at their splice box or secondary compartment terminations.

8. At locations where only one of a pair of meter locations sharing a common side property line will be initially ready to receive service, the objective should be to minimize reexcavation of areas previously trenched. To accomplish this, the trench required initially should also have installed within it those cables that will be needed later. In general, it is more economical to provide an extra length of coiled cable at the end of a trench run than to plan for the installation of a splice. The choice will depend upon the comparative cost of these alternatives considering the physical dimensions involved. In any case, buried cable ends or splices shall be moisture-proofed.

APPENDIX 6A
TECHNICAL REFERENCE DATA

This appendix contains the reference materials that are referred to in the preceding procedure. The data are common to the associated procedures for this type of service to multiple-dwelling development areas and to commercial and industrial development areas. The reference data are presented in Tables 6-12 and 6-13 and in Figs. 6-20 and 6-21.

ABLE 6-12 LOAD ESTIMATING GUIDE: COINCIDENT TRANSFORMER LOAD, kVA

.iving area per home, ft²	*Season*	Number of homes per transformer									
		1	*2*	*3*	*4*	*5*	*6*	*7*	*8*	*9*	*10*
.oad condition 1: All-electric dwellings including electric space heating and central air conditioning											
1200	Winter	12	23	34	44	53	63	72	82	91	100
	Summer	7	12	16	19	22	25	28	30	33	36
1200–1700	Winter	17	32	46	60	73	87	99	112	125	138
	Summer	9	15	20	25	28	32	36	39	43	47
1700–2200	Winter	21	40	59	77	94	110	126	143	159	175
	Summer	11	18	24	30	34	39	43	47	52	56
2200–2700	Winter	26	49	71	93	114	134	153	173	193	213
	Summer	12	21	28	34	40	45	50	55	60	65
2700–3200	Winter	30	58	84	109	134	157	181	204	227	250
	Summer	14	24	33	39	46	51	57	63	69	75
3200–3700	Winter	35	66	97	126	154	181	207	234	261	288
	Summer	16	27	37	44	51	57	64	71	77	84

TABLE 6-12 LOAD ESTIMATING GUIDE: COINCIDENT TRANSFORMER LOAD, kVA (*Continued*)

Living area per home, ft²	Season	Number of homes per transformer									
		1	2	3	4	5	6	7	8	9	10
Load condition 2: Electric base load plus central air conditioning											
1200	Winter	2	3	5	6	7	8	9	10	11	12
	Summer	3	6	9	11	14	17	19	22	25	28
1200–1700	Winter	2	3	5	6	7	8	9	10	11	12
	Summer	3	6	9	11	13	15	17	19	21	22
1700–2200	Winter	2	3	5	6	7	8	9	10	11	12
	Summer	4	8	12	15	18	21	24	27	29	31
2200–2700	Winter	2	3	5	6	7	8	9	10	11	12
	Summer	5	10	15	19	23	26	29	32	35	38
2700–3200	Winter	2	3	5	6	7	8	9	10	11	12
	Summer	7	14	20	25	30	35	39	44	47	51
3200–3700	Winter	2	3	5	6	7	8	9	10	11	12
	Summer	8	16	24	30	36	41	46	50	54	57
Load condition 3: Base load plus central air conditioning plus electric range											
1200	Winter	4	8	10	11	12	14	16	17	18	19
	Summer	6	8	11	13	15	18	20	23	26	28
1200–1700	Winter	5	9	12	14	16	18	20	21	23	25
	Summer	7	11	15	18	21	24	26	29	31	34
1700–2200	Winter	5	9	12	14	16	18	20	21	23	25
	Summer	8	14	18	22	26	30	33	36	39	42
2200–2700	Winter	5	9	12	14	16	18	20	21	23	25
	Summer	9	15	21	26	30	35	39	42	46	49
2700–3200	Winter	5	9	12	14	16	18	20	21	23	25
	Summer	11	19	26	33	39	44	49	54	59	64
3200–3700	Winter	5	9	12	14	16	18	20	21	23	25
	Summer	12	21	29	36	42	48	53	58	63	69

1. Load values indicated in Table 6-12 and Fig. 6-20 allow for load growth, including addition of central air conditioning, if not initially present. Central air conditioning capacity ratings per dwelling are assumed as follows:

Living area, ft²	Capacity, Btu/h
1700	15,000
1700–2199	20,000
2200–2699	30,000
Over 2699	50,000

2. To match estimated transformer loads obtained from Table 6-12 and Fig. 6-20 to transformer capacity rating required, refer to Table 6-13.

Note: Group coincident loads indicated by this table shall be used only for estimating summer season transformer loading and transformer secondary voltage drop. Selection of transformer ratings for supplying homes with electric space heating shall be based on loadings obtained from Fig. 6-20.

Courtesy Long Island Lighting Co.

TABLE 6-13 TRANSFORMER RATING
SELECTION CHART

For estimated load, kVA*		Select rating, kVA**
Winter	Summer	
A. For below-grade transformers		
0–35	0–28	25
35–52	28–42	37½
52–70	42–55	50
70–105	55–86	75
105–140	86–110	100
B. For pad-mount transformers		
0–23	0–23	25
23–50	23–50	37½
50–110	50–80	50
110–135	80–120	75
135–180	120–160	100
180–370	160–270	167

*Load as estimated from Table 6-12 or from Fig. 6-20 (for electric space heating).

**If different ratings are called for by winter and summer estimated loads, install the larger rating.

Courtesy Long Island Lighting Co.

USE OF LOAD-ESTIMATING CURVES FOR RESIDENTIAL LOADS INCLUDING ELECTRIC SPACE HEATING

For URD areas supplying residential consumers with electric space heating, the selection of transformer ratings shall be based upon house service load estimates obtained from the curves in Fig. 6-20. These curves indicate the load conditions that can exist after a long power outage during cold weather. These loads exceed normal levels because, after an outage, heating equipment operates at maximum capacity continuously until normal indoor temperatures are restored. The curves also allow for the demand of some other appliances, such as water heaters and refrigerators, which are also likely to resume operation coincidentally and to continue to operate for longer periods than normal.

To use the curves, it is necessary to know the approximate square footage of all rooms in the house that are used on a regular basis daily (i.e., the main living area), such as bedrooms, living room, dining room, bathrooms, and family room. Where additional heating capacity is provided for areas that are used only occasionally (the supplemental areas), such as garage, basement, and enclosed patio, the total square footage data are then used as illustrated in Example 6-1.

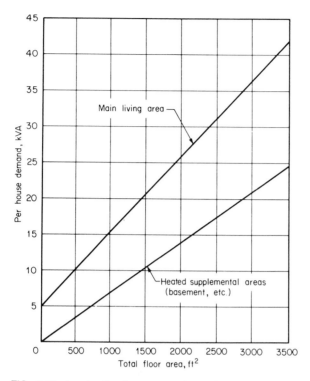

FIG. 6-20 Load-estimating curves for residential loads including electric space heating; see Example 6-1. *(Courtesy Long Island Lighting Co.)*

Example 6-1 Find the load that must be supplied for two houses, each of which has a main living area of 2000 ft^2 and heated supplemental areas totaling 1000 ft^2.

Solution

The demand for each house resulting from the main living area load is obtained by entering the chart at the 2000-ft^2 point on the horizontal axis scale and projecting upward to the intersection with the "main living area" line, and projecting a horizontal line from this intersection to the scale at the left (read 26.25 kVA). The supplemental area load is obtained in a similar manner using the supplemental area line intersection (read 7.00 kVA).

Total load per house is 26.25 + 7.00 kVA, or 33.25 kVA. For the two houses, the total load on the transformer will be 2 × 33.25 kVA, or 66.5 kVA.

This result is used in Table 6-13 to select the required capacity rating of the transformer to supply the two houses.

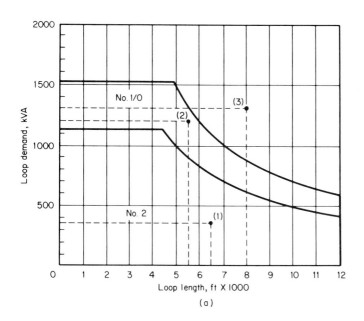

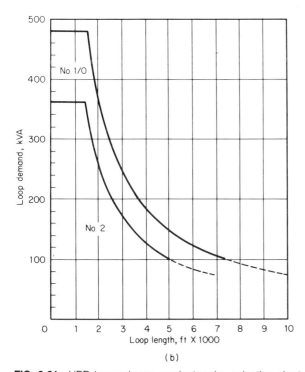

FIG. 6-21 URD loop primary conductor size selection chart. (a) For 13-kV systems. For (1), (2), and (3), see Example 6.2. (b) For 4-kV systems. *(Courtesy Long Island Lighting Co.)*

USE OF THE URD-LOOP PRIMARY CONDUCTOR SIZE-SELECTOR CHART

1. Lay out a tentative URD loop to serve the URD area. Defer the choice of transformer locations and required ratings.

2. Determine the total length of the proposed loop.

3. Determine the normal peak demand (kVA) for the total number of consumers to be supplied from the proposed loop. Refer to the group load-estimating guides in Table 6-12.

4. Enter the chart at the bottom axis at a point corresponding to the total length determined in step 2 and project a vertical line upward from this point (to the length in thousands of feet).

5. Enter the chart at the left-hand axis at a point corresponding to the kVA demand determined in step 3 and extend a horizontal line to the right to the point of intersection with the vertical line drawn in step 4.

6. a. If the intersection falls within the zone labeled no. 2, select no. 2 aluminum as the primary conductor and design the loop as proposed in step 1.

   b. If the intersection falls within the no. 1/0 zone, select no. 1/0 aluminum conductor and design the loop as proposed in step 1.

   c. If the intersection point does not fall within either the no. 2 or the no. 1/0 zone, the loop layout proposed in step 1 cannot be used. Try another layout dividing the area load between the two loops, then repeat steps 1 to 6 for each loop. Depending on the total area load and size, it may be necessary to repeat this procedure using three loops, etc., until the number of loops is adequate to supply the area load level.

   Example 6-2   Refer to Fig. 6-21a.

   1. The loop length is 6500 ft, and the total loop demand is 350 kVA. The intersection point falls in the no. 2 zone. The loop may be designed as proposed, using no. 2 aluminum.

   2. The loop length is 5500 ft, and total loop demand is 1200 kVA. The intersection point falls in the no. 1/0 zone. The loop may be designed as proposed, using no. 1/0 aluminum.

   3. The loop length is 8000 ft, and total loop demand is 1300 kVA. The intersection point falls outside the no. 2 and no. 1/0 zones. Redesign the layout for two loops, dividing the area load between the two loops. (In this example, assume the subsequent layout consists of two loops as follows:

   Loop a—6500 ft, 800 kVA
   Loop b—7500 ft, 500 kVA

   The proposed loops are both acceptable designs, using no. 2 aluminum.)

CHAPTER 7

DISTRIBUTION SUBSTATIONS

The design of a distribution system is affected by the location and design of its supply substation. Indeed, the distribution substation is an integral part of the electrical distribution system.

SITE SELECTION

The availability of land, annual costs, taxes, zoning laws, and environmental and public relations considerations are some of the factors that determine the ultimate location.

The number and locations of the substations may affect the voltage selected for the primary distribution system. The fewer and farther apart the substations, the higher the primary voltage selected and the larger the loads supplied. Also, the length of distribution feeders (and the distance of consumers from the substation) and the number of consumers supplied from a feeder are reflected in the size of conductors, voltage-regulation measures, and, equally important, the losses that may be incurred. Hence, the study of the most economical design of a distribution system must include substation and transmission supply costs as well as the effect of primary voltage on feeders, transformers, equipment, and methods of maintenance and operation.

In addition to the factors cited, other considerations should be taken into account in choosing a site for a substation:

1. It should be located as near as practical to the centers of the loads to be served; the summation of the loads (which are assumed to be concentrated at some points) multiplied by their distances from the proposed substation site should be at a minimum.

2. It should be possible to supply the loads without undue voltage regulation and with available standard equipment.

3. Access for incoming transmission lines and outgoing distribution feeders should be available with a minimum of inconvenience, and should allow for future expansion of such facilities.

4. A shutdown of the substation should not affect an undue number of consumers; the substation location in relation to other, adjacent substations, both present and future, should permit ties to them in event of emergency.

GENERAL DESIGN FEATURES

Since the substation is the link between the transmission and distribution systems, its continuous and uninterrupted operation is of prime importance. For this reason, multiple incoming supply feeders are provided, relays are installed to operate switches and circuit breakers automatically to disconnect faulted feeders and equipment, and spare equipment may be provided for rapid restoration of service in the event of failure.

Equipment Installation

Arrangement Equipment may be connected in various arrangements by means of buses, switches, and circuit breakers as described in Chap. 4. The arrangements are usually such as to insure safety to workers and reliability of operation, the usual arrangement in many substations permitting work on almost any piece of equipment without interruption to the incoming or outgoing feeders. The choice of arrangement is based on economics and the degree of reliability desired; see Fig. 7-1.

Insulation Coordination (BIL)

The feature of coordination of insulation, associated with the reliability of the substation, has been detailed in Chap. 4. The impulse values of the insulation of transformers, circuit breakers, switches, lightning or surge arresters, regulators, bus supports, and other elements are chosen at different levels, determined from a minimum basic insulation level (BIL), usually based on the minimum line-to-ground voltage rating of the surge arresters. Such design assures that potential failures of insulation from overvoltages can be made to occur at the least destructive and most accessible points, where they can be readily found and repairs or replacement of damaged equipment can be made.

Coordination of Protective Devices

The coordination of protective devices, both within the same substation and with devices on the associated transmission and distribution feeders, has been discussed in Chap. 4. Such coordination is essential to ensure that the operation of a particular device does not unnecessarily deenergize remaining unfaulted portions of the distribution system.

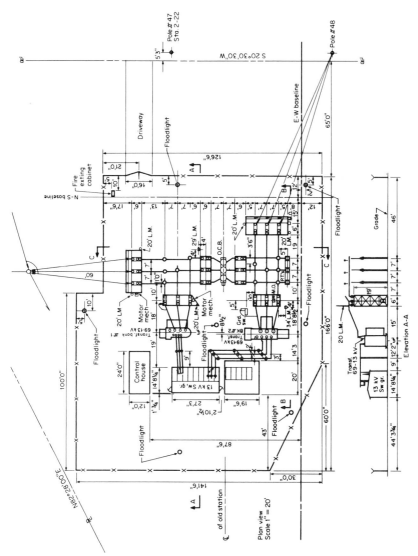

FIG. 7-1 Typical substation arrangement. *(Courtesy Long Island Lighting Co.)*

253

Protection Control

Instrument Transformers   The circuit breakers that serve to deenergize equipment require two sources of power for their proper operation: that which actuates the relays, and that which causes the mechanical operation of the equipment to take place.

The relays receive their actuating power (usually) from instrument transformers which measure the electrical quantities associated with the circuits or equipment they are to protect. These include current transformers, with a standard secondary current rating of 5 A, and, where a voltage input is required, potential transformers, with a standard secondary voltage rating of 120 V.

Auxiliary Circuits   While the instrument transformers furnish the power that actuates the protective relays, a separate source of auxiliary power is provided to operate the trip coils, solenoids, and motors that may be involved. This separate source of power must be as reliable as practical.

In substations in which the power supply bus is sectionalized into two or more parts, transformers supplying station power may be connected to two sections separated from each other, with the transformer supplying the auxiliaries connected to one section and the equipment connected to the other. In other instances, the station transformer may be equipped with a throwover switch, operated manually or automatically, in order to improve the reliability of the supply of auxiliary power.

Where reliability must be of the highest order, storage batteries, "floating" on the line, connected to an ac supply through rectifiers, are also installed to complement other power sources; here, the auxiliary circuit is a dc one.

In almost every instance, one or more of the distribution feeders supplied by the substation are so arranged in the field as to permit them to be energized from adjacent feeders emanating from a different substation. Should the entire substation become deenergized, such feeders may be utilized to reenergize the station power supply, enabling the circuit breakers and other equipment to be operated electrically. Where the substations involved are fed from different transmission or generating sources, care should be taken to avoid the interconnection of transmission and generating sources through the substation buses.

As a further precaution, control and auxiliary wiring systems are ungrounded and are provided with ground-detecting devices that actuate a light or alarm indicating the presence of a fault on the circuit involved.

The auxiliary power supply may also supply some emergency station lighting.

Bus Design

Electrical Considerations   The design of buses in a substation must take into account not only the current-carrying requirements under both normal and short-circuit conditions, but also voltage drops, power losses, and temperature rises (usually a maximum of 30°C above a 40°C ambient). The current-carrying ability of buses, especially for the higher-capacity and higher-voltage circuits, must also take into account the skin effect of the ac flowing through them. Also, where buses are very closely situated (between phases or between circuits), a proximity effect must also be considered which may further distort the distribution of the current flowing in the conductor and may affect the current-carrying ability of the bus.

Voltage drops must also take into account the reactance of the buses, including the self-reactance of the buses themselves (which may be affected by their shapes and composition) as well as the mutual reactance from adjacent buses.

The enclosure of buses, for fire or mechanical protection, may lower the current rating, since the heat generated by the I^2R losses is not as freely dissipated. Further, if the enclosures are metallic, or cause the buses to be spaced farther apart, reactance values may be changed, in turn affecting the voltage drop in the buses. Where individual-phase buses are separately enclosed, no part of the enclosure or any phase may form a loop of conductive material, since such a loop would form a short-circuited turn and would overheat under normal and fault-current flow in the bus.

Mechanical Considerations Buses are usually made of copper or aluminum. Their cross-sectional shapes may include flat bars (single, or several in parallel), tubing, channels, or hollow squares. Sections may be joined together by means of bolts or clamps, or may be brazed or welded together; at intervals, however, expansion-type joints are employed to take care of the expansion and contraction due to load cycles.

The buses must also take into account the forces set up by short-circuit or fault currents that may flow through them (or in adjacent buses), as well as the voltage surges that may result from switching or lightning. The shape of the bus may provide sufficient rigidity and strength. They must also be supported on insulators capable of meeting both the electrical and mechanical requirements. Bus supports specified, therefore, usually include a fairly large safety factor, between 2 and 4.

Current transformers that are connected in these circuits must also withstand both the electrical and mechanical forces imposed on them.

Substation Electrical Grounds

Grounding at a substation is of the greatest importance. Because of the strong alternating magnetic fields set up by the heavy short-circuit currents that may flow in the several elements, voltages of appreciable values may be induced in the metallic structural members, in the equipment tanks and their supporting frames, in metal conduits for power and communication circuits, in metal fences, etc. All of these must be grounded, preferably to a common ground. The ground connections of the surge arresters, as well as the neutral conductors of the feeders, both incoming and outgoing, and of the grounded wye-connected transformers (if any), should also be connected to the common ground.

Draining of the induced voltages to ground prevents dangerously high voltage rises from forming, especially in the vicinity of the substation during fault conditions.

Grounds may consist of a multitude of metal rods driven into the ground at frequent locations and connected together, or of a mesh buried (usually) beneath the substation. The number, spacing, size, and depth of burial of the conductors making up the mesh will depend on the nature of the soil and the ground resistance desired. Both rods and mesh may be installed and connected together.

SUBSTATION CONSTRUCTION

Distribution substations may be constructed entirely indoors, entirely outdoors, or in a combination of the two ways.

Indoor

In purely indoor construction, all the equipment is completely enclosed within a structure, protected from the weather. Measures are taken to ensure that the failure of a piece of equipment does not spread and involve other units. Reinforced concrete fire- and explosion-resistant walls or barriers are installed between major pieces of equipment, such as transformers, circuit breakers, and regulators. Sumps are usually provided beneath oil-filled equipment and connected to waste lines where they exist. The sumps should be of ample size to contain all of the oil in the equipment should its failure result in an oil spill. Control equipment, switchboards, batteries (if any), and other communication facilities may be located in separate fireproof compartments.

Automatic fire-extinguishing systems may be installed to smother any oil fire that may ensue; foam, carbon dioxide, a high-pressure fine spray of water, and other materials may be specified.

Ventilation may be by natural circulation of air, or by means of fans that may operate at peak load periods or when the internal ambient temperature exceeds a predetermined value.

In general, the rating and capacity of the equipment, particularly transformers and circuit breakers, may be lower than for similar units installed outdoors. On the other hand, the units need not provide for inclement weather conditions (e.g., bushings may have shorter creepage distances, tanks need not be watertight, etc.).

Provision, in the form of rails, rollers, or other devices, is made to permit the replacement of the several pieces of equipment. Space around each unit is provided to permit safe access for the maintenance, repair, or replacement of the unit.

The architecture of the exterior should blend with the surroundings, as should the associated landscaping, lawns, or other environmental prerequisites. Incoming and outgoing feeders are installed underground, out of sight.

Outdoor

Outdoor substations have all of the equipment located outdoors within a securely fenced off area. Here, too, provisions are made for the maintenance, repair, and replacement of the pieces of equipment. Sumps may be constructed beneath a unit, in the form of dikes or pits containing coarse gravel or crushed stone, of a sufficient volume to hold possible oil spillage from the unit. Depending on the availability of land and the spacing between units, fire walls between major units may be found desirable.

Transformers, circuit breakers, and other outdoor equipment are designed to operate in all kinds of weather. Tanks are usually hermetically sealed or are equipped with "breathing" devices; bushings and insulators have creepage paths sufficiently long to prevent flashover. Control facilities may be located in sealed compartments either associated with each unit or grouped together in an outdoor-type housing. Small strip heaters may be required to keep condensation from forming in the compartments.

The location of the outdoor substation may present serious environmental problems. These include appearance; sometimes extensive and expensive landscaping is required to conceal the substation partially or entirely from neighboring observers. Another source of objection may be the sound emanating from the transformers; sound barriers then have to be erected around those units to deflect or mitigate the sound emissions in a particular direction.

Combination Indoor and Outdoor

In the combined indoor and outdoor arrangement, the major units, usually the transformers only (with their associated surge arresters), but sometimes circuit breakers also, are located outdoors, and the remaining equipment is housed in a building of some kind. The requirements already outlined should be followed in connection with each portion of the substation. Such substations are located in areas where appearances may not be a major consideration and some equipment, sometimes concealed by landscaping, may be found acceptable.

Unit Substations

Unit substations are small, self-contained, metal-clad units, usually installed in residential areas where larger sites are unobtainable. They are usually well landscaped, and their incoming and outgoing feeders are placed underground. Primary feeders from one unit substation extend to meet those from other, adjacent unit substations so that, when one unit substation is out of service, its load can be picked up by the feeders from the adjacent unit substations. Each unit substation, therefore, must be designed with spare capacity to enable this transfer of load to be made during contingency conditions.

Mobile Substations

Mobile substations, which may be considered portable outdoor-type substations, have been discussed in Chap. 4. Provisions are generally made at the several types of substations to permit a mobile substation to be connected to pick up all or part of the load should a substation be out of service for a prolonged period of time.

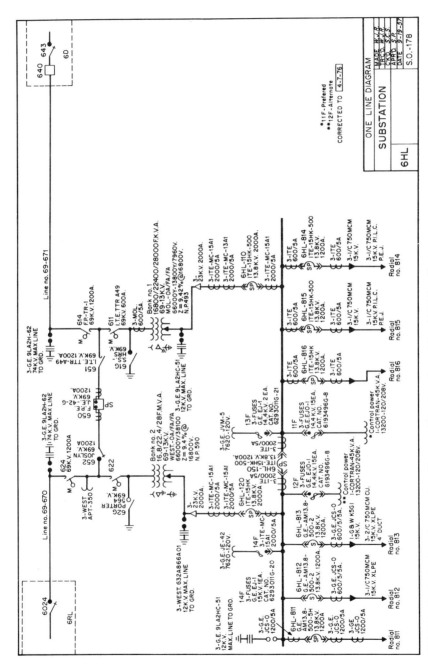

FIG. 7-2 Typical substation one-line diagram. (*Courtesy Long Island Lighting Co.*)

258

Operation

When an operator is in attendance, the speed and smoothness of restoration of service at times of interruption very often depend on the efficient performance of the operator.

Unattended substations may be designed in which some or all of the equipment is to function automatically. They may be operated by remote control from other substations, or from a central dispatching point, or they may be serviced by a "roving" or multistation operator. Automatic stations may, in many instances, shorten the time of interruption occurring from temporary abnormal conditions.

ONE-LINE DIAGRAMS OF CONNECTIONS

The connection of apparatus in a substation is usually drawn in the form of a "one-line diagram" for convenience and simplicity. Inasmuch as all the phase connections in a polyphase system are alike, it is only necessary to show the connections for one phase to indicate the simultaneous operation (as nearly as practical) of all the phases.

There are two types of one-line diagrams. The first, usually referred to as an operating diagram, merely indicates all the major equipment and connections, with such pertinent information as will enable the engineer, field supervisor, and operator to call for and complete the desired switching.

The second type is an elaboration of the first and includes additional information, such as fusing, relaying, and location and rating of instrument transformers and other auxiliary equipment. Such a diagram lends itself to rapid analysis in the event of improper operation of equipment. See Fig. 7-2.

CHAPTER 8

METERING

SCOPE

There are two broad classes of metering that interest the distribution engineer: that used to operate or monitor the several elements of the distribution system, and that mainly used for revenue purposes. With the advent of the digital computer (used for billing, among other applications), the latter may also be employed in furthering the purposes of the first class of metering. Further, the development of microelectronic technology has made possible operations of the electric system heretofore thought impractical if not impossible.

The principle of operation of most meters is that of the interaction of two (or more) magnetic fields tending to produce rotation of one of the elements. They may, however, be thought of as small but accurate motors, with their rotation restrained by springs or other magnetic fields.

Basic electric meters include the ammeter for measuring current in amperes, the voltmeter for measuring electrical pressure in volts, and the wattmeter for measuring power in watts. The ammeter operates on the interaction of a magnetic field, set up by the current to be measured, with the field of a permanent magnet. The voltmeter is essentially an ammeter connected in series with a fixed resistance. A wattmeter is a combination of both the ammeter and voltmeter elements.

OPERATION-MONITORING METERS

Applications

Ammeters are used for measuring loads on substation transformers, distribution feeders and transformers, secondary mains, and services. Voltmeters are used for measuring bus voltages at substations, both on the supply side of power transformers and on outgoing distribution feeders and regulators, at distribution transformers, secondary terminals, and consumer services. In a modified form they serve as control relays for voltage regulators. Both kinds of meters are also used in the setting and maintenance of protective relays and in determining the volt-ampere loads on equipment. Wattmeters measure the power input and output at substations, and at other points where such measurements are desirable.

The three basic meters used in association with each other can determine the power factor of circuits and loads; the three elements may be combined in one instrument to read power factor directly. Likewise, resistance, reactance, and impedance measurements may be obtained from the ammeters and voltmeters used in association with each other, or in one instrument that is made to read such quantities directly.

Extending Ranges

The range of these meters may be extended by means of auxiliary equipment. Ammeters for dc measurements may use a shunt calibrated so that only a predetermined proportion of the current flowing through it is measured by the ammeter; for ac measurements, a current transformer of a predetermined ratio achieves the same results. A voltmeter for dc measurements may use a potentiometer calibrated so that a predetermined proportion of the voltage across the entire potentiometer coil is measured by the meter; for ac measurements, a potential transformer of a predetermined ratio achieves the same results. Shunts and potentiometers are used to extend the range of dc wattmeters, and current and potential transformers are used to extend that of ac wattmeters.

Types

These meters may be of the indicating type or of the recording type; they may also be of the portable or of the station or switchboard type. Indicating ammeters may also be constructed to retain maximum values of current readings; they are sometimes referred to as maximeters.

In ac systems, separate current and voltage measurements are made for each phase of a polyphase circuit, but power and volt-ampere measurements for the sum of all of the phases may be made with one polyphase meter, or may be made using two or more single-phase wattmeters.

Polyphase Power Measurements

Power in a polyphase circuit may be measured by connecting a separate wattmeter in each of the phases and algebraically adding the values obtained. Equal accuracy, however, may be obtained by using one less wattmeter, the remaining ones being properly connected among the phases of the polyphase circuit. This is known as Blondel's theorem for the measurement of power in a polyphase system of any number of wires without regard to any unbalance of currents and voltages that may exist among the phases.

Blondel's Theorem   The theorem may be stated as follows: In any system of N wires, the true power may be measured by connecting a wattmeter in each line but one ($N - 1$ wattmeters), the current coil being connected in series in the line and the potential coil being connected between that line and the line that has no current coil connected in it; the total power is the algebraic sum of all the readings of the wattmeters so connected.

According to the theorem, then, the total power in a three-phase three-wire circuit, either wye- or delta-connected, may be measured by two wattmeters; a three-phase four-wire circuit would require three wattmeters.

If the power factor of the circuit falls below approximately 50 percent, it is likely that the reading of one of the wattmeters may be negative, and it may be necessary to reverse the terminals of the current or potential coil; this change should be noted in connection with any later measurement when the power factor may be greater than 50 percent.

REVENUE METERING

Metering for revenue purposes, in itself, is not the responsibility of the distribution engineer, but the data accumulated by meters so used can serve in optimizing the planning and design as well as the operation of distribution systems. The same computer that translates meter readings into consumers' bills can also be programmed to make selective summaries of such data, simultaneously converting such consumption data into loads and demands on the several elements of the distribution system. The grid coordinate system of mapping is extremely useful in such instances; it is described in Appendix C at the end of this book.

Meters for revenue purposes measure energy in kilowatthours and power demand in kilowatts. Although constructed differently, they are connected in circuits in the same manner as wattmeters, and have both current and potential coils.

Watthour Meters

The watthour meter is essentially a small motor whose rotor, usually a metallic disk, is caused to turn by the torque produced by the reaction of the magnetic fields set up by the current and voltage coils with a magnetic field of the disk resulting from the eddy currents set up in it by those magnetic fields. Its speed of rotation can be made to be proportional to the power flowing in the coils of the meter; a magnetic brake regulates this speed to obtain the desired accuracy. A register which counts the revolutions of the disk acts to integrate the instantaneous values of power over a period of time; the integral is the expression for energy.

Watthour meters may be single-phase or polyphase, although the single-phase units constitute the greatest number in use, measuring the energy consumption of residential consumers.

Transformer Load Monitoring

The consumptions of each of the consumers supplied from one distribution transformer can be totaled to obtain the energy supplied through that transformer over the billing period of time, usually approximately a month. Factors can be applied to this total to convert it into an approximate demand in kW or kVA. In this manner, the loading of that transformer can be monitored. The computer which compiles the consumers' bills for that period of days can be programmed to perform this function. Conversion factors are obtained and kept current by sample testing of small groups of representative consumers.

Similarly, the consumption supplied from each distribution transformer can be summarized for the transformers on each phase of an entire feeder or portions of that feeder. Carrying it a step further, these consumptions can be summarized for all the feeders on a substation bus or on the supply transformers. These values can also be converted to approximate kilowatt demands on the several elements of the distribution system. Comparison of the sum of the individual consumptions on a feeder to such readings on the feeder at the substation (making correction for differences in the time of the readings, if any) can give a rough indication of actual losses on the feeder.

The computer can be programmed to produce periodic readouts of these data together with average demands per consumer, and to identify distribution transformers and other elements that exceed values predetermined to indicate overloads and potential sources of troubles.

The distribution engineer is thus able to initiate actions not only to prevent inconveniences to consumers, but also to avert damage or destruction of equipment and other facilities that could cause long and costly interruptions affecting the safety and well-being of the public. Moreover, the updated values of consumer demands and circuit loadings enable the engineer to do a more effective and economical job of planning and design of the electric distribution system.

Power Factor Correction

For some larger (usually industrial) consumers, the reactive kVA load is also measured in kVA-hours (by shifting the voltage 90° by means of reactors) along with kilowatthour consumption. By properly interrelating these quantities for the period of time covered, a value of average power factor of the consumer's load is obtained. With rate schedules tailored to reflect rewards (or penalties) for power factor, the installation of power factor-corrective measures at the consumer's expense is encouraged; these may be banks of capacitors, synchronous motors, or both. Such power factor improvement enables the distribution engineer to design more economical distribution systems affecting costs to all consumers.

Demand Meters

The measurement of maximum demand for a consumer may be obtained by adding another dial and set of gears to the registers of the watthour meters. Kilowatt demands for consumer billing purposes are obtained basically by measuring the kilowatthour consumption over a 15-min period, multiplying this by 4 to obtain an hourly demand. In some instances, a period of 30 or 60 min is used and the multiplier adjusted accordingly. This is accomplished essentially by a "floating" hand that is pushed by the watthour element that returns to zero every period, after which the action is repeated.

At the billing period, the meter reader reads this floating hand and returns it to zero, eliminating any record of that demand reading. Another system does the same thing, but records the demand on a separate register, the actuating element returning to zero every period; the action is repeated, but the demand value is added to the previous reading. At the billing period, the reader reads the second register, which has "accumulated" the maximum demands, and resets the actuating element to zero. The maximum demand for the period is the difference between the last reading and the previous one on the demand register; hence, a record of the maximum demands for each billing period is maintained.

Demand Control

Rate schedules are designed to encourage the consumer to arrange the operation of loads so as to hold down maximum demand.

Many of the relatively smaller consumers arrange the operation of major pieces of equipment on a predetermined schedule in an effort to hold down their overall maximum demand, usually concentrating on reducing or eliminating short-term peaks.

In large industrial and commercial installations, the demand is continuously monitored by the consumer using impulses generated by the utility's demand meter (impulses that trigger the demand meter's 15-min periodic return to zero). These impulses are fed into the computer that supervises the operation of the consumer's equipment; in some programs, actual shedding of less critical loads within the billing (15-min) interval is done so as not to exceed a predetermined value; see Fig. 8-1.

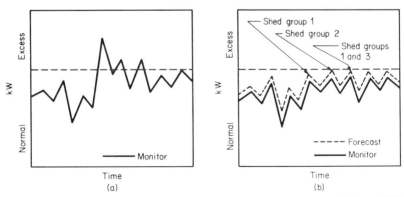

FIG. 8-1 Continuously monitored demand control (a) without demand limit and (b) with demand limit. *(Courtesy Johnson Controls, Inc.)*

In general, the goal is to improve the consumer's load factor, and the first step usually is to obtain the consumer's load profile. If the profile shows only a few 15-min intervals where sharp peaks occur, it is a relatively easy matter to identify the equipment causing those demand peaks and take remedial measures. In most cases, however, a more detailed approach is followed to determine what, if any, demands can be reduced or eliminated. This generally calls for classifying the consumer's major loads into four categories: those which (1) can be rescheduled, (2) can be deferred, (3) can be curtailed or eliminated, and (4) are essential base loads.

Often, this control of consumers' demands is a cooperative effort between the consumers and the utility's distribution engineer. Success of such efforts is beneficial to both the consumer and the utility. While this may reduce the consumer's demand payments and reduce the utility's revenue, the loss in revenue is almost always overshadowed by the utility's reduced carrying charges from the improved usage of fewer facilities and reduced energy costs from the reduction in losses that accompany the reduction of peaks. The reduction of these costs is of increasing concern to the distribution engineer.

WIRING DIAGRAMS

Wiring diagrams for the various meter connections for different types of distribution systems are shown in Fig. 8-2.

ELECTRONIC METERING

The development of so-called microelectronic processing (including solid-state) techniques is making the kilowatthour meter more than just a device for billing purposes; it also provides another tool for use by the distribution engineer and operator. By adding memory and other circuitry to the register of the meter, its functions can be expanded to provide additional data valuable to the distribution engineer. Further, microprocessors can be programmed to process the data and execute predetermined commands. This last feature can be extended to control individual consumer loads in order to restrict consumers' demands and, collectively, reduce the peak demands on the several elements of the electrical system all the way back to the generators. Incidentally, such systems facilitate the remote reading of consumers' meters and instantaneous billing, long the goal of the accounting departments.

Data Procurement

Data that can be obtained from the meter and its electronically enhanced register can include real and reactive power consumption and demands, peak and average demands, power factors, and such other items as the power factor at peak demands of both kW and kVA. Also, calendar information such as time, day, month, and year can be included, and different seasonal, holiday, leap year, and other complex rate schedules can be stored; when such data are employed at predetermined times and conditions, the price in effect and the rate-of-usage information can be readily obtained.

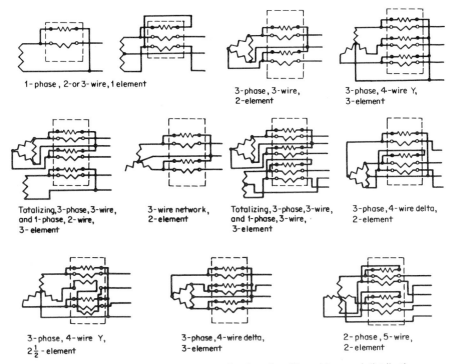

1-phase, 2-or 3-wire, 1 element

3-phase, 3-wire,
2-element

3-phase, 4-wire Y,
3-element

Totalizing, 3-phase, 3-wire,
and 1-phase, 2-wire,
3-element

3-wire network,
2-element

Totalizing, 3-phase, 3-wire,
and 1-phase, 3-wire,
3-element

3-phase, 4-wire delta,
2-element

3-phase, 4-wire Y,
$2\frac{1}{2}$-element

3-phase, 4-wire delta,
3-element

2-phase, 5-wire,
2-element

FIG. 8-2 Schematic diagrams of meter applications for different types of distribution systems. *(Courtesy Westinghouse Electric Co.)*

Operating Functions

By coupling this electronic metering with its sophisticated registers to a communications network (microwave, radio, telephone, carrier, or combinations of these), control of consumers' loads (with prearranged and contractual assent) by the utility operator becomes practical. Such control is directed primarily at restricting demands on the several parts of the *entire* electrical system, starting with elements of the distribution system. Not only is it possible to reduce capital outlays (and related annual carrying charges) for additional facilities, but it is also possible to reduce operating (e.g., fuel) expenses through the reduction of losses in the transmission and distribution systems and in the amount of spinning reserve generation required.

There are several methods which may be used to shift or reduce electrical demands. One simple and "minimum" system limits a consumer's maximum demand during peak periods from a signal actuated by the utility operator. The metering circuit monitors the load automatically with a signal light at the meter and an alarm of some kind on the consumer's premises. Exceeding the limit (after actuating by the operator) triggers the alarm and starts a programmable time delay of (say) 14 min (of the 15-min demand cycle), allowing the consumer time to reduce the demand before service is interrupted; see Fig. 8-3. A second

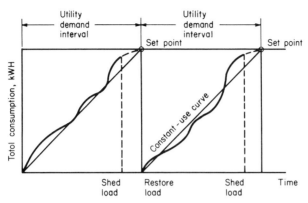

FIG. 8-3 Last-moment method of demand control.

time delay prevents appliance damage to loads such as air-conditioning compressors, when service is restored.

TRANSDUCERS

In many instances, it is desirable to measure nonelectrical quantities, and this may be done by converting them into electrical quantities, which are then metered or measured in the usual manner. This is accomplished by a class of devices generally called *transducers*. Thermocouples, photocells, and microphones are examples of devices which convert heat, light, or sound into electrical quantities; when these quantities exceed predetermined limits, they act to control the equipment with which they are associated. Other devices convert pressure or pressure differences into variations of electrical quantities, which again serve to actuate or control the equipment with which they may be associated.

Transducers may also be applied to convert electrical quantities into nonelectrical quantities.

PART THREE

MATERIALS AND EQUIPMENT

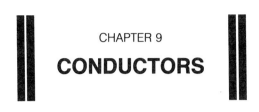

CHAPTER 9

CONDUCTORS

INTRODUCTION

Conductors, as carriers of the electrical energy, are perhaps the most important element of an electric circuit; often they are called wires, and the terms are used interchangeably. They may be installed overhead as bare, covered, or insulated conductors; and, in cable form, they may be installed underground in ducts or buried directly in the ground. Cables are also sometimes installed in trays in substation or industrial installations where immediate access to them is of relatively great importance.

Since a failure of a conductor results in a complete interruption to a circuit, it is imperative that the causes of such failure be minimized. The failure may occur from mechanical causes where the stresses and strains imposed are simply too great and the conductors literally tear apart. More often, however, the cause may initially be an electrical failure which then affects the conductors mechanically. Overloads or short-circuit currents, for example, may cause heating of the conductors to the point where they begin to liquefy (or melt) and ordinary mechanical stresses can no longer be sustained and the conductors pull apart, perhaps vaporizing in the process.

MATERIALS

It is evident, therefore, that both the mechanical and electrical characteristics must be considered in the choice of conductors. In addition, the all-important element of cost limits the number of economically available materials suited for conductors. A brief discussion will highlight the properties of those most commonly used: copper; aluminum; aluminum conductor, steel-reinforced (ACSR);

TABLE 9-1 CHARACTERISTICS OF CONDUCTOR MATERIALS (COMMERCIAL GRADES)

Material	Conductivity, % (pure Cu = 100%)	Weight lb/in³	Weight lb/1000 ft per 1000 cmil	Ultimate strength, lb/in² (× 1000)	Elastic limit, lb/in² (× 1000)	Modulus of elasticity (× 10⁶)	Temperature coefficient of linear expansion per degree (× 10⁻⁶) °C	Temperature coefficient of linear expansion per degree (× 10⁻⁶) °F
Copper—SD	99–100	0.320	3.027	36 to 40	18 to 20	12	17.1	9.5
MHD	98.5–99.5	0.320	3.027	42 to 60	23 to 33	14	17.1	9.5
HD	97–99	0.320	3.027	49 to 67	30 to 35	16	17.1	9.5
Aluminum, plain	61	0.0967	0.920	23 to 27	14 to 16	9	23.0	12.8
Aluminum, steel-reinforced	61	0.147	1.390	44	31	—	19.1	10.6
Steel	8.7	0.283	2.671	45 to 189	23 to 112	29	11.9	6.6
Copper-clad steel—30%	29.25	0.298	2.810	60 to 100	—	16 to 20	13.0	7.2
40%	39	0.298	2.810	60 to 100	—	16 to 20	13.0	7.2

To convert to metric system:
lb/in³ × 0.0277 = kg/cm³
lb/1000 ft × 0.1488 = kg/km
lb/in² × 0.0703 = kg/cm²

Courtesy The Anaconda Co., Wire and Cable Div.

steel; and copper-clad steel. The chief characteristics of these conductor materials are compared in Table 9-1.

Copper

For many years, copper has been the most satisfactory conductor for electrical purposes. Its electrical and mechanical properties, coupled with comparative cost benefits, made it almost exclusively used universally.

Its conductivity is high, surpassed only by that of silver and some other rare metals; indeed, its conductivity is used as a reference for that of other materials.

Degrees of Hardness Although basically soft enough to be handled easily, copper can be increased in strength through annealing processes. It is available in three standard degrees of strength and hardness: hard-drawn, medium-hard-drawn, and soft-drawn. Hard-drawn copper results from the repeated drawing of the wire down to the desired size without annealing, i.e., without deliberate after-heating and cooling. Medium-hard-drawn copper results from drawing the wire down to the desired size; annealing must be made under controlled conditions or the degree of hardness will not be consistent enough to be considered constant for design purposes. Soft-drawn copper results from annealing after drawing is complete; here, too, annealing must be controlled to result in constant values of hardness.

As the names imply, the difficulty in handling copper wire depends on its hardness, as does its strength. Hence, for overhead lines, hard-drawn copper finds greatest application for long spans, medium-hard-drawn for intermediate-length spans, and soft-drawn for short spans and for such other uses as connecting leads, tie wires, and other applications where strength is not a major requirement and flexibility is highly desirable.

For cables, interior wiring, and other uses where mechanical strength is not of paramount importance, soft-drawn copper is almost always preferred because of its flexibility.

Flexibility To obtain additional flexibility, particularly where hard-drawn and medium-hard-drawn copper are involved, conductors are sometimes stranded; i.e., a conductor of a specific size may be made up of a number of smaller-cross-section conductors whose sum is equal to the cross section of the specific conductor. While this makes the conductor more flexible, it does increase the overall diameter of the conductor.

Aluminum

Although the conductivity of aluminum is only some 61 percent that of copper, the weight of aluminum is only about one-third that of copper. For the same conductivity, therefore, an aluminum conductor would weigh only about half as much as copper, although somewhat larger in diameter. The breaking strengths of aluminum and soft-drawn copper are also about the same. The result is that these two materials are economically competitive as conductors.

Aluminum Conductor, Steel-Reinforced (ACSR)

From the viewpoint of mechanical strength, plain aluminum conductors cannot be hardened and, hence, cannot compare with medium- and hard-drawn copper conductors. It is sometimes alloyed, however, with other materials to increase its strength slightly, but more often, its deficiency in strength is overcome by having steel reinforce the aluminum. This is accomplished by having strands of aluminum wire wrapped around a central core of one or more strands of high-strength steel wire. The breaking strength of this type of conductor, 2 or more times that of plain aluminum, is considerably greater than even that of a hard-drawn copper conductor of the same conductivity.

Steel

Although its conductivity is relatively low, yet because of its very high mechanical strength, steel is sometimes used as an electrical conductor. Its use is limited to those few occasions where mechanical strength is of paramount importance. It is usually protected from corrosion by galvanizing or otherwise protected by some covering.

Copper-Clad Steel

The advantages of steel for strength and copper for conductivity are combined into a solid nonstranded-type copper-clad steel conductor. Obviously, this type of conductor is limited to the smaller-size copper conductor equivalents. The layer of copper is continuously bonded to the steel, resulting in a material that is essentially homogeneous. Such conductors are made in two grades of conductivity, 30 and 40 percent copper (as a percentage conductivity compared to the conductivity of hard-drawn copper of the same cross section) and in two grades of strength depending on the grade of steel around which the copper is placed. The copper covering acts to insure the conductors against destructive corrosion. This type of conductor is handled in the same manner as solid-type hard-drawn copper conductors.

Other Materials

From time to time, other materials have been employed as conductors of electricity. Silver, because of its very high conductivity, has sometimes been used in very special situations where strength and cost are not the principal concerns.

Alloys of both copper and aluminum have also found special applications. Brass and bronze, though stronger than copper mechanically, have lower conductivities, and have found application in relatively few special situations.

Similarly, aluminum alloyed with small percentages of silicon, magnesium, iron, and other metals has been used where increased hardness and tensile strength are desired along with light weight, and where conductivity is of secondary importance.

Characteristics of Conductor Materials

The more important characteristics of the several conductor materials mentioned above are given in Table 9-1. Conductivity and weight have already been discussed. While the words *strength* and *tension* have been used, further description is needed; these are quantified in the table in terms of elastic limits and ultimate strengths.

Elastic Limit and Modulus of Elasticity (Sag)   The elastic limit of a metal is the amount of elongation or stretch that the material can make under stress while still being able to return to its original dimensions, i.e., without permanent deformation after the stress is removed. The elasticity of a material can be measured as a ratio between the stress applied (in pounds per square inch or kilograms per square centimeter) and the elongation produced per unit length (say, inches per foot or millimeters per meter). This ratio is known as the *modulus of elasticity* and is a measure of the way a conductor will sag under loading.

A very small percentage elongation of a conductor is accompanied by a comparatively large increase in sag. Tension is approximately inversely proportional to sag; i.e., when the sag is increased, the tension (or stress) in the conductor is decreased. The elongation, up to the elastic limit and including a factor of safety, is taken into account in computing sags.

Elongation beyond the elastic limit can continue until the conductor is at or near the breaking point, which is a measure of the ultimate strength of the conductor. For stranded conductors, the ultimate strength is taken at about 90 percent of the sum of the strengths of the individual strands. The ultimate strength has little importance in the design of electric lines, but is a figure which allows factors of safety to be determined, i.e., a ratio of stress that would ultimately cause the conductor to fail to the allowed stress.

Temperature Coefficient of Linear Expansion   There is still another characteristic of materials that affects the performance of conductors: the temperature coefficient of linear expansion, or a measure of the change in length of the material with temperature. This is of great importance, as a sag installed in a line at a particular temperature, say in summertime, may be considerably different at another temperature, say in wintertime. These variations caused by temperature differences are taken into account in the design and construction of overhead lines.

Conductor Sizes   Since it is impractical to manufacture an infinite number of wire sizes, standards have been adopted for an orderly and simple arrangement of such sizes for manufacturers and users. The American Wire Gauge (AWG), formerly known as the Browne and Sharpe Gauge (B&S), is the standard generally employed in this country and where American practices prevail.

In defining conductor sizes, the *circular mil* (cmil) is usually used as the unit of measurement. It is the area of a circle having a diameter of 0.001 in, which works out to be 0.7854×10^{-6} in^2. In the metric system, these figures are a diameter of 0.0254 mm and an area of 506.71×10^{-6} mm^2.

Wire sizes are given in gauge numbers, which, for distribution system purposes, range from a minimum of no. 12 to a maximum of no. 0000 (or 4/0) for solid-type conductors. Solid wire is not usually made in sizes larger than 4/0, and stranded wire for sizes larger than no. 2 is generally used. Above the 4/0 size, conductors are generally given in circular mils (cmil) or in thousands of circular mils (cmil $\times$ 10^3); stranded conductors for distribution purposes usually range from a minimum of no. 6 to a maximum of 1,000,000 cmil (or 1000 cmil $\times$ 10^3) and may consist of two classes of strandings. These wire sizes and their dimensions are given in Table 9-2.

Gauge numbers may be determined from the formula:

$$\text{Diameter, in} = \frac{0.3249}{1.123^n}$$

or

$$\text{Cross-sectional area, cmil} = \frac{105,500}{1.261^n}$$

where n is the gauge number (no. 0 = 0; no. 00 = -1; no. 000 = -2; no. 0000 = -3).

It will be noted that the diameter of the wire doubles approximately every sixth size (e.g., no. 2 has twice the diameter of no. 8), and the cross-sectional area therefore doubles every third size and is 4 times as great every sixth size (e.g., no. 2 has twice the area of no. 5 and 4 times that of no. 8).

The diameter of stranded wire is approximately 15 percent greater than the diameter of a solid wire of the same cross-sectional area.

The gauge numbers and wire designations apply to conductors of all materials. Usually, however, the equivalent wire sizes are denoted for the several materials in comparison to copper (e.g., 4/0 aluminum is equivalent to 2/0 copper). These are indicated in the tables for such conductors.

Conductor Coverings   Although the practice has essentially been discontinued, in the early days of overhead line construction, conductors were sometimes provided with weatherproof coverings. Such coverings usually consisted of cotton braids impregnated with a water-resistant compound wrapped about the wire. These conformed to no standards, but the various thicknesses of covering used were designated by the number of braids: single-braid, double-braid, and triple-braid. These coverings were used on conductors operating at secondary voltages usually less than 500 V, and at primary voltages under 5000 V. It was felt that such coverings would prove useful in preventing short circuits in stringing new wire, when wires swung together, when they came in contact with tree limbs, or when objects fell or were thrown across the wires. Although the insulating value of such coverings is unreliable, it will be more or less in proportion to the thickness of the covering, the preservative, age, and other considerations. As indicated earlier, this practice has been essentially discontinued, but a great many such installations still exist and will be met in maintenance and reconstruction activities.

TABLE 9-2 CHARACTERISTICS OF SOLID AND STRANDED CONDUCTORS

	Both solid and stranded conductors						Stranded conductor	
	Cross section		Weight, lb/1000 ft*		Resistance, Ω/1000 ft at 20°C		Solid conductor diameter, in	Number and diameter of strands, in	Diameter, in
Size	cmil	in²	Cu	Al	Cu	Al			
—	1,000,000	0.7854	3026.9	921.6	0.010	0.017	—	61 × 0.128	1.150
—	750,000	0.5891	2270.2	691.2	0.014	0.022	—	61 × 0.111	0.998
—	500,000	0.3927	1513.5	460.8	0.021	0.034	—	37 × 0.116	0.813
—	350,000	0.2749	1059.4	322.5	0.030	0.048	—	37 × 0.097	0.681
								19 × 0.136	0.678
—	250,000	0.1964	756.7	230.4	0.041	0.068	—	37 × 0.082	0.575
								19 × 0.115	0.573
4/0	211,600	0.1662	640.5	195.0	0.049	0.080	0.4600	19 × 0.106	0.528
								7 × 0.174	0.522
3/0	167,772	0.1318	507.9	153.6	0.063	0.102	0.4096	19 × 0.094	0.470
								7 × 0.155	0.464
2/0	133,079	0.1045	402.8	122.0	0.078	0.128	0.3648	19 × 0.084	0.418
								7 × 0.138	0.414
1/0	105,625	0.0830	319.5	97.0	0.098	0.161	0.3250	19 × 0.075	0.373
								7 × 0.123	0.368
1	83,694	0.0657	253.3	76.9	0.124	0.203	0.2893	19 × 0.066	0.322
								7 × 0.109	0.328
2	66,388	0.0521	200.9	61.0	0.156	0.256	0.2576	7 × 0.097	0.292
3	52,624	0.0413	159.3	48.4	0.197	0.323	0.2294	7 × 0.087	0.260
4	41,738	0.0328	126.4	38.4	0.249	0.408	0.2043	7 × 0.077	0.232
5	33,088	0.0260	100.2	30.4	0.313	0.514	0.1819	7 × 0.069	0.207
6	26,244	0.0206	79.5	24.1	0.395	0.648	0.1620	7 × 0.061	0.184
7	20,822	0.0164	63.0	19.1	0.498	0.817	0.1443	7 × 0.053	0.167
8	16,512	0.0130	50.0	15.2	0.628	1.030	0.1285	7 × 0.047	0.154

*For PE- and PVC-insulated conductors, add 550 lb per square inch of cross section for every 1000 ft.

To convert to metric system:
in² × 645 = mm²
in × 2.54 = cm

Courtesy The Anaconda Co., Wire & Cable Div.

Tree Wire Where wires are strung among trees with the possibility of abrasion from contacting limbs and resultant short circuits, an especially heavy covering was applied to the wires. Often, this consisted of thick fiber, hemp, or sisal coverings, in some instances applied over some rubber insulation. Again, this practice is virtually obsolete, though some tree wire still remains in place.

Where fear of wires' coming together and tree problems are still considerations to be taken into account, present practice calls for installing fully insulated wire.

Insulated Wire Use of insulated wire on overhead systems was formerly restricted to transformer connections, overhead-to-underground and other jumper connections, and an occasional span or two. Insulation consisted of rubber covered with one or more weatherproof braids. The rubber insulation was subject to deterioration from temperature changes, moisture, sunlight, pollution, and other causes.

The development of plastics, such as polyethylene (PE) and polyvinyl chloride (PVC), has made available insulated conductors in which the plastic serves both as insulation and covering or sheathing. It has completely taken the place of weatherproof covered conductors, tree wire, and other insulated wire mentioned above. Indeed, the availability of such plastic-insulated wire has increased its usage in place of bare wire for overhead and for lead-sheathed underground cables, resulting in generally improved reliability for the former and wider application of underground systems.

One of the features of some of these plastics is their hardness, particularly at low temperatures. They are therefore somewhat more difficult to handle, and the plastic must sometimes be heated before it can be removed by knife or other tool, for making connections or for other purposes.

CABLES

Conductors for use in underground systems must be provided with insulation sufficient to withstand the voltages at which they must operate. Generally, these consisted of stranded copper conductors (except very small-size conductors which were solid) with insulation of rubber, varnished cambric, or special oil-impregnated paper, all contained within a sheath made of lead. In some special circumstances, such as in heavy tree areas or where appearance was an important factor, these cables were installed on overhead systems, usually attached to a messenger wire for support between poles.

Because the rubber insulation sometimes contained sulfur compounds which reacted destructively with the copper, the conductor strands were often tinplated, adding to the complexity and cost of such insulated cables; these were generally limited to secondary voltage applications of 500 V or less and to primary voltages of 5000 V or less.

Single-conductor cables are used for spur or lateral or branch lines of feeders where numerous branch joints or splices are required, for secondary and service supply, for street lighting, and other such purposes, or where the conductor may be very large, making a multiconductor cable impractical. Multiple-conductor cables of two, three, or four conductors are used in the main portion of feeders, where there may be few branch connections required; in such instances they may be more economical, both from material and labor standpoints, than a number of single-conductor cables.

In some instances, where electrolytic or other chemical conditions may cause sheath corrosion or erosion, lead sheaths may be replaced with nonmetallic sheaths or the sheath (and cable) may be covered with nonmetallic materials such as plastics.

In some instances, such lead-sheathed cables may be buried directly in the ground, or laid under water in submarine installations. In these instances, the lead sheath is covered with jute or tar and armor wires of steel are wound around the whole for protection.

Although the installation of lead-sheath cables is a disappearing practice, many miles of such installations exist and will continue to exist for many years to come.

Cables in which the plastic takes the place of both insulation and sheath are now used almost exclusively for distribution circuits, and are more extensively buried directly in the ground. This not only eliminates ducts and manholes, but permits longer sections of cable to be used with fewer splices required. Aluminum often replaces copper as the conductor in these newer types of cables.

SECONDARY MAINS

Secondary mains consisting of two or more conductors were sometimes fastened to insulators mounted on cross arms and more often on insulated racks attached to the sides of the poles. The "live" conductors were generally of weatherproof covered wire, though sometimes they were rubber-insulated with a covering wrapped about the rubber. The "neutral" conductor was generally a bare wire.

More recently, with the advent of plastics in this field, secondary conductors with plastic insulation are wound together or "cabled" with the bare neutral wire, the assembled conductors thus becoming self-supporting; more often, the conductors are made of aluminum, rather than the copper formerly almost universally used. Besides taking up less space on the top or side of the pole, cabled secondaries are easier to install and their reactance (which contributed to a loss or drop in voltage) is less than for the open-type conductors.

SERVICE CONDUCTORS

Overhead services for many years were of the open-wire or multiple-conductor types. Open-wire construction, with separate weatherproof covered wire and bare neutral, held apart on separate insulators, not only was cheaper, but allowed a third or fourth conductor to be added very readily. In multiple-conductor construction, sometimes known as "duplex" or "triplex" cables, the service cable consisted of one or two insulated conductors about which the neutral conductor strands were wound, the whole enclosed by weatherproof braiding. This type of construction made for better appearance, and a slightly better reactance. Many of both open-wire and multiple-conductor services still exist and will continue to do so for an indefinite time.

Services also have employed "cabled" conductors, similar to those used for secondary mains. In some instances, consumers are served from midspan taps; in others, from poles. Economics and appearance are the determining factors.

In many instances, services are placed underground, generally for the sake of appearance. Lead-sheath rubber-insulated cables installed in ducts or conduits, or supplied with armor and buried directly in the ground, have been used. Plastic-insulated conductors requiring no sheathing have largely taken the place of such underground cables. They are buried directly in the ground, as previously described for main-line conductors.

CONNECTIONS

Connections made between conductors, in joining two ends together or in making a tap off the other, should be electrically and mechanically sound. They not only should introduce no additional resistance (and associated heating) at the points of contact, but they should also not be subject to corrosion or conductor stresses or movements.

In earlier times, such connections usually consisted of wires wrapped together and soldered. Later, twisted sleeves were employed in which the two ends of the conductor were inserted in a sleeve and the whole assembly twisted. Stranded conductors had each strand serviced separately before soldering. Many of these connections still exist.

The later development of "solderless" or mechanical connectors made obsolete the wrapped and soldered splices. Parallel-groove clamps, split-bolt connectors, and crimped sleeves made splicing more simple and more uniform, with substantial reduction in labor costs. Some of these are shown in Fig. 9-1. For rapid installation, usually during periods of emergency, the "automatic" splice, employing wedges which, under pressure of the sagging wire, grip the ends of the conductors to be spliced, was also developed; this is relatively more expensive.

When necessary, friction tape, or insulating tape covered with friction tape, is employed to continue the covering or insulation. In present-day applications, the splices are often left bare.

In many instances, where it is desirable to disconnect the connection readily,

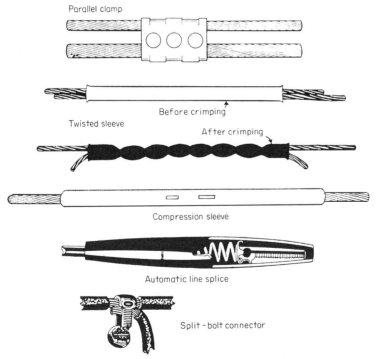

Parallel clamp

Before crimping

Twisted sleeve

After crimping

Compression sleeve

Automatic line splice

Split-bolt connector

FIG. 9-1 Mechanical connectors. *(Courtesy Burndy Corp.)*

special clamps, sometimes known as "live-line" or "hot-line" clamps, are used; these are shown in Fig. 9-2.

The advent of aluminum conductors into a field in which the conductors previously were exclusively copper presents problems where conductors of dissimilar metals need to be connected together. Special care is exercised, since the connection may be affected by chemical interaction between the two metals, especially when wet and in the presence of some pollutants; but even more, because of the different rates of expansion when heated. The uneven expansion and contraction will eventually cause such splices to become loose, and their resistance to increase with consequent abnormal heating, with possible dire results.

Connectors for copper-to-copper conductors are usually made of copper, though bronze is sometimes used for greater strength. Where aluminum or ACSR conductors are to be connected to similar conductors, connectors of aluminum are used. Where the conductors to be connected are of dissimilar metals, connectors are so designed that only surfaces of similar metals come in contact with each other; aluminum clamps with copper bushings, or vice versa, are employed for this purpose. Care is taken to prevent water dripping from copper items, which may contain copper salts, from coming into contact with aluminum items.

While this discussion applies equally to overhead and underground installations, it must be noted that splices on underground cables, especially where lead-sheath cables are involved, are very much more complex. The connector must be smooth so that no corona discharge will pit the metals. The insulation covering the connector is carried over from one cable to the other by means of insulating tapes wound about the connector. The lead sheath is sweated or soldered to the cable sheaths and is usually larger in diameter. The splice may be filled with an insulating compound, which is heated and poured into the splice, where it hardens on cooling. The new plastic-insulated cables are spliced with a connector between the two conductors and plastic tape of the same material as the insulation wrapped about the assembly; the tape tends to become homogeneous with time.

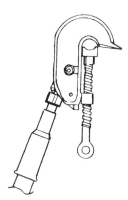

FIG. 9-2 Live-line or "hot-line" clamp. *(Courtesy A. B. Chance Co.)*

OVERHEAD-TO-UNDERGROUND CONNECTION

Connections are often required to be made between overhead and underground conductors. With the advent of plastic-insulated cables, connections have been made directly between the overhead and underground conductors. Live-line clamps furnish a means for easy and rapid disconnection of the conductors involved. Potheads and weatherheads are dispensed with.

In older installations, many of which will continue to exist, special devices have been used. For primary voltages, *potheads* have been used. Here the conductors of the underground cable are connected to terminals in which the conductors are surrounded by poured insulation compound to prevent moisture or air from entering the cable insulation. The overhead wires are connected to the female end of the terminal, enclosed in an insulated cap. The connection is made by placing the cap over the terminal extending from the pothead case. Potheads so described are known as *disconnecting* potheads. Where the connections are made directly to the terminal extending from the pothead in a permanent fashion, the pothead does not carry this distinction.

For lower secondary voltages of 500 V or less, a simpler device was used. Here the conductors of the underground cable are brought out through a preformed insulator, usually of porcelain, in an assembly which inverts the leads so that rain cannot enter the cable. Such devices are known as *weatherheads*. Connection to the overhead wires is made with ordinary connectors of the several types described; see Fig. 9-3.

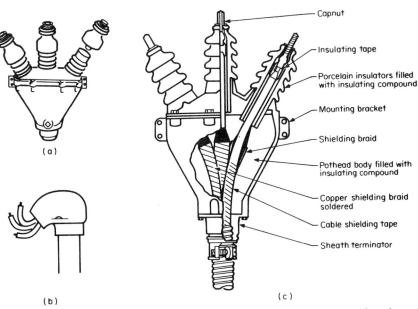

(a)

(b)

(c)

Capnut

Insulating tape

Porcelain insulators filled with insulating compound

Mounting bracket

Shielding braid

Pothead body filled with insulating compound

Copper shielding braid soldered

Cable shielding tape

Sheath terminator

FIG. 9-3 Connection between overhead and underground lines: (a) pothead for primary lines; (b) weatherhead for secondary lines; (c) cross section of pothead. (*Courtesy G&W Electric Specialty Co.*)

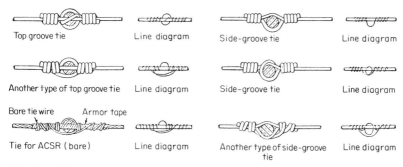

| Top groove tie | Line diagram | Side-groove tie | Line diagram |

Another type of top groove tie — Line diagram — Side-groove tie — Line diagram

Bare tie wire — Armor tape — Tie for ACSR (bare) — Line diagram — Another type of side-groove tie — Line diagram

FIG. 9-4 Wire ties for overhead conductors: handmade wire ties. *(Courtesy EEI Overhead Reference Book.)*

TIES

Ties are pieces of wire used to attach the conductors to the insulators on overhead systems. They should be flexible enough to be handled easily, but must be mechanically strong to prevent the conductor from pulling away from the insulator under stress. For bare copper conductors, this tie wire is usually of soft-drawn copper; for weatherproof covered conductors, bare or weatherproof covered wire is used. For aluminum or ACSR conductors, soft-drawn aluminum wire is used. Wire sizes are optional, but are generally small enough to be flexible but strong enough for the purpose. Often such ties are made from old or discarded conductors of small sizes, no. 6 or no. 8 conductor. Where such ties

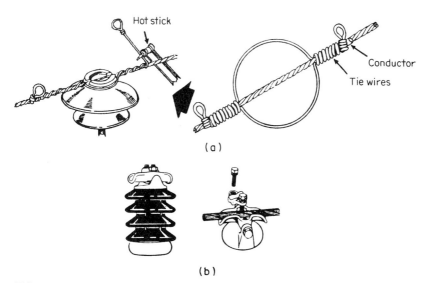

FIG. 9-5 Wire ties for overhead conductors: (a) live-line-type tie; (b) clamp-type tie. *(Courtesy A. B. Chance Co.)*

TABLE 9-3 INDUCTIVE REACTANCE PER SINGLE CONDUCTOR: COPPER OR ALUMINUM, 60 CYCLE, Ω/1000 ft

Wire size,					Spacing between conductors, in							
	$\frac{3}{8}$	$\frac{1}{2}$	1	2	4	6	8	12	18	24	30	36
Cmil or no. stranded												
1,000,000	—	—	—	0.0377	0.0534	0.0628	0.0693	0.0787	0.0881	0.0947	0.1000	0.1042
750,000	—	—	0.0250	0.0419	0.0568	0.0663	0.0727	0.0823	0.0915	0.0982	0.1025	0.1073
500,000	—	—	0.0286	0.0445	0.0604	0.0696	0.0762	0.0855	0.0948	0.1014	0.1065	0.1107
350,000	—	—	0.0354	0.0487	0.0646	0.0738	0.0803	0.0897	0.0990	0.1055	0.1107	0.1149
250,000	—	—	0.0365	0.0524	0.0683	0.0775	0.0841	0.0934	0.1027	0.1093	0.1144	0.1180
4/0	—	—	0.0384	0.0543	0.0702	0.0794	0.0860	0.0953	0.1046	0.1112	0.1163	0.1205
3/0	—	—	0.0411	0.0570	0.0729	0.0821	0.0887	0.0980	0.1073	0.1139	0.1190	0.1232
2/0	—	—	0.0437	0.0596	0.0755	0.0847	0.0913	0.1006	0.1099	0.1165	0.1210	0.1258
1/0	—	—	0.0464	0.0623	0.0782	0.0874	0.0940	0.1033	0.1126	0.1192	0.1248	0.1285
1	0.0276	0.0342	0.0501	0.0660	0.0819	0.0911	0.0977	0.1070	0.1163	0.1229	0.1280	0.1322
Solid												
2	0.0303	0.0369	0.0528	0.0686	0.0845	0.0938	0.1004	0.1097	0.1190	0.1256	0.1307	0.1348
3	0.0329	0.0395	0.0554	0.0713	0.0872	0.0964	0.1030	0.1123	0.1216	0.1282	0.1334	0.1375
4	0.0356	0.0422	0.0581	0.0740	0.0899	0.0991	0.1057	0.1150	0.1243	0.1309	0.1360	0.1402
6	0.0409	0.0475	0.0634	0.0793	0.0952	0.1044	0.1110	0.1203	0.1296	0.1362	0.1413	0.1455
8	0.0462	0.0528	0.0687	0.0846	0.1005	0.1097	0.1163	0.1256	0.1349	0.1415	0.1466	0.1508

Note: For stranded conductors, the inductance is approximately 0.0013 Ω less than for the same size solid conductors.

From *Overhead Systems Reference Book.*

are handled while the conductors remain energized, ties designed with loops that can be handled with so called hot-line tools or hot sticks are employed. Several types of ties are shown in Figs. 9-4 and 9-5.

For special conditions, especially for live-line operations, clamps that are designed to hold the conductors, but are easily opened, are used. The economics of such clamps, however, are such that they are rarely used. Tie wires are almost universally used.

ELECTRICAL CHARACTERISTICS

Resistance values for the various sizes of conductors are assumed to be the same for ac and dc operation; the skin effect prevalent in ac circuits at the values normally encountered in distribution circuits is small and may be neglected; refer to Table 9-2.

Inductive reactance values, for multiwire circuits, depend on the distance between conductors with respect to each other; these are shown in Table 9-3. For distances between conductors not shown, values of inductive reactance may be interpolated from the values shown.

CHAPTER 10

POLES, CROSS ARMS, PINS, RACKS, AND INSULATORS

WOOD POLES

Overhead distribution lines are almost universally supported on poles made of wood, though concrete and metal (steel and aluminum) are also used.

Kinds of Wood

The kinds of wood used frequently reflect the availability of materials in the different areas of the country. Western red cedar has wide application in western and northern regions, while southern or long-leaf yellow pine predominates in the rest of the country. Large amounts of northern white cedar and chestnut, however, still exist among older installations; limited amounts of Douglas fir may be found in the west and redwood in the far west, while cypress is occasionally used locally in some swampy areas of the southeast. Other woods are sometimes also used in the local areas in which they are produced. For special situations where high strength is required, wallaba may be used, though it is imported from northern South America and is comparatively denser and heavier and more expensive than other woods.

Blight has accounted for the rapid disappearance of chestnut, and northern white cedar has become obsolete not only because of diminished supply, but because it is inclined to be knotty and not very straight, making it harder to handle and work on, while it also does not present the best of appearances. Both chestnut and northern white cedar, however, have relatively long natural lives, and many still exist and may continue in service for some time. Most other woods (and particularly long-leaf yellow pine), however, are more susceptible to decay and must be treated with some kind of preservative to attain an economical life span.

Moisture Absorption

Wood, being porous, has a tendency to absorb moisture, and the moisture content depends on the conditions under which the material is used. Wood exposed to the weather can absorb large amounts of water. Since wood is a hygroscopic substance, it tends to give off or take on water vapor until it comes into equilibrium with the surrounding air; changes in atmospheric humidity produce a continual fluctuation in the moisture content of the wood. The relation of the equilibrium moisture content of wood to the relative humidity of the ambient at three different temperatures is shown in Fig. 10-1.

Decay

Much of the difficulty experienced with wood poles is due to decay, particularly at the ground line. Decay is caused by fungi attacking the wood fibers, and the conditions most favorable to the growth of decay fungi are air, moisture, and heat, with the wood acting as their food supply. In the part of the pole below ground, moisture is usually present but air is in short supply; in the part above ground, the reverse is generally true. At and near the ground line, both of these elements exist in relatively substantial quantities and, hence, this particular area is more subject to decay.

Preservatives

To eliminate, or at least retard, this destructive process, wood poles are first heated in a vacuum to drive out moisture (and to kill some of the fungi) and then treated with a preservative under pressure to fill the pores of the wood fiber; this not only poisons the food supply of the fungi (and insects), but also inhibits the absorption of moisture. Many different preservatives have been employed with varying degrees of effectiveness. No penetration process, however, is completely successful, and so periodic inspection and maintenance of the poles is usually scheduled.

Poles are turned on large lathes to improve their appearance before being subjected to treatment with preservatives for their full length. Preservatives may add as much as 25 percent to the weight per cubic foot of the wood.

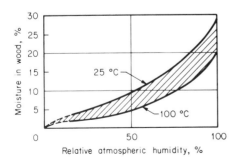

FIG. 10-1 Absorption of atmospheric moisture by wood.

Pole Classification

Pole lengths are standardized and come in increments of 5 ft and range from 20 ft to over 90 ft; for distribution lines, however, lengths are usually limited to from 25 to 55 ft.

The strength of poles depends not only on the material, but on their cross-sectional dimensions. This feature is also standardized and is denoted by "class" numbers, ranging from 00 to 10. For each standard length, the dimensions for each class are defined by standardized circumferences at the top of the pole and at a point 6 ft from the butt or bottom (approximately at the ground line). The top circumference for poles of all of the kinds of wood of the same class is the same; the lower circumference is different for different kinds of wood and determines the taper of the pole. American Standards Association (ASA) standard dimensions are given in Table 10-1 for southern pine, western red cedar, and wallaba poles; also shown are standard depth settings for the several lengths of poles.

Pole Depth Setting

For ordinary soils, depth settings start at 4 ft for 20-ft-length poles and progress to $7\frac{1}{2}$ ft for 55-ft poles; in rock, similar settings range from 3 ft to 5 ft. Pole setting depths for poles over 55 ft increase 0.5 ft for each 5 ft of incremental length, from 8 ft for 60-ft poles to 11 ft for 90-ft poles.

For less resistant soils and other media, deeper settings must be considered, as well as other methods for reinforcing the strength and stability of the pole, such as push braces, cribbing, and guying. Guying is one of the most efficient methods of relieving the pole from some or all of its horizontal load; it has been discussed in Chap. 5.

Pole Strengths

Poles of equal classes will carry equal loads at 2 ft from the top of the pole when set in the ground at standard ASA setting depths. The transverse loads which different class poles will carry are given in Table 10-2.

Some average figures regarding the characteristics of several kinds of wood are shown in Table 10-3.

Pole Framing

In preparing wood poles for use, any cutting or boring is done before the preserving process to eliminate points in which decay may occur. In older poles, a "roof" was cut either in the shape of a gable or at an angle, to prevent snow, ice, and water from accumulating on top and causing decay. Present preserving practices make this unnecessary.

For attaching cross arms, a "gain" cut into the side of the pole, in the form of a channel gouged out to the dimensions of the cross arm, at standard locations on the pole, may be found in older poles. Present practice calls for planing a flat surface on one side of the pole, referred to as a "slab" gain.

TABLE 10-1 ASA STANDARD POLE DIMENSIONS AND DEPTH SETTINGS

Pole length, ft	Depth setting, ft	Wood*	Minimum circumference at 6 ft from butt (approximate ground line), in								
			Class and minimum top circumference, in								
			00 29	0 28	1 27	2 25	3 23	4 21	5 19	6 17	7 15
20	4	P	—	—	31.5	29.5	27.5	25.5	23.5	22.0	20.0
		C	—	—	34.5	32.0	30.0	28.0	25.5	23.5	22.0
		W	32.0	30.0	28.5	26.5	25.0	23.0	21.0	—	—
25	5	P	—	—	34.5	32.5	30.0	28.0	26.0	24.0	22.0
		C	—	—	38.0	35.5	33.0	30.5	28.5	26.0	24.5
		W	35.0	33.0	31.0	29.0	27.0	25.5	23.5	—	—
30	5½	P	—	—	37.5	35.0	32.5	30.0	28.0	26.0	24.0
		C	—	—	41.0	38.5	35.5	33.0	30.5	28.5	26.5
		W	38.0	36.0	33.5	31.5	29.5	27.5	25.5	—	—
35	6	P	45.0	42.5	40.0	37.5	35.0	32.0	30.0	27.5	25.5
		C	49.0	51.5	43.5	41.0	38.0	35.5	32.5	30.5	28.0
		W	40.5	38.0	36.0	33.5	31.5	29.0	27.0	—	—
40	6	P	47.0	44.5	42.0	39.5	37.0	34.0	31.5	29.0	27.0
		C	51.0	48.5	46.0	43.5	40.5	37.5	34.5	32.0	—
		W	42.5	40.5	38.0	35.5	33.5	31.0	28.5	—	—
45	6½	P	50.0	47.0	44.0	41.5	38.5	36.0	33.0	30.5	28.5
		C	54.0	52.0	48.5	45.5	42.5	39.5	36.5	—	—
		W	45.0	42.5	40.0	37.5	35.0	32.5	30.0	—	—
50	7	P	—	—	46.0	43.0	40.0	37.5	34.5	32.0	29.5
		C	—	—	50.5	47.5	44.5	41.0	38.0	—	—
		W	46.5	44.0	41.5	39.0	36.5	34.0	31.5	—	—
55	7½	P	—	—	47.5	44.5	41.5	39.0	36.0	33.5	—
		C	—	—	52.5	49.5	46.0	42.5	39.5	—	—
		W	48.5	46.0	43.0	40.5	38.0	35.0	32.5	—	—
60	8	P	—	—	49.5	46.0	43.0	40.0	37.0	34.5	—
		C	—	—	54.5	51.0	47.5	44.0	—	—	—
		W	50.0	47.5	44.5	42.0	39.0	36.5	33.5	—	—
65	8½	P	—	—	51.0	47.5	44.5	41.5	38.5	—	—
		C	—	—	56.0	52.5	49.0	45.5	—	—	—
70	9	P	—	—	52.5	49.0	46.0	42.5	39.5	—	—
		C	—	—	57.5	54.0	50.5	47.0	—	—	—
75	9½	P	—	—	54.0	50.5	47.0	44.0	—	—	—
		C	—	—	59.5	55.5	52.0	48.5	—	—	—
80	10	P	—	—	55.0	51.5	48.5	45.0	—	—	—
		C	—	—	61.0	57.0	53.5	49.5	—	—	—
85	10½	P	—	—	56.5	53.0	49.5	—	—	—	—
		C	—	—	62.5	58.5	54.5	—	—	—	—
90	11	P	—	—	57.5	54.0	50.5	—	—	—	—
		C	—	—	63.0	60.0	56.0	—	—	—	—

*P = yellow pine; C = western red cedar; W = wallaba.

TABLE 10-2 STANDARD MAXIMUM HORIZONTAL FORCE THAT CAN BE APPLIED TO CLASSES OF POLES*

Pole class	Load resistance 2 ft from top (approximate load that pole must withstand), lb	Increment between class numbers
10	(No specifications)	—
9	900	—
8	1000	100
7	1200	200
6	1500	300
5	1900	400
4	2400	500
3	3000	600
2	3700	700
1	4500	800
0	5400	900
00	6400	1000

*At standard depth settings.

Dielectric Value of Wood

Wood offers a marked resistance to the passage of an electric current and, at least for lower voltages, may be classed as a nonconductor. Its dielectric strength varies with the different species and is greater across the grain than along the grain; changes in temperature affect the dielectric strength substantially, approximately doubling with each decrease of 22.5°F or 12.5°C. The most significant variation, however, takes place with changes in moisture content: as wood dries from fiber saturation to the vacuumed oven-dry condition, its dielectric value approaches infinity. Preservatives, however, have great influence on the dielectric strength of the wood, especially those consisting of chlorides; older poles treated with creosote and its derivatives have been found to experience

TABLE 10-3 CHARACTERISTICS OF VARIOUS WOODS

Wood	Weight, lb/ft^3	Ultimate strength, lb/in^2
Northern white cedar	23	3,600
Redwood	24	4,400
Cypress	29	4,800
Western red cedar	23	5,600
Chestnut	41	6,000
Southern yellow pine	35	7,400
Douglas fir	32	8,000
Wallaba	—	11,000
Locust	—	12,800

little change in dielectric strength as a result of the treatment. As the wood seasons, the effect of still more aging on its dielectric strength diminishes and ceases to be serious in the period of a year or so. Although the wood of poles (and cross arms) aids in the insulation of line conductors from each other and from grounds, it is not to be relied upon where safety to the public and workers is concerned.

Life Expectancy

The life of wood poles, under "normal" conditions of soil and weather, when they are treated and maintained with reasonably effective preservatives, has been estimated to be from 25 years to over 100 years. These figures have been used in making economic studies, in establishing sinking funds for the retirement of poles (and other wood structures), and in designing pole lines calling for considerably larger and stronger poles than initially necessary. In view of the continuing changes in consumer requirements, in civic and traffic requirements, and in the materials and methods employed in providing electric service, the higher longevity considerations (100 years, or even 50 years) appear to be somewhat unrealistic.

CONCRETE AND METAL POLES
General

Concrete and metal poles are at present not used extensively for distribution purposes in the United States. They are, however, used extensively in Europe and other lands where woods suitable for poles are not readily or economically available. Concrete and metal poles are usually used where great strength and appearance are paramount requirements; concrete poles are made in several colors and finishes. Both concrete and metal poles come in cross sections that are circular, square, or polygonal (usually six- or eight-sided). Both allow electrical risers to be installed in the hollow space within them.

Access to Pole-Top Facilities

The problem of access to the facilities at the top of concrete and metal poles has been met by the use of pole steps, whether they are installed permanently, or whether means are provided for their temporary installation. The permanent installation of steps is frowned upon from considerations of safety; their temporary installation is time-consuming, as well as similarly unsafe, as such steps can be installed by unauthorized people. This contrasts with the safety provided by wood poles, where skilled and trained workers, employing climbers, essentially eliminate this unauthorized access by the general public. The widespread use of bucket-type trucks that enable access to pole-top facilities has essentially eliminated the need of climbing, as well as the need for the installation of steps on the pole. Hence, this means has become common to all types of poles.

CONCRETE POLES

Manufacture

Concrete poles are manufactured with hollow cores to reduce their weight, which has been (and still is) a disadvantage, especially when they are handled in the field. Reinforcing steel strands are installed longitudinally for the full length of the pole and prestressed before the concrete is placed; reinforcing steel strands are also installed, essentially at right angles to the longitudinal reinforcing strands, usually as special coils wrapped around and welded to them in a manner to prevent movement during concrete casting. See Fig. 10-2.

In addition to their heavier weight (compared with wood), concrete poles are relatively more expensive, another reason for their lessened usage. Representative data on the characteristics of hollow round and square reinforced concrete poles are contained in Appendix 10A at the end of this chapter.

All concrete poles are tapered, and the square ones have chamfered corners. All provide cable entrance openings and hand holes to permit the installation of electric riser cables in their hollow cores.

Advantages

Concrete poles are not adversely affected by wet or dry rot, birds (especially woodpeckers), fire, rust, or chemicals (such as fertilizers and salt spray). Besides being stronger and more rigid than wood, they are essentially maintenance-free;

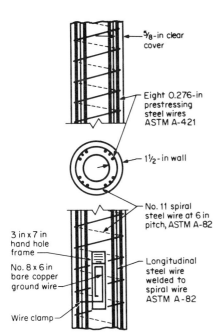

FIG. 10.2 Typical reinforced concrete hollow pole section. *(Courtesy Centrecon, Inc.)*

ground moisture and weather, which work against other types of poles, work in favor of concrete, hardening, toughening, and protecting its integrity. Considering the potential lifespans, concrete claims the lowest cost per year.

METAL POLES

Metal poles are manufactured according to specifications drawn up for particular uses; their length and the thickness of the metal employed depend on their use and the desired strength. They are tapered and come in the same shapes as concrete poles—round, square, and polygons. They may also include provision for pole steps. Though generally of the color of the metal of which they are made, they may be painted in specified colors. They may be buried directly in the ground, set directly in bases of concrete, or bolted to metal base plates permanently set in concrete.

CROSS ARMS

Cross arms are the most common means of supporting distribution conductors on poles. Although they are being used less frequently, their use will persist for some time.

Standard Arms

Standard cross-section dimensions for wood cross arms (width by height) are:

$3\frac{1}{4}$ in by $4\frac{1}{4}$ in

$3\frac{1}{2}$ in by $4\frac{1}{2}$ in

$3\frac{3}{4}$ in by $4\frac{3}{4}$ in

4 in by 5 in

Of these, the first two are most commonly used for distribution purposes, and usually only one of these will be stocked by an individual utility. The larger size finds greater use in the harsher northern and western climates, while the smaller finds use in the south and southwest. The rectangular cross section is slightly rounded or "roofed" on the top surface to shed rain and snow.

The length of the cross arm depends on the number of conductors it is to support and the spacing between them. Standard cross arms include two-, four-, six-, and eight-pin arms, although the four- and six-pin arms, 8 ft in length, are the more widely used. Spacing between pins for the six-pin 8-ft arm is standardized at $14\frac{1}{2}$ in, except for the space between the two center pins, the "climbing space" for the lineman's safety; for primary voltages up to 15,000 V, this climbing space is 30 in, and for voltages above that value it is 36 in (with spacing between the six pins reduced to 13 in). Spacing between pins for the four-pin 8-ft cross arm is 26 in, with a space between the two center pins of 36 in, the climbing space. See Fig. 10-3. Vertical spacing between cross arms is standardized at 2 ft.

Both Douglas fir and southern yellow pine are used for cross arms because

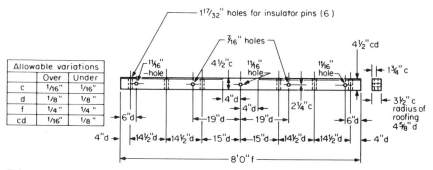

FIG. 10-3   Standard 6-pin 8-ft wood cross arm. *(From Overhead Systems Reference Book.)*

of their comparatively high bending strengths and their durability. Both are treated with preservatives after holes for pins and bolts have been bored in them. Their insulating properties are similar to those described for wood poles.

Steel Cross Arms

Cross arms made of steel are used where stresses are very large and cannot be accommodated by other means such as double arms or arm guys; they do not have the insulating property that wood cross arms have.

Cross-Arm Braces

Galvanized flat steel braces are used to provide strength and rigidity to the cross arm. They are usually attached to the cross arm by means of carriage bolts and to the pole by a lag screw. For heavier loads, a one-piece vee-shaped angle-iron brace is sometimes used.

PINS

Pins are used to hold the insulators to which conductors are fastened. They may be made of wood, usually locust because of its strength and durability (refer to Table 10-3, p. 291), or of iron or steel where greater strength or extra-long lengths are required. Pins fastened to cross arms are made of wood or steel, while those fastened directly to poles are made of steel.

Wood Pins

Wood pins are generally made of yellow or black locust, having an ultimate fiber bending strength of about 10,000 lb/in². They are standardized in the dimensions of the pin and in the threads which accommodate the insulators. Where greater strength is required, galvanized metal pins having an ultimate fiber of 50,000 lb/in² for malleable iron and 60,000 lb/in² for steel are used.

Standard locust pins are 9 in. in length, with a shank $1\frac{1}{2}$ in. in diameter that

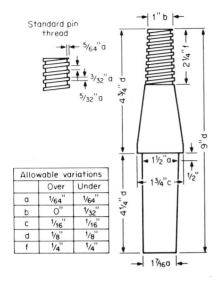

FIG. 10-4 Standard wood pin. *(From Overhead Systems Reference Book.)*

Allowable variations		
	Over	Under
a	$\frac{1}{64}''$	$\frac{1}{64}''$
b	$0''$	$\frac{1}{32}''$
c	$\frac{1}{16}''$	$\frac{1}{16}''$
d	$\frac{1}{8}''$	$\frac{1}{8}''$
f	$\frac{1}{4}''$	$\frac{1}{4}''$

fits into the cross-arm hole. A $1\frac{3}{4}$ in shoulder above this shank tapers to a 1-in-diameter part $2\frac{1}{4}$ in long, which is threaded to receive the insulator; refer to Fig. 10-4.

Steel Pins

Steel pins have many shapes; some are designed to be bolted through the cross arm, others to be clamped around the arm; while others are designed to be bolted directly to the top of the pole. The steel pins may vary in length, and some may also be of hollow core section. Several types of steel pins are shown in Fig. 10-5.

Long steel pins with angle bases, bolted to the side of the pole, serve to replace the cross arm in supporting conductors, making for a more streamlined, stronger, better-appearing line; refer to Fig. 5-13 in Chap. 5.

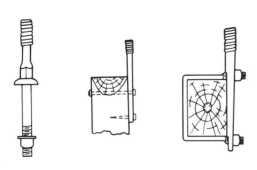

FIG. 10-5 Typical steel pins. *(Courtesy A. B. Chance Co.)*

RACKS

Secondary conductors, for many years supported horizontally on wood cross arms, were relocated to a lower position on the pole and mounted vertically on a rack attached to the pole. They make for a better appearance and are stronger and less costly than cross arms; and the wires being in a vertical plane, service wires running in different directions do not cross each other. With this design the worker on the secondary conductors and services need not approach the primary conductors on the pole, making for a safer work method.

Many secondary-service installations on cross arms and on racks still exist and will continue to do so for a long time.

Multispool Racks

Racks, made of galvanized steel, are attached to the pole by one or more through bolts, some of which may be replaced by lag screws where loadings permit. The racks support insulators, spaced 6 or 8 in apart (although some may have a 4-in spacing), to which the conductors are attached. The conductors may be attached to a number of individual single-insulator racks, or to one rack containing several insulators. The insulators may be of the spool or knob type, as shown in Fig. 10-6.

Single-Spool Racks

Single-spool racks are also manufactured; on these, each insulator is bolted individually to the pole. While spool-type insulators are the more frequently used type, knob-type insulators are also used.

Cabled Secondary Mains and Services

The rack has been supplanted in many instances by a single uninsulated clamp attached to the pole and supporting the secondary mains. Here, the conductors are cabled around a neutral wire, which also acts as a messenger wire carrying the conductors. A similar type of cabled conductor is used for services, with the

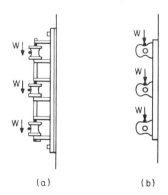

FIG. 10-6 Typical secondary rack: (a) spool type; (b) knob type. (Courtesy McGraw Edison Co.)

(a)   (b)

neutral-messenger supporting the conductors from the pole to the consumer's premises.

INSULATORS

Materials

Insulators, placed between energized conductors and the supporting structures, are now almost universally made of porcelain, although a great many glass insulators are still in service and will be for a long time. Fiberglass, epoxy, and other plastics are now beginning to be used in the manufacture of insulators.

Glass Glass insulators, made in a variety of shapes and sizes, are electrically and mechanically adequate and very economical. Their usage, however, has been generally limited to circuits operating at voltages under 5 kV, principally because of their relatively (compared to porcelain) low resistance to shock and high coefficient of expansion. While the dielectric properties of glass may vary considerably depending on its particular treatment, its mechanical properties are reasonably consistent. Its tensile strength is usually less than 10,000 lb/in^2; its minimum compressive strength is 50,000 lb/in^2; its modulus of elasticity is 10,000,000 lb/in^2; and its coefficient of expansion is 400 to 600 $\times$ 10^{-6} per degree Fahrenheit, or 720 to 1080 $\times$ 10^{-6} per degree centigrade.

Pyrex, a form of glass, partially overcomes the two deficiencies concerning shock and temperature changes, but its relatively high cost restricts its application generally to high-voltage transmission lines.

Porcelain Porcelain is made from white clay to which powdered feldspar and silica are added. The material has greater (compared to glass) ability to resist sudden changes in temperature, and an external glaze protects it from shock. Depending on how insulators are manufactured, the tensile strength of the porcelain may vary from less than 2000 lb/in^2 to as much as 9000 lb/in^2, and the corresponding minimum compressive strengths, from 15,000 to 60,000 lb/in^2; its coefficient of expansion may be as low as 16.6 $\times$ 10^{-6} per degree Fahrenheit, or 30 $\times$ 10^{-6} per degree centigrade. The dielectric strength generally exceeds 16 kV/mm (16,000 kV/m), or some 400 kV/in.

Porcelain insulators are made by two processes: the "wet process," in which the insulation is molded, and the "dry process," in which the insulation is pressed into shape in steel molds. The dry-process insulator has less electrical and mechanical strength, but is also less expensive; it is more apt to puncture under electrical stress rather than to flash over. Dry-process insulators are more economical, however, and their use is generally limited to lower-voltage applications.

The principal types of insulators are described below.

Pin-Type

Pin-type insulators may be constructed in one piece for voltages to about 35 kV and in two or three pieces from that voltage to 69 kV. They are shaped with a

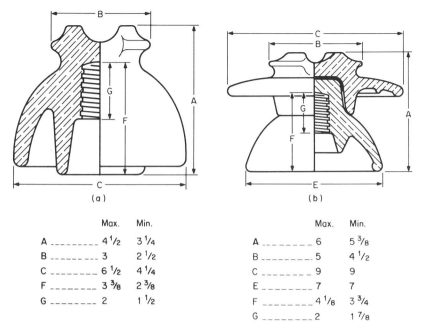

	Max.	Min.
A	$4\,^1/_2$	$3\,^1/_4$
B	3	$2\,^1/_2$
C	$6\,^1/_2$	$4\,^1/_4$
F	$3\,^3/_8$	$2\,^3/_8$
G	2	$1\,^1/_2$

	Max.	Min.
A	6	$5\,^3/_8$
B	5	$4\,^1/_2$
C	9	9
E	7	7
F	$4\,^1/_8$	$3\,^3/_4$
G	2	$1\,^7/_8$

FIG. 10-7 Typical pin-type porcelain insulators: (a) one-piece 11- to 15-kV pin insulator; (b) two-piece 33-kV pin insulator. *(From Overhead Systems Reference Book.)*

groove on top in which the conductor lies and is tied in place; they may also have a groove around the sides in which the conductor is placed and tied. The side groove is generally used where the line turns and the conductor imposes strains on the side. Both glass and porcelain insulators are shaped so as to provide a long path from the line conductor to the point of support where the insulator is screwed to the pin, usually considered a ground. Several ridges on the underside extend this path so that, figuratively, the insulator may be described as having an outside skirt and a number of inner petticoats; see Fig. 10-7. Though constructed in a variety of shapes, colors, and ratings, all types of pin insulators share a common standard in the diameter, shape, and number of insulator threads per inch, a standard that matches that of the pins with which they are usually associated.

Post-Type

The post-type insulator is a variation of the pin-type insulator, in which the porcelain is shaped into a cylindrical block with its surface corrugated horizontally to provide a long path from the conductor to the point of support; grooves for the attachment of the conductor are provided at the top of the insulator; see Fig. 10-8.

FIG. 10-8 Typical post-type porcelain insulator. *(Courtesy Ohio Brass Co.)*

Suspension- or Strain-Type

The suspension- or strain- (or string-) type insulator consists of a porcelain disk contained between a ball-and-socket (or clevis-and-pin, or other) arrangement so that the porcelain is in compression; each disk has certain electrical and mechanical characteristics. The ball-and-socket arrangement permits units to be added to each other in a string, thus accommodating higher voltages. Several such strings can be connected mechanically in "parallel" to achieve necessary strength requirements. Disks vary in diameter from about 5 to 10 in, with standard vertical distances between ball and socket (or other components) of $4\frac{3}{4}$ and $5\frac{3}{4}$ in, and are rated for an allowable tension of 12,000 lb. See Fig. 10-9.

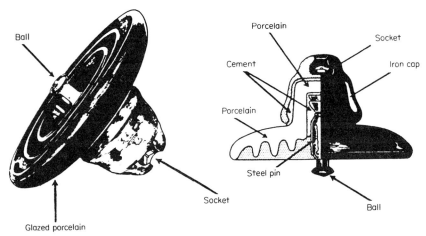

FIG. 10-9 Typical suspension or strain-type porcelain insulator. *(Courtesy Locke Insulator Co.)*

Spool-Type

Spool-type insulators are used with secondary racks or in service brackets to support secondary mains and services; they come in two different sizes and colors for identification purposes. See Fig. 10-10.

Strain-Ball Type

The strain-ball insulator is generally employed in guys to insulate the upper part from the lower part; the lower part is insulated to protect people on the ground, while the upper part is insulated from the ground to protect the lineman on the pole. In some instances involving wye circuits with a common primary and secondary grounded neutral, this insulator is sometimes not inserted in the guy wire. The insulator consists of a ball or block of porcelain in which two transverse holes at right angles to each other contain the guy wires so arranged that the porcelain is always under compression. See Fig. 10-11. This type of insulator is also used to dead-end smaller conductors at the cross arm where the operating voltage is less than 5000 V.

Other Types

Other insulator types include the knob types sometimes used with secondary racks, previously described; and various other shapes for use as bushings, bus and service supports, and other specific purposes. Where porcelain is used, it is always designed to be under compression.

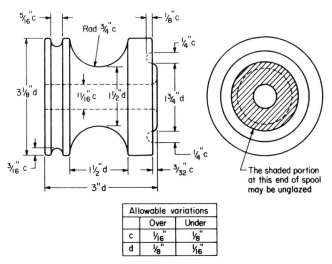

Allowable variations		
	Over	Under
c	$\frac{1}{16}$"	$\frac{1}{8}$"
d	$\frac{1}{8}$"	$\frac{1}{16}$"

FIG. 10-10 Typical spool porcelain insulator for secondary rack. Material: porcelain. Finish: brown glazed. *(From Overhead Systems Reference Book.)*

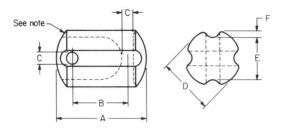

Trade no.	A	B	C	D	E	F	Strength in compression
502	3¼"f	1¾"f	⁷⁄₁₆"c	2½"d	1¾"f	⁵⁄₁₆"c	10,000 lb
504	3¾"f	2¼"f	⁹⁄₁₆"c	2⅞"d	2⅛"f	⅜"c	12,000
506	5¼"f	3⅛"f	¾"c	3⅜"d	2⅜"f	½"c	15,000

Allowable variation		
	Over	Under
c	¹⁄₁₆"	¹⁄₁₆"
d	⅛"	⅛"
f	¼"	¼"

FIG. 10-11 Typical strain-ball-type porcelain insulator. Material: porcelain, wet process. Finish: brown glazed. *Note:* One end of insulator may be unglazed. *(From Overhead Systems Reference Book.)*

TEST VOLTAGES

Nominal ratings of insulators include not only their operating voltage, but flashover values as well. Flashover values depend in general on the leakage distances provided between the conductor and the point of attachment to the support. These values, however, may be subject to other factors, including local atmospheric and environmental conditions, types of support, maintenance programs, etc. Typical minimum flashover voltage values are contained in Table 10-4.

TABLE 10-4 TYPICAL FLASHOVER VOLTAGE REQUIREMENTS

Nominal voltage (between phases), kV	Minimum rated dry flashover voltage of insulators, kV*
0.75	5
2.4	20
6.9	39
13.2	55
23.0	75
34.5	100
46.0	125
69.0	175

*Interpolate for intermediate values.

For current industry recommended values, refer to the latest revision of the National Electric Safety Code.

APPENDIX 10A
CONCRETE DISTRIBUTION POLES:
REPRESENTATIVE SPECIFICATIONS*

SCOPE

These specifications apply to the manufacture of machine-made, pretensioned, prestressed concrete distribution poles, designed in accordance with recommendations of the American National Standards Institute (ANSI).

SHAPE

The poles shall be single, hollow, round or square cross section, with a taper of 1 to 150. They shall have a uniform nominal wall thickness of $1\frac{1}{2}$ in as shown in Fig. 10-12.

DIMENSIONS AND STRENGTH
Round Poles

The dimensions and strength of the round poles shall be as shown in Table 10-5. The poles shall be grouped into seven classes based on the design ultimate strength as given in Table 5-22 in Chap. 5.

Square Poles

The dimensions and strength of the square poles shall be as shown in Table 10-6.

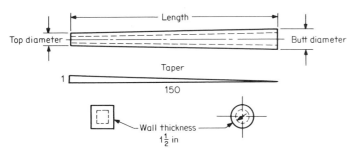

FIG. 10-12 Typical reinforced concrete hollow distribution pole. *(Adapted from Centrecon, Inc.)*

---

*Adapted from data furnished by Centrecon, Inc., and by Concrete Products, Inc.

TABLE 10-5 DIMENSIONS AND STRENGTHS—ROUND HOLLOW CONCRETE POLES

Overall length			Setting depth, ft-in	Top diameter, in	Butt diameter, in	Design ultimate moment, ft·lb	Allowable moment—SF = 2, ft·lb	Nominal weight, lb
m	ft-in	Class						
12	39-4	A	5-11	$7\frac{1}{2}$	$13\frac{13}{16}$	100,530	50,260	2540
13	42-8	A	6-3	$7\frac{1}{2}$	$14\frac{5}{16}$	110,130	55,060	2830
14	45-11	A	6-7	$7\frac{1}{2}$	$14\frac{13}{16}$	119,470	59,730	3130
15	49-2	A	6-11	$7\frac{1}{2}$	$15\frac{3}{8}$	128,800	64,400	3470
16	52-6	A	7-3	$7\frac{1}{2}$	$15\frac{7}{8}$	138,400	69,200	3800
12	39-4	B	5-11	$7\frac{1}{2}$	$13\frac{13}{16}$	84,830	42,410	2480
13	42-8	B	6-3	$7\frac{1}{2}$	$14\frac{5}{16}$	92,930	46,460	2760
14	45-11	B	6-7	$7\frac{1}{2}$	$14\frac{13}{16}$	100,800	50,400	3060
15	49-2	B	6-11	$7\frac{1}{2}$	$15\frac{3}{8}$	108,680	54,340	3380
16	52-6	B	7-3	$7\frac{1}{2}$	$15\frac{7}{8}$	116,780	58,390	3710
9	29-6	C	4-11	$6\frac{11}{16}$	$11\frac{7}{16}$	50,810	25,400	1520
10	32-10	C	5-3	$6\frac{11}{16}$	$11\frac{15}{16}$	57,560	28,780	1770
11	36-1	C	5-7	$6\frac{11}{16}$	$12\frac{7}{16}$	64,130	32,060	2010
12	39-4	C	5-11	$7\frac{1}{2}$	$13\frac{13}{16}$	70,690	35,340	2430
13	42-8	C	6-3	$7\frac{1}{2}$	$14\frac{5}{16}$	77,440	38,720	2710
14	45-11	C	6-7	$7\frac{1}{2}$	$14\frac{13}{16}$	84,000	42,000	3000
15	49-2	C	6-11	$7\frac{1}{2}$	$15\frac{3}{8}$	90,560	45,280	3310
16	52-6	C	7-3	$7\frac{1}{2}$	$15\frac{7}{8}$	97,310	48,650	3640
9	29-6	D	4-11	$6\frac{11}{16}$	$11\frac{7}{16}$	41,780	20,890	1490
10	32-10	D	5-3	$6\frac{11}{16}$	$11\frac{15}{16}$	47,330	23,660	1720
11	36-1	D	5-7	$6\frac{11}{16}$	$12\frac{7}{16}$	52,730	26,360	1960
12	39-4	D	5-11	$7\frac{1}{2}$	$13\frac{13}{16}$	58,120	29,060	2390
13	42-8	D	6-3	$7\frac{1}{2}$	$14\frac{5}{16}$	63,670	31,840	2670
14	45-11	D	6-7	$7\frac{1}{2}$	$14\frac{13}{16}$	69,070	34,540	2960
15	49-2	D	6-11	$7\frac{1}{2}$	$15\frac{3}{8}$	74,460	37,230	3260
16	52-6	D	7-3	$7\frac{1}{2}$	$15\frac{7}{8}$	80,010	40,000	3570
9	29-6	E	4-11	$6\frac{11}{16}$	$11\frac{7}{16}$	33,870	16,930	1470
10	32-10	E	5-3	$6\frac{11}{16}$	$11\frac{15}{16}$	38,370	19,180	1700
11	36-1	E	5-7	$6\frac{11}{16}$	$12\frac{7}{16}$	42,750	21,370	1930
12	39-4	E	5-11	$7\frac{1}{2}$	$13\frac{13}{16}$	47,130	23,560	2360
13	42-8	E	6-3	$7\frac{1}{2}$	$14\frac{5}{16}$	51,630	25,810	2640
14	45-11	E	6-7	$7\frac{1}{2}$	$14\frac{13}{16}$	56,000	28,000	2930
15	49-2	E	6-11	$7\frac{1}{2}$	$15\frac{3}{8}$	60,380	30,190	3220
16	52-6	E	7-3	$7\frac{1}{2}$	$15\frac{7}{8}$	64,880	32,440	3530
9	29-6	F	4-11	$6\frac{11}{16}$	$11\frac{7}{16}$	27,100	13,550	1460
10	32-10	F	5-3	$6\frac{11}{16}$	$11\frac{15}{16}$	30,700	15,350	1680
11	36-1	F	5-7	$6\frac{11}{16}$	$12\frac{7}{16}$	34,200	17,100	1920
12	39-4	F	5-11	$6\frac{11}{16}$	13	37,700	18,850	2160
13	42-8	F	6-3	$6\frac{11}{16}$	$13\frac{1}{2}$	41,300	20,650	2410
14	45-11	F	6-7	$6\frac{11}{16}$	$14\frac{1}{16}$	44,800	22,400	2680
15	49-2	F	6-11	$6\frac{11}{16}$	$14\frac{9}{16}$	48,300	24,150	2960
9	29-6	G	4-11	$6\frac{11}{16}$	$11\frac{7}{16}$	21,450	10,720	1460
10	32-10	G	5-3	$6\frac{11}{16}$	$11\frac{15}{16}$	24,300	12,150	1670
11	36-1	G	5-7	$6\frac{11}{16}$	$12\frac{7}{16}$	27,080	13,540	1900
12	39-4	G	5-11	$6\frac{11}{16}$	13	29,850	14,920	2150
13	42-8	G	6-3	$6\frac{11}{16}$	$13\frac{1}{2}$	32,700	16,350	2400
14	45-11	G	6-7	$6\frac{11}{16}$	$14\frac{1}{16}$	35,470	17,730	2660
15	49-2	G	6-11	$6\frac{11}{16}$	$14\frac{9}{16}$	38,240	19,120	2930

Courtesy Centrecon, Inc.

TABLE 10-6 DIMENSIONS AND STRENGTHS—SQUARE HOLLOW CONCRETE POLES

Overall pole length, ft	Pole size—tip/butt, in	E.P.A., ft² Concrete strength, lb/in² 6000, standard	7000, when specified	Ultimate ground-line moment, ft·lb	Breaking strength load 2 ft. below tip, lb	Deflection per 100 ft·lb, in	Deflection limitations, ft·lb	Pole weight, lb
25	7.6/11.65	43.3	45.3	84,000	4540	0.03	4100	2260
30	7.6/12.46	38.7	40.7	132,000	5740	0.03	5000	2880
35	7.6/13.27	33.8	36.2	147,000	5350	0.03	5900	3600
40	7.6/14.08	29.4	31.7	163,000	5090	0.03	6800	4370
45	7.6/14.89	25.7	28.1	178,000	4880	0.06	3850	5225
50	7.6/15.70	22.3	24.7	193,000	4710	0.06	4300	6160
55	7.6/16.51	19.1	21.7	209,000	4590	0.06	4750	7270

Glossary of Terms

E.P.A. Effective projected area, in square feet of transformers, capacitors, streetlight fixtures, and other permanently attached items which are subject to wind loads.

Concrete strength This is a reference to the compressive strength of the concrete in pounds per square inch as measured by testing representative samples 28 days after casting.

Ultimate ground-line bending moment This is the bending moment applied to the pole which will cause structural failure of the pole. This is the result of multiplying the load indicated in the column Breaking Strength by a distance 2 ft less than the pole height (i.e., 2 ft less than the length of pole above ground). Figures under Ultimate Ground-Line Moment assume embedment of 10 percent of the pole length plus 2 ft. The figures in this column on technical charts are maximum moments expected to be applied to the pole. Appropriate safety factors should be used by the designer.

Breaking strength This is the approximate load which, when applied at a point 2 ft below the tip of the pole, will cause structural failure of the pole.

Ground line The point at which an embedded pole enters the ground or is otherwise restrained.

Deflection The variation at the tip of the pole from a vertical line resulting from the application of loads such as equipment, wind, ice, etc.

Ground-line bending moment The product of any load applied at any point on the pole multiplied by its height above ground line.

Dead loads This refers to the load on a pole resulting from the attachment of transformers and other equipment permanently.

Live loads These are loads applied to the pole as a result of wind, ice, or other loads of a temporary nature.

Ultimate Strength

The design ultimate strengths are determined to meet the working strength requirement of wood poles in accordance with the ASA *Specifications and Dimensions for Wood Poles*. The factor of safety of 2 shall be used for prestressed concrete poles.

COLORS AND FINISHES

The poles shall be furnished in colors and finishes specified by the buyer. Colors shall be white, gray, buff, green, brown, and black. Finishes shall be Mold Finish; Natural Aggregate, exposed or polished; or Terrazzo Aggregate, exposed or polished. Other finishes will be furnished on request.

MATERIALS

Cement

The cement used for the concrete shall be portland cement, type I, II, or III, conforming to ASTM designation C-150.

Aggregates

The concrete shall be made with fine and coarse aggregates conforming to ASTM designation C-33.

The fine aggregates shall be graded from $\frac{3}{8}$ in to no. 100 sieve, with 100 percent passing the $\frac{3}{8}$-in sieve and not more than 10 percent passing the no. 100 sieve.

The maximum-size coarse aggregate shall not exceed one-fifth of the minimum dimension of the member nor three-fourths of the clear spacing between the prestressed wires. The maximum size of coarse aggregate to be used shall conform to the smaller of the two limitations mentioned above.

Water

The water used for mixing concrete shall be clean and free of injurious quantities of substances (acid, alkalines, oil, and vegetable matter) deleterious to concrete or to prestressing steel.

Admixture

The admixture used for the concrete shall conform to ASTM C-494, and shall not contain more than a trace of calcium chloride.

Reinforcement

The prestressing steel wire shall conform to ASTM A-421 with the strengths as noted in the following table.

Diameter	Yield strength	Tensile strength
9 mm (0.354 in)	192,000 lb/in^2	220,500 lb/in^2
7 mm (0.276 in)	177,000 lb/in^2	206,200 lb/in^2

The reinforcing steel wire used for main reinforcement and spiral reinforcement shall conform to ASTM A-82.

Embedded Materials

The embedded materials, such as sleeves for through holes, sockets for step bolts, and other inserts, will be corrosion-resistant and weatherproof.

GENERAL REQUIREMENTS
Design Ultimate Strength

The design ultimate strength for each class of pole shall be as given in Table 5-22 (in Chap. 5) for round poles and in Table 10-6 for square poles.

Through Holes

The through-hole spacing and sizes for framing shall be as shown in Fig. 10-13a, b, and c for round poles and Fig. 10-14 for square poles.

Step Bolts

Each pole may have provisions (if specified) for installing step bolts to gain access to, and perform work on, the poles as shown in Fig. 10-13a, b, and c as well as Fig. 10-14.

Grounding Wire Exits

The grounding wire exits shall be as shown in the specifications. Fish wire shall be attached to poles at the time of manufacture to facilitate ease in installation of grounding wire.

Pole Cap

Each pole shall have permanently attached to its top an insulating pole cap. This cap shall be constructed of a nonconductive material that will not become conductive during its service life.

Butt Cap

Each pole manufactured under this specification shall have permanently attached thereto a concrete butt cap.

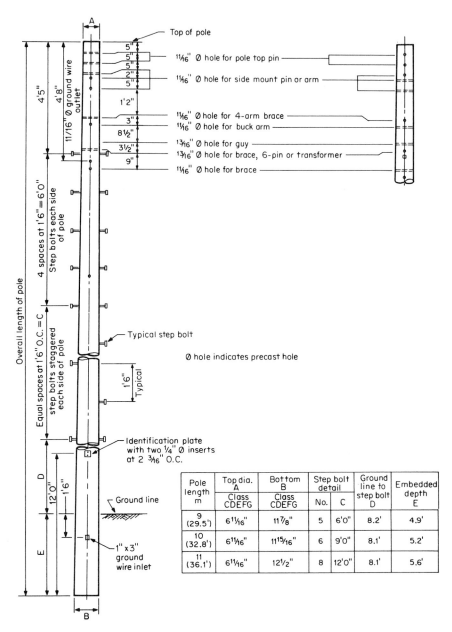

FIG. 10-13 Reinforced concrete round hollow distribution pole. Pole steps optional. (a) Use of top third of upper part of pole. (Courtesy Centrecon, Inc.)

Pole length m	Top dia. A Class CDEFG	Bottom B Class CDEFG	Step bolt detail No.	C	Ground line to step bolt D	Embedded depth E
9 (29.5')	6¹¹⁄₁₆"	11⅞"	5	6'0"	8.2'	4.9'
10 (32.8')	6¹¹⁄₁₆"	11¹⁵⁄₁₆"	6	9'0"	8.1'	5.2'
11 (36.1')	6¹¹⁄₁₆"	12½"	8	12'0"	8.1'	5.6'

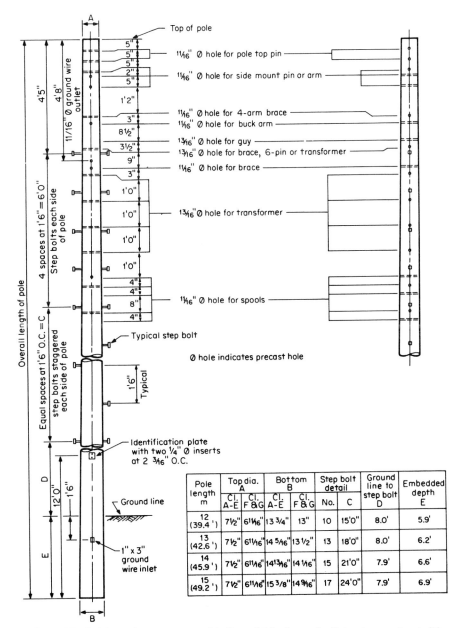

FIG. 10-13 Reinforced concrete round hollow distribution pole. Pole steps optional. (*b*) Use of top two-thirds upper part of pole. *(Courtesy Centrecon, Inc.)*

Pole length m	Top dia. A		Bottom B		Step bolt detail		Ground line to step bolt D	Embedded depth E
	Cl. A-E	Cl. F&G	Cl. A-E	Cl. F&G	No.	C		
12 (39.4')	7½"	6¹¹⁄₁₆"	13¾"	13"	10	15'0"	8.0'	5.9'
13 (42.6')	7½"	6¹¹⁄₁₆"	14⁵⁄₁₆"	13½"	13	18'0"	8.0'	6.2'
14 (45.9')	7½"	6¹¹⁄₁₆"	14¹³⁄₁₆"	14¹⁄₁₆"	15	21'0"	7.9'	6.6'
15 (49.2')	7½"	6¹¹⁄₁₆"	15³⁄₈"	14⁹⁄₁₆"	17	24'0"	7.9'	6.9'

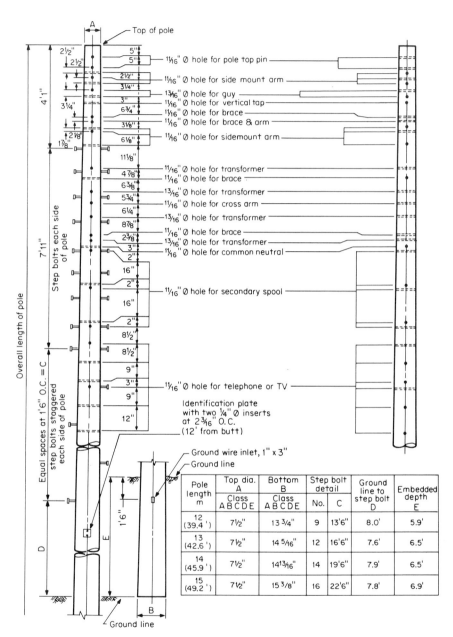

FIG. 10-13 Reinforced concrete round hollow distribution pole. Pole steps optional. (c) Use of top full upper part of pole. *(Courtesy Centrecon, Inc.)*

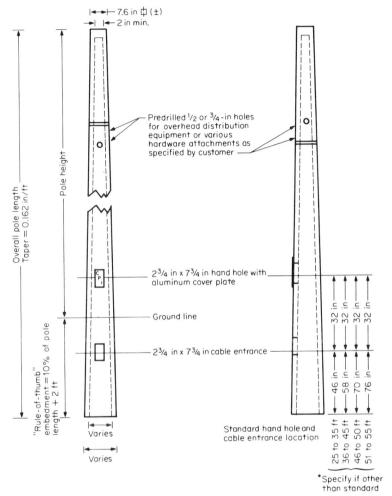

FIG. 10-14 Reinforced concrete square hollow distribution pole. *(Courtesy Concrete Products, Inc.)*

Cable Entrance

These poles shall be furnished with cable entrance opening, hand hole, and hollow core or PVC raceway for use with underground wiring.

MANUFACTURE
Placement of Steel Reinforcement

All prestressing steel and reinforcing steel shall be accurately positioned and satisfactorily protected against the formation of rust or other corrosion prior to placement to the concrete.

All prestressing steel shall be free of loose rust, dirt, grease, oil, or other lubricants or substances which might impair its bond with the concrete.

All unstressed reinforcing steel shall be free of loose rust, dirt, grease, oil, or other lubricants or substances which might impair its bond with the concrete.

Spiral reinforcement shall have a minimum size of gauge no. 11 with 6-in pitch, or equivalent steel ratio in volume, and shall be mechanically welded to the unstressed reinforcement.

The amount of concrete cover on the outside of the major steel reinforcement shall be not less than $\frac{5}{8}$ in.

Embedded Items

Sleeves, sockets, inserts, or other embedded items shall be accurately set in the molds and secured to prevent movement during the concrete placing process. They shall be cast in the concrete so that the axis of the insert is normal to the outside surface of the pole. Particular care shall be used to insure proper cover on all embedded items.

Mixing Concrete

The proportions of water to cement shall produce a concrete after steam curing having a minimum compressive strength of 3500 lb/in². A minimum 28-day compressive strength of 6000 lb/in² after atmospheric curing shall be required.

Concrete shall be mixed until there is a uniform distribution of the materials and shall be discharged completely before the mixer is recharged.

Placing Concrete

Round Poles   The poles shall be formed and compacted by centrifugal force in a machine of suitable type so designed that the molds may be revolved at the required speed to insure distribution and dense packing of the concrete without the creation of voids behind the reinforcing steel.

Metal molds shall be used which shall be adequately braced and stiffened against deformation under pressure of the wet concrete during spinning. They shall also be sufficiently rigid to take the prestressing force without allowing deformation, which will reduce the spinning speed.

Filling the mold and spinning shall be a continuous operation and the spinning shall take place before any of the concrete in the mold has taken an initial set. Excess water forced to the center of the mold during the spinning cycle must be drained in a suitable manner.

Square Poles   Square poles shall be cast in steel molds which are true to shape and line.

Concrete shall be poured in one continuous operation while being vibrated with high-frequency vibration to achieve consolidation and insure high density.

Applying Prestressing Force

The intial prestressing force shall be the smaller of 70 percent of tensile strength or 80 percent of yield strength. Tensioning shall be carried out in such a manner that the initial prestressing force shall vary no more than ±5 percent from design value.

Concrete Curing

The poles shall be cured with low-pressure steam for as long as necessary to reach the strength required by the design for transfer of prestressing force.

Prior to the application of heat, a minimum initial setting period of 2 h is required. The rate of rise of temperature shall not exceed 20°F per hour during the second 2 h of the curing cycle nor 60°F per hour during the third 2 h. The maximum temperature shall not exceed 175°F. Fresh concrete shall be protected so as to be free from rain, hot sun, or wind and rapid loss of moisture prior to the start of the steam-curing cycle. Temperature and moisture for each steam-curing cycle shall be continuously controlled and recorded to ensure the adequate curing of the concrete.

Concrete Testing

Samples for strength tests of concrete shall be taken not less than once each day in accordance with ASTM C-172. Standard 4- × 8-in test cylinders shall be used to insure the required strength of products. Standard cylinders for acceptance tests shall be molded and laboratory-cured in accordance with ASTM A-39. Each strength test shall be performed at 1, 7, and 28 days after casting. The strength level of the concrete will be considered satisfactory if the averages of all sets of two consecutive strength test results equal or exceed the required strength and if no individual strength test result falls below the required strength by more than 500 lb/in².

Bending Test

A bending strength test of a pole shall be performed in order to assure that the pole meets the minimum structural strength requirements in accordance with specifications.

Manufacturing Tolerance

1. Length: $\pm\frac{3}{8}$ in per 10 ft of length
2. Outer dimension: $+\frac{3}{16}$ in, $-\frac{1}{8}$ in
3. Wall thickness: $-\frac{1}{8}$ in
4. Deviation from straight line: not more than $\frac{1}{8}$ in per 10 ft of length

5. Location of reinforcement: main reinforcement cover, $\pm\frac{1}{8}$ in; spacing of spiral, $+\frac{1}{4}$ in

6. Location of embedded items: $\pm\frac{1}{8}$ in

Workmanship

All poles shall be unpolished but free of burrs, chips, holes, excess cement, and other surface irregularities. All poles shall present a straight and symmetrical appearance after erection. Crooks, curvatures, and twisting edges shall be prohibited. All raceway openings to cable entrance and hand holes shall be free of burrs, cement, or aggregate. Any surface roughness or obstructions that would injure or harm the insulation of electrical cables under normal installation or operating procedures will be prohibited.

Quality Control

Manufacturing procedures and quality control shall follow the recommendations of the Prestressed Concrete Institute.

CHAPTER 11

TRANSFORMERS, CUTOUTS, AND SURGE ARRESTERS

TRANSFORMERS

The transformer, one of the most efficient pieces of electrical apparatus (usually better than 98.5 percent efficient), having no moving parts, transfers electric power from one circuit to another magnetically; in the process, it enables changes to be made in the voltage (and current) output. It is this latter characteristic that has enabled the development of economically feasible ac systems.

While output voltage values may thus be stepped up (at power plants and transmission substations), they can also be stepped down to values suitable for distribution purposes. The first such transformation takes place at the distribution substation, where incoming transmission line voltages are stepped down to primary feeder voltage values. A second transformation takes place at the distribution transformer connected to the primary feeder; here, the voltage is further reduced to values suitable for utilization at the consumer's premises.

Main Parts

Transformers consist essentially of two windings, insulated from each other, wound around a common core usually made of steel and having excellent magnetic properties. The assembly is contained in a steel tank and submerged in an insulating coolant, usually an oil of high dielectric strength. The electric leads connecting the coils to their respective circuits are passed through the tank by bushings made of porcelain or plastics. The dielectric values of the insulations of the several components are deliberately so scaled between them that voltage surges which may cause damage will cause failure to occur at the bushings, a point which can be readily discovered and repaired or replaced at relatively minor cost. More detailed discussion of this insulation coordination is contained in Chap. 4.

Core Types Cores may be of the *shell* type or *core* type. In the first, the steel core surrounds the windings; though providing a better magnetic circuit, the units become larger and heavier, and are usually restricted to the larger-size transformers generally found in substations and in underground network units; they are usually polyphase units.

In the core type, the windings surround the steel core, resulting in a more compact, smaller unit, generally more suitable for the sizes of transformers associated with overhead and radial primary distribution systems; these are almost always single-phase units.

Windings Windings are usually designed to have the requisite number of turns take up the minimum amount of space, to sustain the forces set up when large or short-circuit currents flow in them, and to allow space for the cooling medium to transfer the heat created to the tank and atmosphere. They are also designed electrically to produce an optimum combination of I^2R losses and IZ impedance drops (IR resistance and IX reactance voltage drops).

In core-type transformers, the low-voltage winding is usually placed next to the core and the primary winding outside and next to the secondary winding, requiring only one high-voltage insulation between the two windings and the core. In shell-type transformers, the primary and secondary windings consist of a number of pancake-shaped coils connected in series, but the primary and secondary "pancake coils" are placed alternately around the core; the arrangement reduces the reactance between windings and provides better cooling paths for heat dissipation.

Polarity In a transformer, the vector relationship between the primary and secondary voltages depends on the way the windings are wound in relation to each other. To identify this relationship, if both windings are connected in series, as in an autotransformer, the voltage in the secondary can add to or subtract from the primary voltage. Where the primary and secondary voltage vectors are in opposite directions, the transformer is said to have *additive* polarity; where the vectors are in the same direction, the transformer is said to have *negative* or *subtractive* polarity. These polarities are shown in Fig. 11-1.

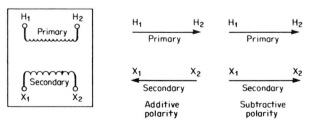

FIG. 11-1 Polarity of a transformer.

Since transformers are connected in parallel with each other, or in polyphase banks, it is essential that the terminals be connected so that the proper voltage relationships ensue. The polarity marks or designations provide the means of accomplishing this. Primary terminals are marked H_1, H_2, H_3, etc., while the secondary terminals are marked X_1, X_2, X_3, etc. See Fig. 4-6a through k in Chap. 4 for transformer connections using these markings. In these standardized designations, the H_1 (primary) and X_1 (secondary) terminals in a transformer having subtractive polarity will be on the *left* when the observer faces the low-voltage bushings; for an additive-polarity transformer, the X_1 (secondary) terminal only will be on the *right* when the observer faces the low-voltage bushings. Standards of the ASA call for distribution transformers rated up to 200 kVA and voltages below 8600 V to have additive polarity, and those above these values to have subtractive polarity; the polarity is usually specified on the transformer nameplate.

DISTRIBUTION TRANSFORMERS

Distribution transformers may be installed on poles, on the ground on pads, and under the ground directly or in manholes and vaults. The transformers used in these types of installations differ mainly in their packaging, as the internal operating features are very much the same.

Overhead Transformers

The overhead type of distribution transformer is mounted directly on a pole by means of two lugs, welded to the transformer tank, that engage two bolts on the pole, as shown in Fig. 11-2a; this is known as *direct mounting*, in contrast to older methods in which the transformer was bolted to a pair of hanger irons that were hung over a cross arm. Where more than one transformer is required, as in power banks, the transformer lugs engage studs on a bracket which is bolted, like a collar, around the pole; the units form a cluster around the pole, from which the term *cluster mounting* is derived; see Fig. 11-2b.

Where the load (weight) of the transformer or transformers may be too great for the pole, they may be placed on a platform erected between two or more poles in a structure, or they may be placed on a protected ground-level pad.

Pad-Mounted Transformers

Transformers may be mounted on concrete pads at, or slightly below, ground level within an enclosure or compartment that may be locked for protection. These are generally installed as part of so-called underground residential distribution (URD) systems, where appearance is a major consideration; refer to Fig. 6-19 in Chap. 6.

The transformers may have their energized terminals exposed when the compartment is open, or the terminals may be mounted behind an insulating barrier and connections from the cables made through bayonet-type connections on insulated elbows which are plugged into jacks connected to the terminals;

these units are referrred to as *dead-front* units and provide an additional margin of safety. Refer to Figs. 6-18 and 6-19 in Chap. 6.

Underground Transformers

In the underground type of transformer, also called the subway type, the tank is not only hermetically sealed for water tightness, but its walls, bottom, and cover are made thicker to withstand higher internal and external pressures; the cover is bolted to the tank (with intervening gaskets) by a relatively large number of bolts, and in some instances, welding is used. These units are designed to operate completely submerged in water.

In larger units, where cooling of the tank itself is not sufficient, radiator fins are welded to the tank to provide additional cooling surface, or pipes are welded to the tank for the circulation of oil through them; in the latter case, the additional surface of the pipes as well as the circulating oil is useful for cooling.

Connections to the supply cables are made by means of watertight wiped joints between a fluid-tight bushing and the cable sheath. Another means provides for the making of connections in a chamber attached to the transformer tank in which the primary-voltage transformer windings are brought out in fluid-tight bushings. In some units, this chamber also houses high-voltage disconnecting and grounding switches.

Where these units supply low-voltage secondary networks, they also house the network protector in another watertight compartment, usually situated at the opposite end of the transformer tank from the primary connection and switch chamber. Refer to Fig. 6-4 in Chap. 6.

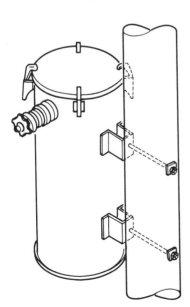

FIG. 11-2 (a) Direct pole mounting of a transformer *(Courtesy Westinghouse Electric Co.)*

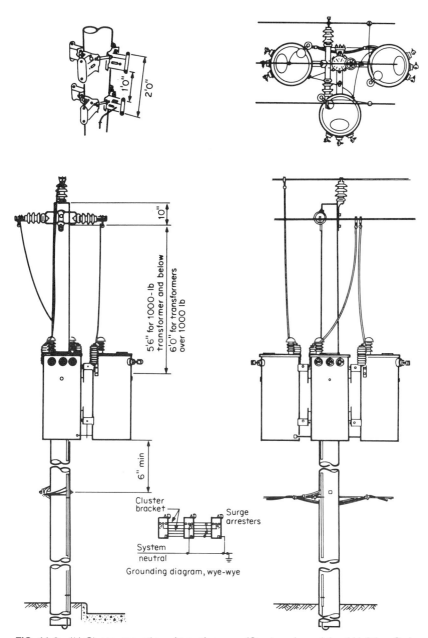

FIG. 11-2 (*b*) Cluster mounting of transformers. *(Courtesy Long Island Lighting Co.)*

Within the figure:

2'0"

1'0"

10"

5'6" for 1000-lb transformer and below

6'0" for transformers over 1000 lb

6" min

Cluster bracket

Surge arresters

System neutral

Grounding diagram, wye-wye

Ratings

Transformers are rated by voltage class and in kVA for capacity. Standard voltage classes are 2400, 4160, 7620, 13,200, 23,000, 34,500, 46,000, and 69,000 V for single-phase and three-phase units, with secondary voltages of 120/240, 120/208, 240/480, and 277/480 V. Standard sizes may include 5, 10, 15, 25, 37½, 50, 75, 100, 167, 250, 333, and 500 kVA for overhead and pad-mounted units, and in addition to these, underground units may also include 667-, 1000-, 1500-, and 3000-kVA units. The units may be both single-phase and three-phase, although the larger sizes are almost always three-phase units.

Transformer Protection

Distribution transformers generally fall into two broad classifications: "conventional," in which the protective devices are mounted externally to the unit; and completely self-protected (CSP), in which protection devices are included inside the transformer tank.

Protection is provided the transformer against overload or failure, and from voltage surges (lightning, switching) that may damage the insulation.

Conventional Transformers   Associated with the conventional transformer are a fused cutout which separates the transformer from the primary in the event of overload or the failure of some component of the transformer, and a lightning or surge arrester which bleeds the high-voltage surge to ground before it has a chance to damage the insulation or other parts of the transformer. Fused cutouts and surge arresters will be discussed later in this chapter.

Completely Self-Protected Transformers   In the completely self-protected transformer, the primary fuse is situated within the case, and because its characteristics are different from those of the external primary fuse, it is referred to as a "weak link." Additional protection from overloads is provided by means of circuit breakers on the secondary side, which coordinate with the weak link; i.e., in the event of overload or fault on the secondary main, the secondary circuit breakers open before the weak link blows. The surge arrester is mounted outside the tank, but connected to its primary terminal. For single-phase units on a wye-grounded system, therefore, only one connection is required to be made to the primary line; this makes for a simpler, neater-appearing, and more economical mounting of transformers. Moreover, these units are also equipped with a warning light which may be seen from the ground and which indicates the unit has been thermally overloaded.

Grounds   In a grounded wye system employing a common neutral for both the primary and secondary systems, the secondary neutral is connected to the transformer tank and the tank grounded. For safety purposes, a second visible ground conductor is connected between the neutral terminal and the neutral conductor of the secondary mains.

Similar protection is provided for ground-level or pad-mounted transformers. Surge protection is usually not required for transformers installed underground, although exceptions are made where such transformers are installed close to underground risers connected to overhead systems; surge arresters may then be installed at the transformer or at the pole, or both.

Single-Phase Transformers

Most of the distribution transformers in service are single-phase units supplying single-phase loads directly or supplying polyphase loads in banks of two or three units. The secondary winding is usually divided into two equal parts, each part having a voltage of 120 V between its terminals. The two parts may be connected in parallel for two-wire 120-V operation, in series for two-wire 240-V operation, or in series for three-wire 120/240 V operation; refer to Fig. 4-6a in Chap. 4. In older units, the four leads from the two parts of the secondary winding were brought outside the tank through insulated bushings, and connections were made outside the tank; in more recent construction, the connections of the parts are made inside the tank and only those leads are brought out that the circuit requires. In many cases, the middle or neutral lead is not brought out through an insulated bushing, but is connected to a stud on the tank which also serves as a means of grounding the transformer tank.

Leads from the primary winding are brought out of the tank through insulated bushings usually made of porcelain of sufficient dimensions to accommodate the intended primary supply voltage. Where the primary supply is a delta-connected or ungrounded wye circuit, the leads employ two bushings. Where the primary supply is a grounded wye circuit, only one lead is brought out through a porcelain bushing; the other lead is connected to the tank and brought out by means of a stud, which may also be connected to the secondary neutral lead, and which also serves for making connection to the distribution-circuit neutral conductor.

(It is most important that care be exercised when connecting or energizing single-bushing transformers, or any transformer with one end of the primary coil connected to the transformer tank, to make *absolutely* certain that the ground connections on the tank of the transformer are made to the system neutral *first*. Failure to do so could lead to energizing the transformer tank at line voltage, jeopardizing the safety of anyone working at or near that transformer.)

Single-phase units may also be used to supply two-phase and three-phase loads from a polyphase primary supply; they may also be used as boosting or bucking transformers in single-phase primary supply circuits; connections are shown in Fig. 4-6a through k in Chap. 4.

The primary windings may also be furnished with taps which permit changes in the transformation ratio to accommodate the need for a fixed raising or lowering of the secondary voltage, or to provide for phase transformation from three-phase to two-phase (or vice versa) as noted previously.

Polyphase Transformers

In polyphase (usually three-phase) transformers, the connections between the phase windings on both the primary and secondary sides are made within the tank of the transformer, and the leads are brought out through porcelain bushings. Whether connected for delta or wye and whether on the primary or secondary side, three leads are brought out on each side. In most instances where a grounded wye circuit is involved, the neutral is brought out through a stud connected to the transformer tank, as mentioned previously for single-phase units. In the relatively few instances where ungrounded wye circuits are involved, the "neutral" or common junction point is brought out through a fourth porcelain bushing.

Taps on the primary side are also furnished to permit the same changes in ratio of transformation as were described for single-phase transformers.

Transformer Cooling

Most distribution transformers, whether overhead, pad-mounted, or subway types, have their cores and coils submerged in insulating oil. The heat produced by the iron and copper losses is carried by convection currents in the oil to the tank and there dissipated into the atmosphere. Excessive temperatures and the formation of hot spots are thus prevented, avoiding damage to the insulation and conductors.

Moisture and sludge formed from the effect of the oxygen in the air on hot oil tend to reduce the dielectric quality of the oil.

Where the use of oil is undesirable, principally because of fire hazard, air-cooled or askarel-cooled units may be employed, although these are generally limited to larger sizes and to transformers installed in vaults. Askarels, although nonflammable, are heavier, less effective coolants, and more expensive than oil; they are also environmentally undesirable, containing carcinogenic PCB compounds, and are extremely irritating to eyes and skin.

Transformer Impedance

Transformer nameplate data also include the impedance of the transformer expressed as a percentage of its nominal rated voltage. This information is important when transformers are connected in parallel, or in polyphase banks, as units of different impedances will not share the load equitably and will cause circulating currents producing unnecessary heating when connected in banks; refer to Chap. 4 for additional discussion.

FUSE CUTOUTS

Distribution transformers (of the conventional type) are usually connected to the primary supply lines through a fuse cutout. The cutout contains a fusible element (a link) whose melting automatically disconnects the transformer from the primary to prevent damage from overloads; it also disconnects the line from the transformer in the event of a fault in the transformer, not only preventing spread of damage in the faulted transformer, but also preventing interruption to the primary supply circuit that would affect other transformers and consumers served from that circuit.

Fuse cutouts are also used to disconnect faulted or overloaded parts of a primary circuit from the remaining unfaulted portion of the circuit.

The size of the fuse is generally based on the size of the transformer or load on the primary section it is to protect.

For voltages of 5000 V or lower, the fuse element, enclosed in a fiber tube, is mounted inside a porcelain box, engaging the two contacts at the top and bottom of the box. The fuse is usually mounted on the door of this box, which is hinged at the bottom; in some models, the melting of the fuse causes the door to drop open as an indication that the fuse has blown. See Fig. 11-3a.

For voltages above 5000 V, the fuse element is mounted in the open between two contacts at either end of an insulating porcelain support, as shown in Fig. 11-3b. The tube containing the fuse drops when the fuse melts, indicating that the fuse has blown. For obvious reasons, this type is known as an open-type cutout.

The design of fuse cutouts should take into consideration the voltages which will be applied to them and the currents that will flow through them. As the currents that may flow under fault conditions may be of rather large magnitude, the fuse cutouts should be strong enough to withstand the resultant forces without damage to themselves or to surrounding objects.

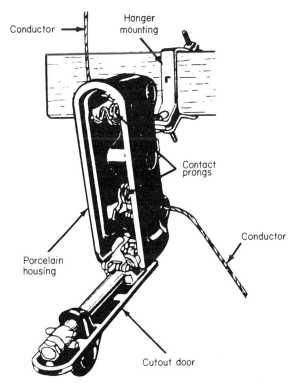

FIG. 11-3 (a) Door-type fuse cutout. *(Courtesy A. B. Chance Co.)*

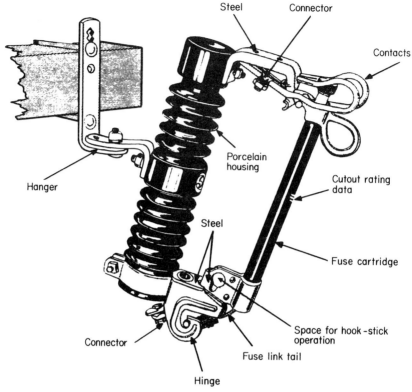

FIG. 11-3 (b) Open-type fuse cutout. (Courtesy A. B. Chance Co.)

Fuses

A fuse is a fusible link that melts when excessive current flows through it, the time of melting varying with the magnitude of the current flowing through it. Further description and discussion of fuse operation are given in Chap. 4.

Fuses for different applications are of several types; those used in connection with the primaries of distribution feeders and transformers are of the expulsion type. These contrast with the open or link types used with secondary voltages and the filled-type fuses of large capacity generally used with higher voltages such as are found on incoming supply feeders at substations.

Expulsion-Type Fuses The fuse link consists of a fusible element with a cap on one end and a flexible stranded cable on the other. The fuse fits into an insulating tube, usually of fiber, with the cap holding it in place on one end and making contact with the nut or fitting at that end; the flexible cable is held taut by a spring or by being wound around a stud and held in place by a knurled nut. The assembly is fitted between the stationary terminals of the cutout.

The minimum time required to melt the link is determined by the composition and dimensions of the link. At the primary voltages involved, however, an arc will persist between the ends of the melted link. The rapid expansion of the surrounding air confined in the tube will tend to blow out the arc; the heat acting on the fiber will also produce a gas under pressure, aiding in extinguishing the arc. The mounting or enclosure of the fuse, therefore, will determine the length of time to extinguish the arc. The clearing time of the fuse will be the sum of these two times, the melting time and the arc-extinguishing time. Hence, the fuses are rated on their voltage class, their normal current-carrying capacity, and a time-current relationship expressed as fast or slow (e.g., type K, type T, etc; refer to Chap. 4).

In the open-type cutout, the expulsion tube may be made to vent at both ends by having the cap at the top made deliberately weak to blow off when high currents produce high gas pressures. Thus, this type of cutout permits the expulsion tube to vent at only one end when fault currents are low, and to "double-vent" when fault currents are high, allowing a higher interrupting rating for the fuse and cutout.

Three-Shot Fuses

When fuses are used to sectionalize a primary feeder, it is sometimes desirable to reenergize the portion disconnected by the blowing of a fuse, e.g., after a fault of a temporary nature. Three fuses are mounted in parallel, but only one fuse is connected in the line. When that fuse blows, as it swings open it trips a latch that places the second fuse in service. Should this fuse also blow, the operation is repeated for a third fuse. If the third fuse should blow, the line remains deenergized.

Safety Considerations

Besides providing adequately for normal and fault currents and for the voltage at which a cutout is to be operated, designs provide for the mounting or cutout to withstand severe climatic conditions, including snow, ice, rain, and severe temperature differences. Protection is also provided against decay, insects, birds, rodents, snakes, etc. Metal parts are made noncorrosive, not only from rust but from corrosive atmospheres. All nonenergized metal parts are bonded together and connected to one or more grounds. The cutouts are also installed out of reach of the general public and are operated with auxiliary devices such as an insulated switch stick handled from the ground.

Oil Fuses

Where very large interrupting capacity is required in a fuse, it is submerged in oil in a hermetically sealed steel tank. Here, the hot gases are not expelled, but are allowed to expand into sealed explosion chambers at the top of the cutout; the remnants of the fuse remain under the relatively cool oil. The magnetic field set up by the large current flowing through the fuse in turn sets up magnetic fields within the tank that aid in snuffing out the arc that forms, the insulating lining of the tank preventing the arc from contacting the tank.

The fuse link is fitted on an insulating frame that fits between the contacts inside the tank. The assembly is so shaped that it can only be inserted into an open position, preventing a premature blowing of the fuse before the assembly is locked in the closed position. Stationary contacts inside the tank are brought out by means of porcelain bushings.

The relatively high cost of oil fuses generally restricts their use to transformers in vaults or underground manholes, and to where voltages are less than 15 kV.

SURGE ARRESTERS

Lightning or surge arresters serve to bleed a high-voltage surge to ground before it reaches the line or equipment which they are to protect. They do this by presenting a lower-impedance path to ground than that presented by the line or equipment. The voltage surge breaks down the insulation of the arrester momentarily, allowing the surge to go to ground and dissipate itself; the insulation of the arrester then recovers its properties, preventing further current from flowing to ground, and returning the arrester to a state ready for another operation.

Types of Arresters

Lightning or surge arresters consist basically of an air gap in series with another element which has the special characteristic of providing a relatively low resistance or impedance to the current produced by a high-voltage surge, and a high resistance or impedance to the flow of power current at the relatively low operating voltage of the distribution line to which it is connected. In some later units, the air gap may be omitted.

Pellet Type In the pellet type of arrester, the second element is made up of a tube full of lead pellets. The lead pellets are actually lead peroxide coated with

TABLE 11-1 STANDARD RATINGS OF SURGE ARRESTERS FOR DISTRIBUTION VOLTAGES

Voltage reference class, kV	Crest voltage, kV*	Voltage reference class, kV	Crest voltage, kV*
1.2	30	15	110
2.5	45	23	150
5.0	60	34.5	200
8.7	75	46	250
12.0	95	69	350

*Basic insulation level (BIL) in kilovolts with a standard 1.5- by 40-μs wave.

Courtesy McGraw Edison Co.

lead oxide. The pellets normally act as insulation preventing current from flowing to ground. When a high-voltage surge is impressed on them, a current will flow that heats them and turns the lead oxide (a poor conductor) into lead peroxide (a good conductor). After the surge is discharged to ground, the surface of the pellets is changed by the discharge current back to lead oxide and restores the arrester to its original condition. Although rapidly becoming obsolete, a great many of this type of arrester exist and will for a long time.

Valve Type   In the valve type of arrester, the second element may be made of some particular substance such as ceramic material containing conducting particles, such as metal oxides (Thyrite and Granulon are commercial names), or other substances having characteristics under surge-voltage conditions similar to those described above. Many of these are built in modular units, several connected in series to accommodate the line voltage impressed on them.

Expulsion Type   The expulsion type of arrester may or may not employ a second air gap enclosed in a tube made of fiber in series with a fixed air gap. As with fuse holders made of fiber, when a high voltage occurs creating an arc across the gap, the heat acting on the fiber gives off a nonconducting gas under pressure that blows out the arc, interrupting the flow of surge current and restoring the arrester to its original condition.

Installation

Surge arresters are installed as close as possible to the equipment or line to be protected so that the resistance of the connection to ground may be held to a minimum. The ground is of the utmost importance, as the arrester will not operate without one. If possible, the arrester should have its own ground, in addition to connections to other grounds.

Since the arrester is to protect the insulation of the line or equipment associated with it, its insulation should be coordinated with that of the line or equipment. This is discussed in some detail in Chap. 4.

Rating

Standard arresters are rated not only on the nominal voltage class of the line to which they are to be connected, but also as to the crest voltage (the basic impulse insulation level) they can withstand; refer to Chap. 4 for more detail. Table 11-1 lists some standard ratings for surge arresters associated with distribution circuits of various voltage classes.

CHAPTER 12

REGULATORS, CAPACITORS, SWITCHES, AND RECLOSERS

VOLTAGE REGULATORS

A voltage regulator is used to hold the voltage of a circuit at a predetermined value, within a band which the control equipment is capable of maintaining and within accepted tolerance values for distribution purposes. Regulators may be installed at substations or out on distribution feeders on poles, pads, or platforms or in vaults.

Voltage regulators are essentially autotransformers, with the secondary (or series) portion of the coil arranged so that all or part of its induced voltage can be added to or subtracted from the line or incoming primary voltage (across which the primary or exciting portion of the winding is connected). The voltage variations are accomplished by changing the ratio of transformation automatically without deenergizing the unit.

Types

There are two types of voltage regulators in use in distribution systems: the induction regulator and the tap-changing-under-load (TCUL), or step-type, regulator. The first is usually limited to circuits operating at 5000 V or less and is being rapidly replaced by the latter, employed where relatively larger amounts of power and higher voltages are involved.

Induction Regulator In the induction type of voltage regulator, the primary (high-voltage) winding and the secondary (or series) windings are so arranged that they rotate with respect to one another (Fig. 12-1). The primary coil is usually the stator and the secondary coil the rotor, the direction of rotation

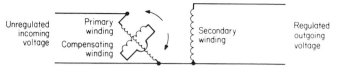

FIG. 12-1 Induction regulator windings.

generally depending on whether the incoming voltage is to be raised or lowered. The voltage induced in the secondary or series winding will depend on the position in relation to the primary winding. Depending on the position, the induced voltage can add to or subtract from the input voltage to obtain the outgoing voltage.

During the rotation of the primary coil, the moving magnetic field can cause a large reactance voltage drop in the secondary. To dampen (or cancel) this effect, a third coil is mounted at right angles to the primary coil on the movable core and short-circuited on itself. The moving primary coil will induce a voltage in the third coil which will, in turn, set up a moving magnetic field of its own, which will tend to oppose that set up by the motion of the primary coil. The reactance of the regulator unit is thus kept essentially constant.

Step-Type (or TCUL) Regulator The TCUL, or step-type, regulator is also essentially an autotransformer, and is connected in the circuit in the same manner as the induction regulator. This type does not employ rotation of one of the coils, but changes voltages by means of taps in the primary coil, as shown schematically in Fig. 12-2. The portion of the coil with taps is a separate part of the primary coil with arrangements included for a reversal in its connection so that the voltage within that portion of the primary coil can be added to or subtracted from the voltage in the rest of the primary coil.

Each tap is changed by the opening and closing of an associated "selector" switch. To avoid disconnecting the transformer from the line each time a tap is changed, the taps are so arranged that two adjacent taps are connected through a small autotransformer each time the tap change is in progress. The midpoint of this "preventive" autotransformer is connected to the primary coil, as illustrated in Fig. 12-3. A small air gap is inserted in the core of the autotransformer to reduce the size of the magnetic field, which could cause an excessive voltage drop in the coil.

Small circuit breakers, known as *transfer switches,* make and break the circuit under oil. The selector switches are always closed while the corresponding transfer switch is open, and opened while the transfer switch is closed. In this design,

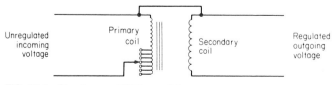

FIG. 12-2 Step-type or TCUL regulator.

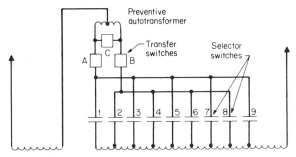

FIG. 12-3 Operation of step-type (TCUL) regulator. *(Courtesy Westinghouse Electric Co.)*

three transfer switches and one selector switch for each tap are required for operation. The transfer switches are often contained in a separate compartment attached to the main tank of the regulator so that they may be maintained without the necessity of draining the oil from the entire regulator unit. The oil in this compartment may be more readily contaminated because of the frequency of the switches' operation.

The sequence of operation of both the selector and transfer switches is shown in Table 12-1. The switches are operated in proper sequence by a motor-operated mechanism which may be controlled manually or automatically. A time-delay device prevents short-duration dips from operating the control relays.

Control

The rotation of the primary coil in the induction regulator and the tap changing in the step-type (TCUL) regulator are controlled by a voltage-regulating relay (earlier known as a *contact-making voltmeter*) connected to the output side of the

TABLE 12-1 SEQUENCE OF OPERATION OF TCUL REGULATOR SWITCHES

	Position								
Switch	**1**	**2**	**3**	**4**	**5**	**6**	**7**	**8**	**9**
Transfer switch A	X X		X X X		X X X		X X X		X X
B		X X X		X X X		X X X		X X X	
C	X	X	X	X	X	X	X	X	X
Selector switch 1	X X								
2		X X X							
3			X X X						
4				X X X					
5					X X X				
6						X X X			
7							X X X		
8								X X X	
9									X X

Courtesy Westinghouse Electric Co.

regulator. Associated with it is a *line-drop compensator,* which is essentially a miniature reproduction of the electric circuit to be regulated and determines the voltage applied to the voltage-regulating relay. A more detailed description of these devices' operation is contained in Chap. 4.

Rating

Both induction and step-type regulators are rated on their nominal voltage classification and their plus-minus percent of voltage regulation. Their capacity, or kVA rating, as a percentage of the volt-amperes transformed is the same as the percent voltage transformed using the incoming primary voltage as a base. This is the same rating used for autotransformers.

For example, if the regulator boosts (or bucks) the voltage 10 percent, it transforms only 10 percent of the load in kVA. If the load to be served is 1000 kVA, the size of regulator required is 100 kVA.

CAPACITORS

Capacitors are also used to improve voltage regulation on distribution circuits, but their operation differs from that of induction and step-type regulators. By introducing capacitive reactance in the circuit, they effectively counteract the inductive reactance of the circuit, affecting its impedance. This, in turn, may cause a voltage drop or rise. It will also tend to improve the circuit power factor, thereby decreasing the current required for a given load and reducing losses in the circuit.

Construction

Capacitors are usually constructed of sheets of aluminum foil separated by liquid-insulating-impregnated paper or other material, wrapped in bundles. The bundles are connected together electrically in series-parallel circuits, and the number of such bundles determines the voltage and capacitance rating of the units. The bundles are contained in a steel tank through which the leads are brought out by means of porcelain bushings.

Operation

Sufficient numbers of units are assembled in a bank of capacitors to provide the required capacitance or reactive power. Capacitors should be installed as close as possible to the inductive load (equipment) so that the current supplied over the circuit will be as nearly as possible in phase with the voltage (unity power factor).

In utility practice, however, capacitors are installed at intervals on the circuit to counter the effect of the inductance of the circuit itself as well as that of the inductive loads connected to it. Capacitor banks may be installed outdoors on poles in racks, on platforms, or on pads. They may also be installed indoors in vaults or other enclosures, and in substations.

Since the loads on a circuit change almost continuously, means are sometimes provided to switch on or off some or all of the units in a capacitor bank, so that the capacitance (approximately) cancels out the inductance of the system. This may be accomplished by time clocks, overvoltage or undervoltage relays, or remote-control devices.

Protection   When installed outdoors, capacitor banks are usually protected by fuse cutouts and surge arresters, somewhat in the way transformer installations are protected. In indoor installations, the banks may be protected by switches or fuse cutouts. Large banks of capacitors, such as may be found connected to substation buses, are usually connected by means of circuit breakers of ample capacity and short-circuit duty to accommodate the switching on and off of such banks, either in their entirety or in part. The employment of capacitors is discussed in greater detail in Chap. 4.

To protect against the failure of one unit in a bank affecting the entire bank, each capacitor unit in the bank is usually fused individually.

Discharge of Capacitors   Some arrangement of resistors or reactors connected to the terminals of the capacitors is often provided for discharging the potentially dangerous energy that remains in a capacitor that has been charged, even after it is disconnected from the line. This phenomenon is known as *dielectric absorption*. Where the capacitors are connected between a phase of the primary feeder to which distribution transformers are connected and ground, discharge facilities may not be required, since the charge in the capacitor will drain off through the transformer winding. The rapidity at which this occurs will depend on the distance and the size and configuration of the conductors between the capacitor and the transformer.

Series Capacitors

Capacitors may be connected in series with the line to compensate for reactive voltage drop in the circuit and provide an instantaneous and almost perfect voltage regulation. Their use is generally limited to low-voltage heavy-current application, such as welders and furnaces. The units so employed may be the same as those used for shunt operation.

Rating

Capacitor units are rated according to their nominal voltage class (e.g., 2400 or 7620 V) and their kVA rating: 15, 25, 50, 100, 150, 200, and 300 kVA. Banks of such units, both for single-phase and three-phase operation, are assembled to provide the required capacitance.

Maintenance

Capacitors require little maintenance beyond occasional inspection for blown fuses, cracked bushings, leaking tanks, and the accumulation of dirt or other pollutants (severe smoke, salt fog, chemical fumes, etc.) on insulating surfaces. Maintenance is often accomplished by replacing only the unit in the bank requiring attention.

SWITCHES

In addition to the fuse cutouts and potheads mentioned in Chap. 9 and 11, other devices are used for connecting or disconnecting circuits or portions of them.

Disconnects

Disconnects are switches designed *not* to be opened when any amount of load current flows through them. Their use is generally limited to places where no load current or only a small charging current is to be interrupted. They are usually installed in a circuit so as to enable a line or a piece of equipment to be isolated from the energized portion of the circuit; they provide a visual break in the circuit as a positive safety measure for the worker. On the other hand, when used as normally open devices, they may be closed to energize a line or piece of equipment where only charging current will flow; in some instances, some relatively large load current may also flow when the disconnect is closed. In either case, the disconnect should be closed firmly to prevent the possibility of arcs forming between the blade and terminal clip as the blade approaches the (energized) clip.

Disconnect switches, air-break switches, and oil switches are sometimes gang-operated; i.e., the three single-pole disconnects (for a three-phase system) are mechanically connected and operated together.

Air-Break Switches

Air-break switches are generally used to interrupt relatively small amounts of load current, such as in sectionalizing primary feeders or interrupting the exciting current of large transformers or a group of smaller transformers. The switches may be closed to pick up loads, but extreme care must be used in opening a circuit carrying load.

Air-break switches are essentially disconnect switches equipped with so-called arcing horns. These are metal rods attached at an angle to the stationary terminal of the switch. As the blade of the switch is opened under load, the arcing horns remain in contact with the blade until after the main contacts have separated. As the blade continues to travel, an arc will form between it and the arcing horn. The arc will lengthen as the blade continues to travel, until it can no longer sustain itself and becomes extinguished. Pitting or burning from the arc occurs on the horn and part of the switch blade, where it can be tolerated.

Auxiliary devices are sometimes employed to increase the load-interrupting capability of such switches. One scheme interposes a low-capacity high-interrupting-duty fuse between the stationary contact and the moving element or blade; the fuse interrupts the circuit safely while the switch blade is still in place.

Construction   Air-break switches are built rather ruggedly not only because they may have large fault currents flowing through them, but also because they are exposed to the weather. They are designed to operate under severe weather variations, with special attention paid to prevent contacts and hinges from "freezing" in position in cold, snowy, and icy weather.

Oil Switches

Operation   Where load currents are to be interrupted relatively often, oil switches not designed to interrupt fault currents are used. Here, the switch is opened under oil, the oil serving to quench the arc that forms. Usually, however, a fuse is connected in series with the switch to clear the circuit should the switch fail to interrupt the circuit. Oil switches of this type, such as double-throw switches, are often used to transfer a primary service from one feeder to another and may be operated manually or automatically.

While the oil switch is more dependable than the air-break switch, it is more costly and its use is generally limited to those applications where other means are unsatisfactory.

Rating   Like disconnects and air-break switches, oil switches are rated on their voltage classification and their normal current-carrying capacities. Inasmuch as fault current may flow through them, their fault-current-carrying capabilities are often included in their specifications.

CIRCUIT BREAKERS

Types

A circuit breaker generally is an oil switch but is built more ruggedly to enable it to interrupt not only the relatively large load currents but also the much larger fault currents which may occur on a circuit. Some circuit breakers, however, are designed to open in air, with special provisions for handling the arc that follows when the contacts are opened. These are known as air circuit breakers, and their interrupting capability is usually much lower than that for oil circuit breakers. Other circuit breakers are designed to have their contacts open in a vacuum or in an ambient of inert gas such as sulfur hexafluoride (SF_6), with improved interrupting capability claimed for these latter types.

Operation

All types of circuit breakers are designed to operate automatically in opening the circuit under fault conditions, or to be opened or closed manually when desired.

Circuit breakers are generally installed at substations on the ends of primary feeders. They may be used elsewhere, however, where heavy short-circuit currents must be interrupted and cannot be handled by other means such as fuses. Such locations, for example, are at-large consumers served at primary voltages.

Circuit breakers are usually actuated by overcurrent- or fault-sensing relays which serve to open them. This is done by tripping a compressed spring. Opening may be achieved in from 2 to 60 or more cycles, including the extinguishing of the arc, from the time the relay contacts are made. Relays also control their reclosure, which is accomplished by means of solenoids or electric motors. Circuit breakers may be set for a single operation, or for multiple reclosing operations before "locking out."

To insure their operation, the operating power source for the operation of the relays and the circuit breakers may be storage batteries floated on the line through a rectifier so that they are continuously fully charged.

Further discussion of circuit breakers may be found in Chaps. 4, 7, and 13.

Rating

In addition to their voltage classification and current-carrying capacity, the interrupting capability is a most important part of the circuit breaker rating.

RECLOSERS

Reclosers are essentially circuit breakers of relatively lower normal current- and short-circuit current-carrying capability. Their overcurrent- and fault-sensing devices and reclosing controls are a part of the unit and are contained within it.

Operation

The coils operating the closing mechanism obtain their power from the source side of the recloser. These coils also operate a mechanism that compresses the springs, which, when tripped, open the circuit breaker contacts rapidly. Another coil, in series with the coil operating the closing mechanism, trips a latch that releases the compressed springs to open the circuit breaker. A relay reenergizes the coil operating the closing mechanisms, which automatically close the circuit breaker. A time delay inserted in the relay circuit permits the reclosing of the breaker a number of times (usually three) before locking it open. The reclosing feature may be switched out of operation when only a single operation is desired (e.g., when workers may be working on the circuit). The unit may also be operated manually.

Reclosers may be installed out on one or more branches of a circuit so that a fault on those branches need not affect the entire circuit. Temporary faults (such as a tree limb falling on the line) will not cause a lengthy outage to service on the branch, but only a momentary interruption.

Reclosers may sometimes be installed at substations as circuit breakers on distribution feeders where loads are relatively small and likely short-circuit or fault current low enough not to exceed the rating of the recloser. Such conditions may often be found in rural areas.

Rating

Recloser units may be single-phase or polyphase, and are rated according to voltage class, normal carrying current, and fault-current interrupting capability. Additional discussion may be found in Chap. 4.

CHAPTER 13

DISTRIBUTION SUBSTATION EQUIPMENT

EQUIPMENT

The principal equipment usually found in a distribution substation includes power transformers; circuit breakers and their associated protective relays and control devices; high-voltage fuses; air-break and disconnect switches; surge or lightning arresters; voltage regulators; storage batteries; measuring instruments; and, in some instances, capacitors and street lighting equipment. Associated with all of these are cables and buses and their supports.

Many of these items have been discussed previously in connection with their functions in the planning and design of distribution systems and in their construction and installation; this discussion may generally be found in Chaps. 2, 4, 5, 6, 7, 11, and 12. Discussion in this chapter will focus on those characteristics of these same items which were not discussed previously but which are associated with the activities of the distribution engineer.

TRANSFORMERS

Substation transformers function in a manner similar to distribution transformers, but have significant differences in their construction and operation. Features shared by both categories include the usual employment of oil (sometimes air or askarels) for insulating and cooling purposes, taps for changing the ratio of transformation, and insulation coordination together with basic insulation levels. Transformers may be of the single-phase or three-phase types.

Core

Substation transformers, and especially the polyphase units, are usually constructed with shell-type cores which surround the windings, as compared with the usual distribution transformer, in which the windings surround the core.

Polarity

Substation transformers are usually wound for additive polarity, in accordance with EEI, NEMA, and other standards, as contrasted with the subtractive polarity of distribution transformers.

Bushings

The bushings of substation transformers on the low-voltage side are usually made of porcelain and are similar to the primary bushings found on distribution transformers, but of greater current-carrying capacity. The high-voltage-side bushing, however, in addition to the greater current-carrying capability, depending on the magnitude of the voltage, may consist of a solid porcelain cylinder (with petticoats) as insulation for voltages up to about 35 kV, or an oil-filled hollow porcelain cylinder for values up to about 69 kV; for 69 kV and higher voltages, the hollow porcelain cylinder may contain layers of oil-impregnated paper insulation with metal foil inserted at several locations among the layers, forming a series of capacitors which serve to even out and equalize the electrostatic stresses set up within the bushing; see Fig. 13-1. Other high-voltage bushings may be filled with an inert gas such as sulfur hexafluoride (SF_6).

Like distribution transformers designed for line-to-ground operation, single-phase transformers may have only one bushing on both the high- and low-voltage sides; three-phase transformers may have only three high- and three low-voltage bushings, with a neutral stud as a common terminal for both high- and low-voltage windings.

Oil

Substation transformers may also show evidence of greater precaution taken in keeping air and moisture from the oil. In some units, an inert gas, such as nitrogen, fills the space above the oil, and the transformer tank is sealed. A "relief diaphragm" is sometimes installed in a vent in the sealed transformer which ruptures when the internal pressure exceeds some predetermined value, indicating possible deterioration of the insulation. In some units, a pressure relay is installed to give an indication of pressure rise in the tank.

Some outdoor units are equipped with a tank on top of the transformer, called a *conservator*, in which the expansion and contraction of the oil takes place. The tank is sometimes open to the atmosphere through a breathing device. The condensation of moisture and the formation of sludge take place in the tank, which is also provided with a sump from which the condensation and sludge may be drawn off. See Fig. 13-2.

Cooling

Substation transformers may be equipped with fins or radiators to enhance the ability of the transformers to dissipate the heat generated by their copper and iron losses. Both the fins and the radiators increase the surface area transferring heat to the atmosphere. The radiator, in addition, increases the natural circulation of oil by the convection currents set up within the unit.

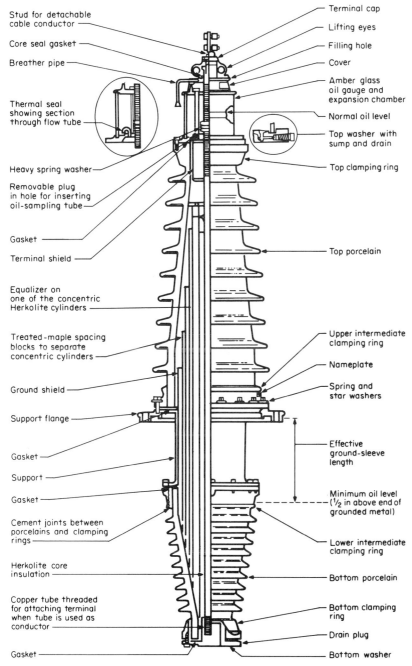

Stud for detachable cable conductor

Core seal gasket

Breather pipe

Thermal seal showing section through flow tube

Heavy spring washer

Removable plug in hole for inserting oil-sampling tube

Gasket

Terminal shield

Equalizer on one of the concentric Herkolite cylinders

Treated-maple spacing blocks to separate concentric cylinders

Ground shield

Support flange

Gasket

Support

Gasket

Cement joints between porcelains and clamping rings

Herkolite core insulation

Copper tube threaded for attaching terminal when tube is used as conductor

Gasket

Terminal cap

Lifting eyes

Filling hole

Cover

Amber glass oil gauge and expansion chamber

Normal oil level

Top washer with sump and drain

Top clamping ring

Top porcelain

Upper intermediate clamping ring

Nameplate

Spring and star washers

Effective ground-sleeve length

Minimum oil level (½ in above end of grounded metal)

Lower intermediate clamping ring

Bottom porcelain

Bottom clamping ring

Drain plug

Bottom washer

FIG. 13-1 Typical oil-filled bushing for a 69-kV transformer. *(Courtesy General Electric Co.)*

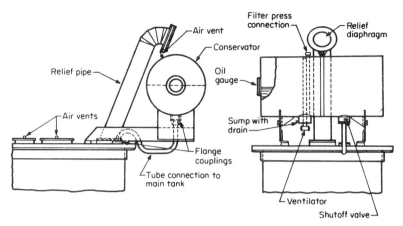

FIG. 13-2 Diagram showing main features of a conservator. *(Courtesy General Electric Co.)*

This cooling capability may be further increased by fans blowing against the fins and radiators, increasing the rate of heat transfer to the surrounding atmosphere. Further cooling may be obtained by pumps forcibly increasing the rate of circulation of oil in the radiators. Both means are often used together. In some cases, resort is had to water cooling by means of pipes installed within the transformer tank through which water is circulated, or by means of an external heat exchanger through which the hot oil and water are separately circulated with the aid of pumps.

Rating

In addition to the voltage classification, substation transformers may have several ratings expressed in kVA, each rating associated with the type of cooling employed: a normal rating with no added means of cooling; a higher rating with forced air or forced oil circulation; and a still higher rating when both means are used in combination; for example, 10,000 kVA; 12,000 kVA-FA; 15,000 kVA-FA/FO. Ratings may also include permissible noise levels at maximum loads, expressed in decibels at standard distances from the unit.

Impedance

The impedance of substation transformers is usually higher than that of distribution transformers in order to hold down the current that may flow during a fault on the system connected to its low side.

CIRCUIT BREAKERS AND PROTECTIVE RELAYING

Protective devices have been described in Chaps. 4, 7, and 12. The same considerations that apply to the types of bushings for the several levels of voltage for transformers on their high-voltage side also apply to circuit breakers. Current transformers for relaying, metering, or other purposes may also be included in the bushing.

Circuit breakers may be single-phase or three-phase; these may consist of a single pole within an individual tank, or the three poles may be contained in a single tank.

Ratings of circuit breakers include their voltage classification in kV, their normal current-carrying capacity in amperes, their short-circuit- or fault-current interrupting capability in amperes or kVA, and, in some instances, their speed of opening, in cycles.

Metal-Clad Switch Gear

Modular metal-clad switch gear is constructed to incorporate circuit breakers, disconnecting devices, interlocks, measuring instruments, current and potential transformers, relays, and buses into a single, compact, factory-assembled unit or module for each circuit; a number of these units are capable of being assembled together.

Beside the compactness, the modular construction provides for flexibility in switching arrangements and for ready, inexpensive future expansion. With no live parts exposed and with interlocks to prevent energized switch gear from being drawn out, protection for both worker and equipment is enhanced. Connections into and out of this type of switch gear can be made via underground cables, making for neat appearance.

Ratings of metal-clad switch gear are determined by its current-breaking capability:

Voltage classification	Normal current capacity	Fault-current interrupting duty
5 kV	2000 A	50,000 kVA
15 kV	2000 A	500,000 kVA
34.5 kV	2000 A	1,500,000 kVA

Units above these ratings may be tailor-made for specific installations.

FUSES

Fuses associated with distribution transformers and primary circuits have been described in Chaps. 4, 5, and 11. Their time-current characteristics and their coordination with each other and with reclosers and circuit breakers have been discussed.

High-Voltage Fuses

In some smaller or rural substations, where possible short-circuit currents may be limited to relatively low values, a high-voltage fuse may be substituted for the incoming circuit breaker. While its operation, including coordination with other protective devices on the same circuit, is the same as for distribution fuses,

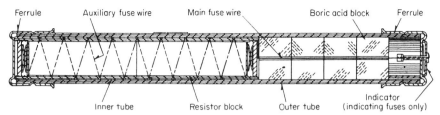

FIG. 13-3 Cross section of boric acid-type expulsion fuse. *(Courtesy Westinghouse Electric Co.)*

its construction is greatly different from that associated with door- and expulsion-type cutouts. The high-voltage fuse links are usually enclosed in a container, which tends to suppress and confine the arc and vaporizing metal, making the fuse capable of interrupting moderate short-circuit currents at the higher voltages.

Liquid-Filled Fuses In liquid-filled construction, the fuse link is enclosed in a tube that is filled with a fire-extinguishing fluid such as carbon tetrachloride. A spring holds the fuse under tension so that, when it blows, the resultant arc is quickly lengthened and quenched in the fluid; the gas formed is inert and aids in blowing out the arc. If the pressure becomes excessive, provision is made to have the cap on the "outgoing" side of the fuse blow off, preventing the rupture of the tube and confining damage to the fuse itself.

Solid-Filled Fuses In solid-filled construction, the material surrounding the fuse element may be sand or powdered boric acid. In the sand-filled type, the heat and gases generated when the fuse melts are absorbed by the sand, which tends to squelch the arc. In the boric acid type, the heat produced decomposes the boric acid, which forms steam under pressure that acts to squelch the arc. See Fig. 13-3.

Ratings

High-voltage fuses for this type of application are rated not only for their voltage classification and normal current-carrying capacity, but also as to their short-circuit-current interrupting capability.

DISCONNECT AND AIR-BREAK SWITCHES

Disconnects are rated by voltage classification in kV and normal current-carrying capacity in amperes. They are not rated to interrupt any current. These have been discussed in Chaps. 4, 5, 7, and 12.

Also discussed in these chapters have been air-break switches. They may be single-pole or three-pole gang-operated. They are also rated by their voltage classification and normal current-carrying ability, but, in addition, have a current-interrupting capability expressed in amperes.

SURGE OR LIGHTNING ARRESTERS

Surge arresters may be of the valve or expulsion type and have been described in Chap. 6. They are rated not only on their normal voltage classification in kV, but also on their crest voltage capability in kV at a standard 1.5×40-μs wave (or other specified wave), and their discharge-current capability in amperes or thousands of amperes (kA). For high-voltage application, surge arresters may consist of a number of unit-value valve arresters connected in series in one overall unit, as shown in Fig. 13-4 for a 69-kV arrester.

VOLTAGE REGULATORS

Voltage regulators may be of the induction type or the TCUL or step type, and have been described in Chaps. 4, 7, and 12. A cross section of an induction regulator is shown in Fig. 13-5. Regulators are rated by voltage classification in kV and the plus-minus percent voltage-regulation capability. In step-type regulators, the percent voltage of each step and the number of steps is often also

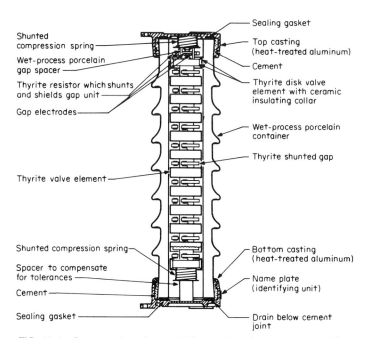

FIG. 13-4 Cross section of a 69-kV Thyrite (valve) surge arrester. *(Courtesy General Electric Co.)*

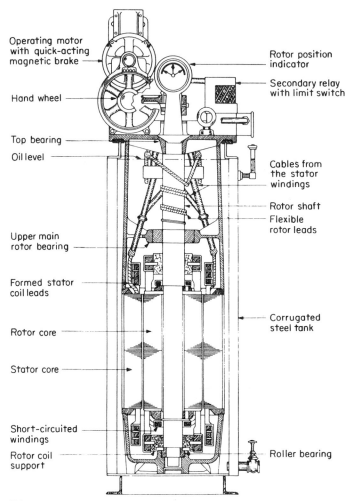

Operating motor
with quick-acting
magnetic brake

Rotor position
indicator

Secondary relay
with limit switch

Hand wheel

Top bearing

Oil level

Cables from
the stator
windings

Rotor shaft

Flexible
rotor leads

Upper main
rotor bearing

Formed stator
coil leads

Corrugated
steel tank

Rotor core

Stator core

Short-circuited
windings

Rotor coil
support

Roller bearing

FIG. 13-5 Cross section of typical induction voltage regulator. *(Courtesy Westinghouse Electric Co.)*

specified. The current or kVA capability of the regulators is the same as the maximum voltage change times the full load rating of the circuit in which they are connected.

STORAGE BATTERIES

Storage batteries constitute a reliable source of dc energy to the control systems in substations. Such control systems may operate at nominal voltages of 6, 12, 24, 48, or 120 V, although the latter three are the preferred values. Since the batteries are made up of a number of individual cells, each producing approximately 2 V, a number of them are connected in series-parallel circuits to provide the required voltage and capacity. These have been mentioned in Chap. 7 in connection with the operation of circuit breakers and their associated relays.

The most frequently used batteries are the so-called lead-acid type and the nickel-alkaline type (sometimes also called the Edison battery). The first type employs plates made of lead and sulfuric acid as the electrolyte; the second, plates made of nickel oxide and iron or cadmium, with potassium hydroxide as the electrolyte. Separators are made of treated wood, glass wool, porous rubber, or plastic material. Containers for the lead-acid cells are usually made of glass, rubber, plastic, or asphaltic compositions, with ports or vents for exhausting of gases and renewal of water or acid. Containers for the nickel-cadmium alkaline cells are nickel-plated sheet steel, hermetically sealed, but with valves to allow the venting of gases; the electrolyte needs no replacement.

Storage batteries are usually rated in ampere-hours and their capacity on the basis of an 8-h rate. Battery efficiencies run from 85 to 90 percent, with the lower efficiencies experienced at the more rapid rates of discharge and charge.

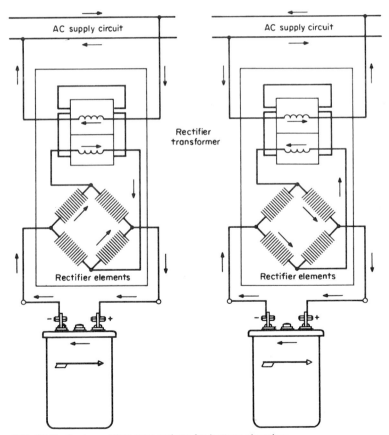

FIG. 13-6 Basic rectifier connections for battery charging.

Batteries are permanently connected to a dc source usually supplied through small copper oxide or selenium rectifiers from an ac supply, and connected as shown in Fig. 13-6.

MEASURING INSTRUMENTS

Measuring instruments usually include ammeters, voltmeters, watthour meters, and kilowatt demand meters; wattmeters, reactive volt-ampere, and power factor meters may also be found in some installations. These have been generally discussed in Chap. 8, and as they are universal types of instruments they will not be described further. Associated current and potential transformers have been described elsewhere.

CAPACITORS AND STREET LIGHTING EQUIPMENT

Capacitors and lighting equipment have been described in Chaps. 4, 5, and 12 and need no further discussion.

BUSES AND BUS SUPPORTS

Buses and their supports have been discussed in Chap. 7 and need no further discussion.

ALL SUBSTATION EQUIPMENT

For current information and further details on substation equipment, reference should be made to manufacturers' instruction books and catalogs. Consultation with manufacturers' and suppliers' representatives and engineers will prove most valuable.

PART FOUR

OTHER DESIGN CONSIDERATIONS

CHAPTER 14

NONTECHNICAL CONSIDERATIONS

INTRODUCTION

The maintenance of good employee, consumer, and community relations cannot be divorced from the planning and design of distribution systems. Indeed, along with the other functions connected with the utility's business, these must be considered an integral part of the distribution engineer's responsibility. The distribution system involves not only the utility's employees, but contact with every consumer at some stage in the process of providing and maintaining electric service. Moreover, electricity is intimately connected with the life of a community as a whole. The distribution engineer, therefore, must be ever conscious not only of the obligations of a fellow employee and a good neighbor, but also those of a good citizen.

The technology and expertise which the distribution engineer brings to the problems of safety, quality of service, and economy must also be extended to include those involving nontechnical considerations. While the previous chapters deal with the resolution of technical questions employing the natural laws of science and mathematics, the solution to nontechnical problems requires the employment of heavy dollops of common sense.

SAFETY

Safety to the employees and the public (who are, after all, the distribution engineer's neighbors) has been previously discussed in terms of electrical and mechanical technical considerations (e.g., selection of utilization voltages, adequate strength of poles and conductors, etc.). In addition, however, several other considerations should be taken into account.

Accessibility

As much as practical, the utility's facilities have to be both accessible (to the workers) and inaccessible (to the public). For example, a distribution pole line should be so located that free and easy access to the facilities is available at all times, yet it should not interfere with pedestrian and vehicular traffic, nor intrude into areas (such as playgrounds) where its presence may constitute a particular hazard. This accessibility and inaccessibility should prevail even under adverse or contingency conditions.

Moreover, the conductors and equipment on the poles should be so situated that they can be handled safely by the people working on them. This not only implies providing sufficient working space, but includes considerations of how the work may be performed safely. Safe methods include the use of protective equipment (such as rubber gloves and sleeves, line hose, etc.), the use of live-line tools and equipment, and the deenergization and grounding of the facilities on which work is to be performed. This last, in turn, may include the design of additional facilities to maintain service, as much as practical, to the consumers involved.

The continual development of new and better tools and equipment is also reflected in the design of distribution lines. For example, the development of bucket trucks and plastic insulation resulted in "armless" construction wherein cross arms are eliminated and insulated conductors are spaced closer together chiefly to obtain a more desirable appearance.

Horizontal and vertical clearances between the power line and adjacent structures must take into account the possibility of contact with television and radio antennas, flagpoles, and other encumbrances that may exist on, or may later be added to the structures rather than being limited only by technical or code requirements.

Clearances are specified in the National Electric Safety Code, and in local rules and regulations. These, it must always be remembered, are only recommendations for minimum values. The judgment of the distribution engineer can, and must, supersede them when other factors make exceeding these values desirable, if not imperative.

Maintainability

Each item selected, and its place in the distribution system, must be viewed in light of its possible failure or malfunction for whatever reason. Provision, therefore, must be made in the design for adjustment, repair, revamping, or even replacement, should that occasion arise. The design of the distribution system, therefore, should take into consideration the method of maintaining each of the several elements making up the distribution system.

The Distribution System In general, the safest means is to be able to deenergize and ground the particular item requiring maintenance, preferably without affecting the remainder of the circuit. Circuits, both primary and secondary, are arranged so that small sections may be deenergized by interconnecting the remaining portions to other sources, by means of some sort of switches. Smaller pieces may be deenergized by means of hot-line or live-line clamps.

Where deenergization is not practical, work may be carried out by insulating the worker. This is accomplished by protective gear, such as rubber gloves, sleeves, blankets, line hose, insulator hoods, and other, similar devices. Another method insulates workers from ground by having them work from insulated platforms or insulated buckets mounted on line trucks. Still another means involves the handling of the energized facilities by tools having sufficient insulation properties which, properly handled, enable the worker to accomplish required maintenance by essentially a remote operation of the tools; this is referred to as live-line, or hot-line, maintenance. To facilitate such operations, appropriate details and modifications are included in the design of distribution systems, e.g., wider spacing of conductors; hot-line ties that hold conductors to the insulators; "unnecessary" extensions of primary and secondary mains so that mains butt each other, permitting their temporary connection during contingencies by means of jumpers; the arrangement of terminals at substations to accommodate portable substations; and the limitation of buried cables in URD systems to enable their bypass by an emergency length of cable laid on the ground during cable failures.

Equipment Design for maintainability also extends to individual pieces of equipment installed on the distribution system. For example, in both substation and distribution transformers, provision is made so that taps can be changed through a hand hole in the cover of the transformer (while it is deenergized), bushings can be repaired or replaced, or oil changed, without the necessity of removing the transformer to the shop for dismantling.

QUALITY OF SERVICE

There is more to the quality of electric service to the consumer than the electrical and mechanical considerations described in previous chapters. The selection and maintenance of proper voltage, of materials and equipment of ample capacity, and the maintenance of frequency within very rigid limits all contribute to the quality of electric service rendered the consumer. Equally, if not more, important, however, are service continuity and environmental considerations, which also play a large part in the high standard of quality established by the power suppliers in this country.

Continuity of service may be affected by the nature of the area served. For example, if the facilities are to be installed in a heavily treed area, initial and periodic tree trimming may be required, and armless design with insulated conductors, or the installation of an overhead sheathed cable, may also be specified in order to have a minimal effect on the trees. Indeed, if the aesthetic value of the trees is prized by the consumers in the area, good consumer and community relations may dictate consideration of underground construction.

Similarly, areas exposed to severe storm or wind conditions, or to lightning displays, may require additional considerations, as will areas of atmospheric pollution such as concentrations of salt spray or chemical contamination. Also, the proximity of airports and military bases may introduce additional sources of hazard that may require relocation or undergrounding of facilities.

Occasionally, underground installations may also require remedial or preventive considerations. For example, chemical action from minerals in the soil, and from electrolytic action caused by stray direct currents from railroad return circuits or by those generated by the presence of dissimilar elements (usually metals) either in contact with each other or immersed in soil chemicals acting as electrolytes, may be such as to require special considerations. Some have been mentioned in the previous chapters, but relocation of facilities or their installation overhead may be the individual solutions to such problems.

Moreover, both overhead and underground systems in particular areas may be subject to the effects of hurricanes, tornadoes, floods, earthquakes, landslides, and other destructive acts of nature. The incidence of their occurrence may substantially affect the planning and design of distribution systems in those areas.

In addition to the environmental problems brought about by nature, other environmental problems, usually concerning appearance, may affect final plans and designs. These may arise from the desires of single consumers, from groups of consumers, or from the community as a whole. In areas where property values may be adversely affected by the appearance of utility facilities, alternate plans and designs may need to be considered. For example, installation of overhead facilities along rear-lot property lines to improve appearance is sometimes considered, although access to the facilities, while not restricted, may be hampered. Armless construction has also been discussed. Other measures may include the substitution of concrete or metal poles for wood poles, and the coloring of poles (wood, concrete, or metal) and visible equipment (transformer and capacitor tanks, insulators, etc.) to blend with the surroundings, or with the sky.

Community pride in government buildings, civic centers, historical structures, and landmarks may call for special attention and treatment, which the distribution engineer must recognize; the engineer might also seek to participate in the determination of preservation measures.

The interests and desires of members of a community are often reflected in ordinances and regulations promulgated by their representatives on appropriate boards and councils of authority. In many cases, the statutes specify the community's relationship with the electric utility serving it and, more specifically, the design of facilities to be installed. Revenues from the community may or may not justify their demands, but the distribution engineer has little choice but to modify the designs to meet the requirements.

ECONOMY
Operating Modifications

In the planning and design of distribution systems, the economics of the cost of installation (including materials, equipment, labor, engineering, and overhead) to serve a given load are generally determined by Kelvin's law; i.e., that combination of designs and equipment whose annual carrying charge is equal to the annual cost of losses sustained therein usually indicates the economic preference among the several designs under consideration. The designs are often modified by the distribution engineer to provide spare capacity to serve future load growth. The amount of additional expenditures that can be justified may be calculated by determining the future worth of present expenditures (or vice versa). This, however, may be based on many assumptions of which the distribution engineer may be unaware; therefore the engineer must use discretion and experience in selecting the proper ones for study.

As indicated above, the conclusions reached from economic considerations may be substantially modified because of special conditions that interpose themselves for geographic, seasonal, or environmental reasons.

Standardization

Even after all of these factors have been taken into account, the use of standardized sizes, kinds, and types of materials and equipment again skews the economics and may seriously affect the cost-effectiveness of the final design and construction.

Standardization, however, permits substantial economies to be realized. It enables quantity purchase of materials at lower unit costs as well as the purchase of fewer items and smaller overall amounts to be stocked; it also allows fewer and more specialized tools to be employed. It makes for the interchangeability of materials, equipment, and tools; and because such standardization leads to standardized construction and maintenance practices, it also permits the interchangeability of personnel.

Indeed, such standardization of materials and work practices also tends to result in greater safety to and productivity from the work force. Training can be more effective if concentrated on fewer and more repetitive tasks. In addition to humanitarian considerations, these contribute to more efficient and economical operations.

The utilization of such standards may disturb the economic optimization sought, as its effect on cost-effectiveness may be negative for an individual project. On overall balance, however, such standardization results in positive improvement in the economics of distribution design.

Civic Improvement

Another instance in which the discretion of the distribution engineer is exercised is that in which projects of reinforcement, rebuilding, or modernization of distribution facilities cannot be ordinarily justified economically. Advantage is taken, however, of other activities, usually associated with civic improvements such as road widenings, sewer, water-line, and subway construction, beautification programs, and similar projects, which may necessitate the handling, making safe, or temporary relocation of the distribution facilities. Here, logic would indicate the undertaking of changes that would otherwise not have been justified from a strictly economic viewpoint.

CONCLUSION

These are only a few examples in which recommendations based solely on tangible technical considerations may be overridden by nontechnical considerations influencing the design of the distribution finally selected.

The distribution engineer, therefore, must not only be aware of new and improved materials and equipment, but must be cognizant of changes in codes, regulations, and construction and maintenance standards and practices. Many of these are brought about by the development of new methods as well as new materials and equipment; many are also influenced by the changing public expectations resulting from changing social values and economic conditions. Indeed, the engineer must not only keep abreast of such changes, but should be an active contributor to them.

In selecting each item and designating its place in the distribution system, the engineer must examine step by step its impact on safety, service quality, and economic results before making a final determination and issuing orders to the field.

CHAPTER 15

OPERATING CONSIDERATIONS

INTRODUCTION

There are yet other requirements which the design of distribution systems must meet, other than those of meeting the consumer's and community's needs and desires. The additional requirements, in the main, have arisen from the changing national economic and energy situations. Collected under the general subject of operating requirements are the installation and arrangement of facilities to achieve a better quality of service, but also a more efficient distribution system and a more economical overall electric system from the generating plant to the consumer's premises.

The operations may be classified into four specific functions and may be listed under simplified headings:

1. Quality of service
2. Load shedding
3. Cogeneration
4. Demand control (or peak suppression)

These are somewhat interrelated, and all involve the distribution system.

QUALITY OF SERVICE

Operations involving the improvement of quality of service to the consumer have been discussed in previous chapters, and include measures to:

1. Isolate faults and restore service to the unfaulted portion of the distribution system

2. Transfer loads between phases or between circuits to relieve overloads or potential overloads, and improve voltage conditions

3. Switch on and off capacitors installed out on the distribution feeders (and in the substations) to improve power factor, reducing the value of current flowing and releasing capacity of distribution facilities, with resultant voltage improvement

4. Enable portions of the distribution system, including the substation portion, to be deenergized for construction and maintenance purposes without affecting the remaining portion of the circuit

Designs of distribution systems include provisions for carrying out these operations by means of suitable circuit extensions and switches. With the development of electronic and miniaturized systems of control and communication, many of these operations, generally performed manually and sometimes with a significant lapse of time, may now be performed automatically almost instantaneously.

LOAD SHEDDING
Why

The need for load shedding stems from two general causes, usually unforeseen:

1. Lack of sufficient power supply

2. Lack of sufficient transmission or distribution load-carrying ability

These conditions may come about from:

1. Load growth faster than the construction of new facilities can be accomplished

2. Abnormally high unforeseen demands that are created by unusual seasonal changes or by some special events that cause a significant loss in diversity of consumers' loads

3. Failure or overload in some element or elements of the supply facilities; e.g., transmission line failure, substation transformer failure, etc., for a prolonged period.

How

In this context, load shedding implies decreasing the load on the substation bus, substation transformer, or incoming transmission line. This may be accomplished in two basic ways:

1. Voltage reduction

2. "Brownout," or periodic disconnecting of feeders for relatively short periods of time on a predetermined schedule

In rare instances is resort had to the employment of both of these methods at the same time.

Voltage Reduction   Voltage reduction may be accomplished by manipulating voltage regulators at the substation on individual feeders, or on the substation bus, where the substation's voltage is so regulated. Voltages may be reduced by steps according to the amount of load shedding required. In circuits that supply essentially lighting or unity power factor loads, a 1 percent voltage drop results in almost a 1 percent drop in load.

Steps may be 1 or 2 percentage points each to a maximum of about 8 percent; more often, however, voltage is lowered in two steps, (say) 5 percent and 8 percent.

Lowering voltage beyond this 8 percent value may prove self-defeating as light output from incandescent lamps decreases to the point where additional lighting may be turned on. Fluorescent lighting is also affected as maintenance of the electron flow in the fluorescent tube becomes tenuous. Power loads usually continue to operate satisfactorily at the lower voltage, drawing more current, so that they have little effect on load reduction. This increase in current, however, may cause overheating, loss of torque, and other undesirable conditions to take place.

In some instances, the lowering of voltage may take place on the transmission or subtransmission incoming supply circuit or circuits when they are equipped with voltage regulators at the sending ends. In this event, the voltage regulators on the distribution feeders (or on the distribution feeder bus) at the substation may have to be blocked in a fixed or neutral position so as not to negate the effect of the reduced voltage on the incoming supply.

Regulators installed in the field on portions of primary circuits, for practical reasons, are usually left alone and allowed to travel to their maximum position if necessary. The relatively small amounts of load that can be shed by attending to these units is often not worth the effort necessary to adjust them to nonautomatic operation at the start and to return them to automatic operation at the end of a usually short period of time.

Distribution System Problems   Where distribution feeders are deliberately designed to accommodate such lowering of voltage below normal values, it may be necessary to reinforce or provide for some consumers, usually at the ends of primary circuits. These may have a sufficiently low normal voltage that the additional drop may cause damage to some of their connected loads. In most cases, provision involves the shortening of secondary mains and the more closely spaced installation of transformers; also, taps may be set for a lower (or the lowest) ratio of transformation. Sometimes, the installation of a booster transformer in those farthest sections of the primary circuit will accomplish the purpose. In any event, this feature requires investigation and the taking of necessary measures preferably before the need for such voltage reduction to shed load is placed in effect.

Where the need for voltage reduction stems only from some deficiency in the distribution system, that deficiency should be identified and removed as quickly as possible; such a need might occur in the case of overloads in an outgoing underground cable supplying a feeder where the cable carries too great a load under normal conditions, or under contingency conditions where it may be called upon to carry the load of an additional circuit or circuits, or parts of them.

Low-Voltage Network Where the feeders supply a low-voltage secondary network, extreme care must be exercised in lowering the voltage on the supply primary feeders. Operation of the regulators must be coordinated so that the voltage on each of the feeders is lowered as simultaneously as practical to prevent opening of network protectors on the feeder if its voltage alone is lowered. In turn, the additional load picked up by the feeder or feeders whose voltage is not lowered may cause "hunting" between the protectors of the several feeders, or may cause feeders to trip from overloads in a cascading effect that would shut down the network. In some instances, the overcurrent relays on the feeder circuit breakers may be blocked to prevent feeders from tripping from temporary overloads until all the regulators have been adjusted and locked in at the desired lowered voltage level. If the voltage lowering operation is to be a relatively frequent occurrence, relay settings on the network protectors in the field may be made sufficiently insensitive to keep the protectors from hunting and the feeders from cascading out.

Brownout

Brownout is a procedure in which feeders are taken out of service for a relatively short period of time on a predetermined basis, usually one at a time, to reduce the demand on the substation supply transformers or on the facilities back to the generating station. Critical loads, such as hospitals and military bases, are usually supplied from two sources with double-throw facilities to accomplish a switchover to an energized source, either manually or automatically. Where this arrangement does not exist, it may be necessary to sectionalize that portion of a feeder supplying such critical loads, connecting it to an energized feeder when its normal supply feeder is deenergized. In some instances, where more than one critical load may exist on a feeder, that feeder may be exempt from the brownout procedure. Feeders supplying low-voltage networks are usually exempt as the load dropped by them in such an operation would be picked up by the others supplying the network, and the net reduction in load would be very nearly zero.

COGENERATION
Why

Basically, cogeneration involves the interconnection of consumers' generation to the utility's distribution system. Changing energy and economic conditions have made such interconnections feasible in many instances, and, in some areas, mandated by law (e.g., Texas). Large users of steam and hot water have found it economically desirable to generate electricity and use the "waste" heat to meet their steam and hot water requirements. The electricity so produced that they do not use themselves is sold to the utility, usually at an advantageous rate. This is because regulatory guidelines tend to favor a rate paid for power purchased by a utility (the avoidable cost to produce power) to be based on the utility's *least* efficient power plant, whereas the cost of power to a utility consumer is usually based on an *average* cost to produce power.

Paralleling the Systems

Paralleling the consumer's generation facilities with those of the utility requires that additional protection equipment be installed at the cogenerator's facilities. The principal features of this additional protection include:

1. Automatic synchronizing of the generator output with the utility

2. Relaying to prevent the closing of the circuit breaker to the utility system until the cogenerator's generator is open, for protection of that generator

3. Relaying to open the circuit breaker to the utility system on loss of power in the utility system

4. Relaying to open the circuit breaker to the utility system on a ground fault on the utility system

5. Relaying to control the cogenerator's generator circuit breaker to provide generator overcurrent protection, phase current balance protection, reverse power protection, under- and over-frequency protection, and under- and overvoltage protection

6. Control of engine governor equipment for speed, generator phase match, and generator load

The electrical connections and indicated protection are shown on the one-line diagram in Fig. 15-1.

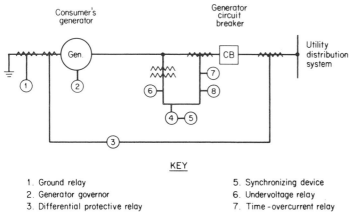

KEY

1. Ground relay	5. Synchronizing device
2. Generator governor	6. Undervoltage relay
3. Differential protective relay	7. Time-overcurrent relay
4. Frequency meter and relay	8. Directional power relay

FIG. 15-1   One-line diagram showing protection relaying for consumer cogeneration unit.

Modes of Operation

There are several load relationships that may exist between the cogenerator and the utility:

1. The cogenerator always supplying power at a constant rate; i.e., the cogenerator supplying a part of the utility's base load

2. The cogenerator always supplying power in variable amounts, fluctuating with the consumer's needs; i.e., the cogenerator supplying only a marginal part of the utility's load

3. The cogenerator and the utility both supplying the consumer's requirements on a normal or contingency basis

4. The utility supplying all of the consumer's requirements on a contingency basis

The Distribution System

The variation in the modes of operation not only may affect the settings applied to the protective relays, but may influence the design of the utility's distribution system. Obviously, the distribution facilities required will vary with the mode of operation, and the problem of maintaining voltage within acceptable limits as conditions change (e.g., from mode 1 to mode 4 above) may tax the engineer's ingenuity.

The wide variations that may take place in voltage profiles and current distribution in the distribution circuit to which the cogenerator is connected may require rearrangement of the circuit configuration to maintain safe and acceptable standards of electricity supply. These may require the installation of additional switching facilities to achieve desired sectionalization and rearrangement; preferably the switches should be automatically operated. Moreover, as generators are usually under control of one operating group while the distribution system is controlled by a separate group, some difficulties in coordination may arise.

DEMAND CONTROL (OR PEAK SUPPRESSION)

From an economic viewpoint, it is desirable to hold down the peak load or maximum demands on all the parts of the electric system—generation, transmission, and distribution. This has the desirable effects of reducing plant investment and at the same time reducing operating expenses (fuel) because of reduced I^2R losses. This has been touched upon in Chap. 8.

Load shedding as described above is a form of demand control or peak suppression, but is associated with contingencies usually of a temporary nature; demand control applies to a permanent reduction in maximum demands as a normal condition.

Basic Concept

For a utility, the most efficient use of its facilities, from an investment point of view, is to use them throughout their lifetimes at maximum loads. By definition, if this were done, the load factor for the facilities would be 100 percent. This does not occur, because some consumer loads are not always required and are turned off. The closer the utility can approach 100 percent load factor, however, the better the investment can be utilized, and the lower a unit of output can be priced.

The load factor concept in supplying a given load applies equally well whether it is applied to the utility itself or to a single consumer. For example, it is not unusual for a utility to have an annual load factor of 50 to 60 percent because of seasonal air-conditioning loads; this implies that a great part of the facilities used to meet summer peak demands will remain idle the rest of the year.

Typical load factors range from less than 20 percent for some residences to over 90 percent for some industrial plants (like some manufacturing plants having large air-filtering installations). In between, office buildings typically may have load factors of 20 to 30 percent and large, three-shift manufacturing plants 70 to 80 percent. Small to medium industries may have load factors ranging from 20 to 70 percent.

Conservation

Demand control is also a conservation measure, as it will substantially reduce losses, in both the utility's and the consumer's facilities. These reductions will be reflected in fuel consumption at the utility's generating plants and in both demand and energy charges (including fuel adjustment charges) in the consumer's bill.

Although the total overall consumption by a consumer may remain the same, the leveling of demands will decrease the *maximum* current flow, though the *reduced* current flow may continue for a longer period of time. As losses (I^2R) vary as the square of the current, the lower current should result in a substantial reduction in the total energy requirements of the consumer.

Experience has also indicated that reviews for reducing demand often discover unnecessary operation of some equipment, and result in elimination of some operations and improved methods of operation—all of which result in decreased energy consumption.

Load Management

Basically, to reduce the demand on its facilities, the utility must seek to reduce the individual consumers' demands. Preferably, the individual consumers' demands should also be coordinated so as to achieve a minimum coincident demand. This latter feature is more difficult of attainment than the first, as it involves cooperation not only between the utility and the consumer, but among a number of consumers as well. Methods for reducing demands differ for large consumers, such as industrial plants, and for small consumers, such as residences; both, however, employ rate incentives.

Large Consumers   For large consumers this subject was touched upon in Chap. 8, in connection with the demand metering of large consumers and their role in schemes employed for reducing consumer demands. This is important to the consumer, as utility rates include demand charges based on the registered maximum demand. The timing impulses from the demand meter are used in several schemes to hold down demands to predetermined values.

To reduce demand requires that nonessential load peaks be reduced or eliminated. Loads are analyzed into several categories:

1. Those that are essential, that cannot be turned off without affecting safety and operations.

2. Those that may be curtailed or turned off for relatively short periods of time without being noticed (e.g., 10 min out of each hour); they may be programmed to be shut off sequentially for predetermined periods of time.

3. Those that may be deferred, put off to some random off-peak time which may differ from day to day (or other period).

4. Those which may be conveniently rescheduled regularly to off-peak periods.

Typical examples of such categories for large industrial consumers are shown in Table 15-1.

Load Cycling Cycling involves the turning off and on of individual loads or groups of loads. How long they can be turned off and how often is predeter-

TABLE 15-1 TYPICAL CATEGORIES OF LARGE CONSUMER LOADS

Category	*Examples*
Essential	Lighting (some)
	Elevators
	Production equipment (some)
	Ventilators (some)
	Pumps (some)
Curtailable	Air conditioners
	Heaters
	Ventilators
	Refrigerators
	Water pumps
	Ovens
Deferable	Coolers
	Air compressors
	Water heaters
	Equipment testing
Reschedulable	Electric furnace
	Process ovens
	Incinerators
	Trash compactors
	Battery chargers

mined, and the off-on times are staggered so that a minimum number of such loads are on at any one time to achieve the smallest practical maximum demand.

Most automatic demand-control systems employ load cycling and involve some technique of demand forecasting to determine when loads should be turned off and on. All are based on the consumer's actual consumption, and its rate compared to some predetermined ideal rate. Several methods of obtaining the comparison have been devised yielding different degrees of accuracy and precision; these will not be further detailed as they are not within the scope of this book. Almost all employ pulses obtained from the utility's demand meter for matching the demand under consideration with that being recorded by the demand meter. Demand-control equipment also acts to control "average" maximum demands, leaving the utility still needing to meet the actual or peak maximum demand.

Demand control usually leads to increased off-on switching of equipment. Care must be taken, therefore, to ensure that thermal overloads and mechanical failure of switching devices do not occur because of short cycling of equipment.

Small Consumers   Most small consumers are residential consumers whose demand is not usually metered. Efforts by the utility to hold down their demands are limited almost to promotional rates which, for practical purposes, cannot be policed. More and more, however, utilities are installing demand meters or clocking devices connected to the larger loads, such as hot water heaters, dish and clothes washers, etc. Consumers are thereby encouraged to install small computer-actuated devices for controlling the turning off and on of such loads; these include supplementary time delays so that the initial inrush currents on the various appliances will not (because of possible loss of diversity) all occur simultaneously, causing service interruption and possible damage to the appliance.

Thermal devices have been developed for storing cold and heat during off-peak periods to be used during periods of electrical peaks. Such devices are also beyond the scope of this book.

Load Management Control

A number of utilities have assumed control of devices that will switch off loads on their system automatically when undesirable demand levels are being reached. This involves the identification of noncritical loads that can be placed under the control of the utility with agreed-upon constraints regarding maximum time to be left off. Agreements with consumers include favorable "interruptible" rate schedules. These may be controlled by means of signals transmitted by radio, carrier, or telephone communication. (Similar means of communication are also employed in small residential consumers, mentioned earlier.)

Costs for such demand-control installations are sometimes shared with the (large) consumer, where the same equipment is also used to control the consumer's maximum demand. This demands an almost continuous review to ascertain that a target value set up by the consumer will not run afoul of that set up by the utility.

Utility Problems

The utility generally must meet seasonal and daily load peaks, peaks of relatively short duration. Generation, transmission, and distribution facilities must be provided to meet these demands.

Generation   Utilities generally have three levels of power generation equipment; base-load units, usually the newest and most highly efficient (as well as the most expensive); midrange units, which are less efficient and are perhaps the most recent of older generation facilities; and units generally operated as peak units, the least efficient and usually the oldest, but sometimes new units specifically designed for this short-term purpose. Controlling system demands, therefore, results in lessened need to operate the least efficient units and, in the long term, defers the addition of the most expensive newest units.

Transmission   Transmission lines are the bulk carriers of electrical energy, and fall into the same category as generators as far as investment is concerned. Here, too, the older and less efficient transmission lines, usually those operating at lower voltages than the newest lines, constitute part of the transmission network, including those facilities needed to ensure continuity of adequate service in contingencies. Controlling system demands will not only result in lowered I^2R losses, as indicated previously, but will also defer, if not make unnecessary, the installation of new and expensive transmission lines.

Distribution   The same general observations concerning transmission lines also apply to distribution circuits. Additional distribution facilities (including substations) and conversions or construction at higher voltages can be deferred if the system demands can be controlled.

In the case of distribution facilities, however, a problem arises as to the rating of equipment, particularly transformers. The rating of such equipment, though nominally predicated on its current-carrying ability, is actually based on the allowable temperature at which insulation may be subject to failure. This temperature is a function not only of the heat caused by the losses (copper and iron) developed within the unit, but also of the duration of the developed heat. The control of demands on the distribution system will, on the one hand, reduce the maximum values of heat applied (as noted previously), but on the other, may reduce the thermal margin of safety resulting from the duration of the heating cycles. In some instances, no doubt, the net overall effect may be negligible; in others, especially during periods of prolonged hot ambient temperatures, such as exist in summertime, the net overall effect may be appreciable. Units so affected should, therefore, be closely monitored.

CONCLUSION

The need for conservation of investment capital and use of energy resources indicates the desirability of operating practices that no longer separate the distribution system from other parts of electrical system operations. Moreover, electronic and miniaturization advances make practical the integration and automatic management of such operations. These may be listed as:

1. Load management for conservation and improved overall system efficiency. This may include the direct control of consumer loads (and energy storage where practical), and other means to alter the shape of the utility system load curve.

2. Alternative energy resource management to control utility- or consumer-owned generation which will integrate with the distribution system.

3. Distribution system management during normal and contingency operation, including reporting of interruptions, sectionalizing and reconfiguration of circuits, remote operation of equipment, load monitoring and reactive power (VAR and capacitors) control, and even remote meter reading for revenue billing purposes.

4. Integrated systemwide control to provide for overall system operation during periods of widespread emergencies.

The distribution engineer, to a greater degree, must take into account the impact of these integrated operations on planning and design practices.

APPENDIXES

CIRCUIT ANALYSIS METHODS

INTRODUCTION

In the analysis of circuits to determine current distribution and voltage drops in the individual parts of the circuits, several basic principles are employed. These call for the reduction of the circuitry into successively simpler forms until a single loop circuit results. Computation of current and voltage values can then be made, essentially reversing the order of simplification, until all the individual parts of the circuit have been analyzed. These procedures are especially applicable to network-type circuits.

Kirchhoff's Laws

Kirchhoff's laws encompass two fundamental simple laws which apply to both dc and ac circuits, no matter how complex they may be:

1. The current flowing away from a point in the circuit (where three or more branches come together) is equal to the amount flowing to that point. Expressed another way, the vector or algebraic sum of all the currents entering (and leaving, a negative entry) a point is zero.

2. The voltage acting between two points in a circuit acts equally on all the paths connected between the two points. Expressed another way, the vector or algebraic sum of all the voltage drops (or rises, negative drops) around a closed loop is zero.

Applying Kirchhoff's laws to the parts of a circuit, a number of equations between the unknowns can be drawn. The number of independent equations which can be written from the first law is 1 less than the number of junction points; from the second law, the number is equal to the number of branches less the number of junction point equations. The equations can be algebraically solved for all the unknowns. In practice, however, these laws are more often used to check results obtained by other means.

CIRCUIT TRANSFORMATIONS
Wye to Delta
Refer to Fig. A-1.

$$I_a + I_b + I_c = 0$$

$$I_a = I_B - I_C$$

$$I_b = I_C - I_A$$

$$I_c = I_A - I_B$$

$$\Delta = Z_a Z_b + Z_b Z_c + Z_c Z_a$$

and

$$Z_A = \frac{\Delta}{Z_a} = Z_b + Z_c + \frac{Z_b Z_c}{Z_a} \qquad Z_B = \frac{\Delta}{Z_b} = Z_c + Z_a + \frac{Z_c Z_a}{Z_b}$$

$$Z_C = \frac{\Delta}{Z_c} = Z_a + Z_b + \frac{Z_a Z_b}{Z_c}$$

Delta to Wye
Refer to Fig. A-1.

$$I_A = \frac{I_c Z_B - I_b Z_C}{Z_A + Z_B + Z_C} = \frac{(I_c Z_c - I_b Z_b) Z_a}{\Delta'} \qquad Z_a = \frac{Z_B Z_C}{Z_A + Z_B + Z_C}$$

$$I_B = \frac{I_a Z_C - I_c Z_A}{Z_A + Z_B + Z_C} = \frac{(I_a Z_a - I_c Z_c) Z_b}{\Delta'} \qquad Z_b = \frac{Z_C Z_A}{Z_A + Z_B + Z_C}$$

$$I_C = \frac{I_b Z_A - I_a Z_B}{Z_A + Z_B + Z_C} = \frac{(I_b Z_b - I_a Z_a) Z_c}{\Delta'} \qquad Z_c = \frac{Z_A Z_B}{Z_A + Z_B + Z_C}$$

where $\Delta' = Z_A Z_B + Z_B Z_C + Z_C Z_A$.

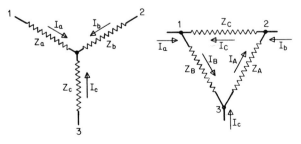

FIG. A-1 Delta-wye transformations.

Changing Bases

To convert Z in ohms at a voltage E to Z', the equivalent value on a voltage base E',

$$Z' = Z \left(\frac{E'}{E}\right)^2$$

To convert Z in percent on a kVA base U to Z', the equivalent value on a kVA base U',

$$Z' = Z\frac{U'}{U}$$

To convert Z_p in percent on a kVA base U to Z_z in ohms on a voltage base E,

$$Z_z = \frac{3Z_pE^2}{U \times 10^5}$$

and, conversely

$$Z_p = \frac{U \times 10^5}{3E^2}$$

where E = line-to-neutral voltage and U = total three-phase kVA.

Paralleling Two Impedances

Refer to Fig. A-2.

$$Z = \frac{Z_aZ_b}{Z_a + Z_b}$$

where Z is the equivalent impedance; and

$$I_a = I \frac{Z_b}{Z_a + Z_b} \qquad I_b = I \frac{Z_a}{Z_a + Z_b}$$

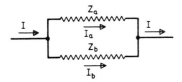

FIG. A-2 Paralleling two imped-ances.

SUPERPOSITION THEOREM

In a network containing several voltage sources, the current in the several branches may be found by replacing all but one of the voltage sources by their particular resistances (dc) or impedances (ac) and determining the current contributed by the one source in each of the branches. The process is repeated with each of the other voltage sources, and separate current distribution in the several branches from each of the voltage sources is again determined. The vector or algebraic sum of all of the currents in each branch (as determined above) gives the value of the current in that branch with all of the voltage sources in place.

SYMMETRICAL COMPONENTS

General

The method of symmetrical components is essential to the analysis of unbalances between phases of a polyphase circuit.

A balanced three-phase system is represented by three vectors of equal magnitude with a phase displacement of 120°. Normal current and voltage conditions assume balanced phases, allowing the magnitude of the current or voltage of one phase to represent each of the three phases. Calculation of the three-phase performance then involves the analysis of one phase by a single-phase calculation with the result rotated through 120° and 240° to complete the three phases.

The unbalanced three-phase system is represented by vectors whose magnitude and phase displacement are not equal. The equivalent single-phase circuit represented by one phase cannot be used for calculations, as it does not include a measure of the conditions on the other two phases.

For ease of calculation, the method of symmetrical components can be used to separate the unbalanced three-phase system into three components, two being balanced three-phase components, the third a uniphase component. Each of these component systems can be represented by an equivalent single-phase system. Thus, the solution of the component systems separately by an equivalent single-phase calculation and then the addition of the three components in proper phase relation simulates the unbalanced system, the solution of which is desired.

Symmetrical Component Systems

The unbalanced system of vectors can be divided into three components called the positive-phase sequence components, negative-phase sequence components, and zero-phase sequence components (see Fig. A-3). Positive-phase sequence components are balanced three-phase vectors, whose time sequence of maxima occur in the order E_a, E_b, E_c. Being balanced, the vectors have equal amplitudes and are displaced 120° relative to each other; and

$$E_{a1} = E_{a1}$$
$$E_{b1} = e^{j240}E_{a1} = a^2E_{a1} \qquad (1)$$
$$E_{c1} = e^{j120}E_{a1} = aE_{a1}$$

in which a is the unit vector, $e^{j120} = -0.5 + j0.866$, that is, an operator which indicates that a vector to which it is attached has been rotated through 120° in a positive or counterclockwise direction; and a^2 is the unit vector $e^{j240} = -0.5 - j0.866$, that is, an operator which indicates that a vector to which it is attached

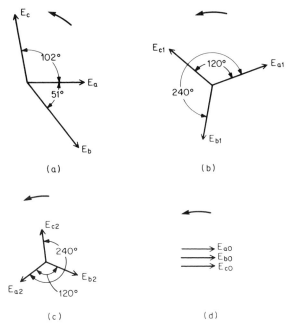

FIG. A-3 Unbalanced vectors and symmetrical components. (a) Three-phase system with unbalance. (b) Positive-phase sequence component. (c) Negative-phase sequence component. (d) Zero-phase sequence component.

has been rotated through 120° twice or 240° in a positive direction; see Fig. A-4 and Table A-1.

The negative-phase sequence components are balanced three-phase vectors whose time sequence of maxima occur in the order E_{a2}, E_{c2}, E_{b2}, as shown in Fig. A-3c.

$$E_{a2} = E_{a2}$$
$$E_{b2} = e^{j120}E_{a2} = aE_{a2} \qquad (2)$$
$$E_{c2} = e^{j240}E_{a2} = a^2E_{a2}$$

The zero-sequence components are three-phase vectors of equal magnitude without phase displacement, as shown in Fig. A-3d.

$$E_{a0} = E_{a0}$$
$$E_{b0} = E_{a0} \qquad (3)$$
$$E_{c0} = E_{a0}$$

In each of the three systems, fixing the phase position and magnitude of one vector immediately determines the other two vectors. Phase sequence should not be confused with the rotation of the vectors. Standard convention calls for coun-

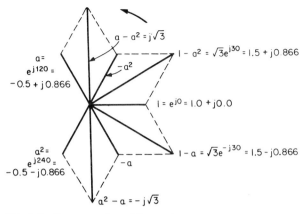

FIG. A-4 The properties of the vector a.

terclockwise rotation for all vectors; hence, positive and negative vectors will rotate in the same direction.

Combination of Sequence Quantities to Form Phase Quantities

In all three systems of the symmetrical components, the subscripts denote the components in the different phases. The total voltage is then equal to the sum of the corresponding components of the different sequences in that phase. The three unbalanced vectors in Fig. A-3a may be equated:

$$E_a = E_{a0} + E_{a1} + E_{a2} \tag{4}$$

$$E_b = E_{b0} + E_{b1} + E_{b2} \tag{5}$$

$$E_c = E_{c0} + E_{c1} + E_{c2} \tag{6}$$

TABLE A-1 PROPERTIES OF THE VECTOR a

$$1 = e^{j0} = 1.0 + j0.0$$

$$a = e^{j120} = -0.5 + j0.866$$

$$a^2 = e^{j240} = -0.5 - j0.866$$

$$a^3 = e^{j360} = 1.0$$

$$a^4 = e^{j480} = a$$

$$1 + a + a^2 = 0$$

$$1 - a = \sqrt{3}e^{-j30} = 1.5 - j0.866$$

$$1 - a^2 = \sqrt{3}e^{j30} = 1.5 + j0.866$$

$$a - a^2 = \sqrt{3}e^{j90} = j\sqrt{3}$$

$$a^2 - a = \sqrt{3}e^{-j90} = -j\sqrt{3}$$

or, by substituting their equivalent values:

$$E_a = E_{a0} + E_{a1} + E_{a2} \tag{7}$$

$$E_b = E_{a0} + a^2E_{a1} + aE_{a2} \tag{8}$$

$$E_c = E_{a0} + aE_{a1} + a^2E_{a2} \tag{9}$$

The unbalanced system is thus defined in terms of three balanced systems.

Resolution of Three Vectors into Their Symmetrical Components

The magnitude of the symmetrical components in an unbalanced system can be found by graphical methods, by analytical methods, or by meters with suitable connections. Each of these solutions depends upon the addition of the unbalanced three-phase vectors in a manner that cancels two of the components and adds the third component. Graphically, the positive-sequence component E_{a1} is found by rotating phase B $+120°$ with respect to phase A, rotating phase C $-120°$ ($+240°$) with respect to phase A, and adding the two rotated phases to phase A and dividing by 3 to correct for the addition, as shown in Fig. A-3b. Analytically, the positive-sequence component may be obtained by multiplying Eq. (8) by a and Eq. (9) by a^2 and adding them to Eq. (7):

$$E_a + aE_b + a^2E_c$$

$$= E_{a0} + E_{a1} + E_{a2} + aE_{a0} + a^3E_{a1} + a^2E_{a2} + a^2E_{a0} + a^3E_{a1} + a^4E_{a2}$$

$$= (1 + a + a^2)E_{a0} + 3E_{a1} + (1 + a^2 + a)E_{a2}$$

Solving for E_{a1} and recalling that $1 + a^2 + a = 0$,

$$E_{a1} = \frac{E_a + aE_b + a^2E_c}{3} \tag{10}$$

This is the procedure, outlined under the graphical solution, above, expressed as an equation.

Either of these operations (graphical or analytical) eliminates the negative and zero components from the resultant.

The negative-sequence component is found graphically by rotating phase B $+240°$ ($-120°$) in respect to phase A, rotating phase C $+120°$ in respect to phase A, and adding the two rotated phases to phase A and dividing by 3 to correct for the addition; refer to Fig. A-5c.

Analytically, the negative-sequence component may be found by multiplying Eq. (8) by a^2 and Eq. (9) by a, and adding the results to Eq. (7):

$$E_a + a^2E_b + aE_c$$

$$= E_{a0} + E_{a1} + E_{a2} + a^2E_{a0} + a^4E_{a1} + a^3E_{a2} + aE_{a0} + a^2E_{a1} + a^3E_{a2}$$

$$= (1 + a^2 + a)E_{a0} + (1 + a + a^2)E_{a1} + 3E_{a2}$$

Solving for E_{a2},

$$E_{a2} = \frac{E_a + a^2E_b + aE_c}{3} \tag{11}$$

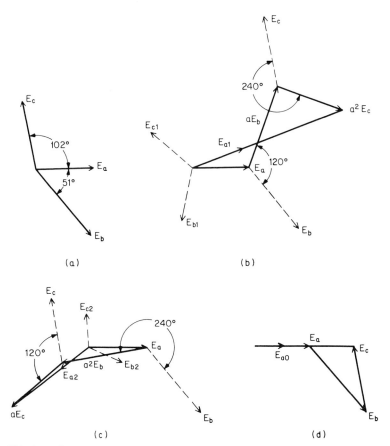

FIG. A-5 Graphical construction for the determination of sequence components. (a) Original vectors. (b) Positive sequence. (c) Negative sequence. (d) Zero sequence.

These operations eliminate the positive- and zero-sequence components from the resultant.

The zero-sequence component is found graphically by adding the phase B and phase C vectors to phase A without rotation, then dividing by 3 to correct for the addition; refer to Fig. A-5d. Analytically, the zero-sequence component may be found by adding the three equations, Eqs. (7), (8), and (9):

$$E_a + E_b + E_c$$
$$= E_{a0} + E_{a1} + E_{a2} + E_{a0} + a^2E_{a1} + aE_{a2} + E_{a0} + aE_{a1} + a^2E_{a2}$$
$$= 3E_{a0} + (1 + a^2 + a)E_{a1} + (1 + a + a^2)E_{a2}$$

Solving for E_{a0},

$$E_{a0} = \frac{E_a + E_b + E_c}{3} \tag{12}$$

Suppose the given voltages are balanced; e.g.,

$$E_a = E_a \quad \text{and} \quad E_b = a^2 E_a \quad \text{and} \quad E_c = a E_a$$

Then

$$E_{a0} = \frac{E_a + a^2 E_a + a E_a}{3} = \frac{1 + a^2 + a}{3} E_a = 0$$

$$E_{a1} = \frac{E_a + a^3 E_a + a^3 E_a}{3} = \frac{1 + a^3 + a^3}{3} E_a = E_a$$

$$E_{a2} = \frac{E_a + a^4 E_a + a^2 E_a}{3} = \frac{1 + a + a^2}{3} E_a = 0$$

In a balanced three-phase system, therefore, the zero- and negative-sequence components disappear, leaving only the positive-sequence components; i.e., the three voltages E_a, E_b, and E_c, as measured, are the positive-sequence components. While the above equations have been stated in terms of voltage, they apply equally well to currents.

The zero-sequence component, being a uniphase component, can exist in a three-phase system only when a path to ground exists which allows the flow of ground current (zero-sequence current times 3) or the establishment of a definite potential in relation to ground. It is apparent, therefore, that in three-phase three-wire systems and in lines feeding delta connections, where the currents and voltages must add to zero, the zero-sequence component is not present. The negative-sequence component, not being present in balanced systems, indicates an unbalance whenever it is present in a system.

SEQUENCE FILTERS

If a device is obtained which will perform the rotations and additions for a particular component, the resultant can be applied on a meter which will indicate the magnitude of the component desired. The measurement of sequence components requires that a current flow through the measuring device which is directly proportional, but not necessarily equal, to the component being sought, and therefore that the effects of the other two components be eliminated. If the rotation of vectors were performed exactly in accordance with Eq. (10) or (11), the measuring device would be more complicated than necessary. Refer to Fig. A-6, which shows the positive- and negative-sequence components for both the wye and delta voltages of the unbalanced three-phase voltages (Fig. A-6a).

It will be noted that the positive- and negative-sequence components of the delta voltages bear definite relations to the positive- and negative-sequence components, respectively, of the wye voltages, and the zero-sequence component of the delta voltages is always zero. These relations follow:

$$E_{a1} = \frac{E_{ab1}}{\sqrt{3}} e^{j150} \qquad E_{ab1} = \sqrt{3} E_{a1} e^{-j150}$$

$$E_{a2} = \frac{E_{ab2}}{\sqrt{3}} e^{-j150} \qquad E_{ab2} = \sqrt{3} E_{a2} e^{j150} \tag{13}$$

$$E_{a0} = \text{indeterminate} \qquad E_{ab0} = 0$$

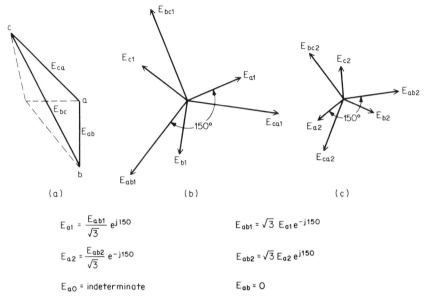

$$E_{a1} = \frac{E_{ab1}}{\sqrt{3}} e^{j150}$$

$$E_{a2} = \frac{E_{ab2}}{\sqrt{3}} e^{-j150}$$

$$E_{a0} = \text{indeterminate}$$

$$E_{ab1} = \sqrt{3}\, E_{a1} e^{-j150}$$

$$E_{ab2} = \sqrt{3}\, E_{a2} e^{j150}$$

$$E_{ab} = 0$$

FIG. A-6 Positive- and negative-sequence components: (a) unbalanced voltages; (b) positive-sequence components; (c) negative-sequence components.

It is obvious that if the delta voltages are used in metering positive- or negative-sequence components, the zero-sequence component is automatically eliminated, and it is only necessary to eliminate the effect of one component.

Positive-Sequence Voltage Filter

Assume that it is desired to measure the positive-sequence voltage of a three-phase four-wire system. Using phase a as a reference, then:

$$E_{a1} = \frac{E_{ab1}}{\sqrt{3}} e^{j150} = \frac{\sqrt{3}}{9} (E_{ab} + aE_{bc} + a^2 E_{ca})e^{j150} \tag{14}$$

and

$$E_{ab0} = \frac{E_{ab} + E_{bc} + E_{ca}}{3} = 0 \tag{15}$$

From Eq. (15),

$$E_{bc} = -E_{ca} - E_{ab} \tag{16}$$

Substituting Eq. (16) in Eq. (14) gives

$$E_{a1} = \frac{\sqrt{3}}{9}(1 - a)E_{ab} + (a^2 - a)E_{ca}e^{j150} \tag{17}$$

Recalling that $1 - a = \sqrt{3}e^{-j30}$ and $a^2 - a = \sqrt{3}e^{-j90}$, then

$$E_{a1} = \frac{\sqrt{3}}{9}(\sqrt{3}\, E_{ab}\, e^{-j30} + \sqrt{3}\, E_{ca}\, e^{-j90})e^{j150}$$

$$= \frac{E_{ab}}{3}\, e^{j120} + \frac{E_{ca}}{3}\, e^{j0} \tag{18}$$

Recalling that $e^{j120} = a$ and $e^{j0} = -a^2$, then

$$E_{a1} = \frac{aE_{ab} - a^2E_{ca}}{3} \tag{19}$$

The coefficient of E_{ab}, namely, a, indicates a rotation through 120°. Likewise $-a^2$ indicates a rotation of E_{ca} through 60°. The difference in angle is 60°, so that the current through a measuring device will be proportional to the positive-sequence voltage, if it consists of a component proportional to E_{ab} and another bearing the same relation to E_{ca} except lagging 60°; refer to Fig. A-7a. The current I_m is therefore proportional to the positive-sequence voltage E_a, but lags it by 120°. If the impedance of the measuring coil is negligible, such a current will obviously be obtained by the connection shown in Fig. A-8a. The resistor R across the voltage E_{ab} allows the current I_r proportional to E_{ab} to flow through the measuring device; while the reactor, which has a phase angle of 60° or equals Re^{j0}, across the voltage E_{ca} allows the current I_x, which is proportional to E_{ca} and lagging it by 60°, to flow through the metering device. The total current through the metering device is, therefore, proportional to the positive-sequence voltage E_{a1}. That the current is proportional to the positive-sequence voltage, even when the impedance of the metering device is not negligible, can be shown from Fig. A-8a.

$$E_{ab} = I_r(R + Z_m) + I_xZ_m \tag{20}$$

$$E_{ca} = I_x(Re^{j0} + Z_m) + I_rZ_m \tag{21}$$

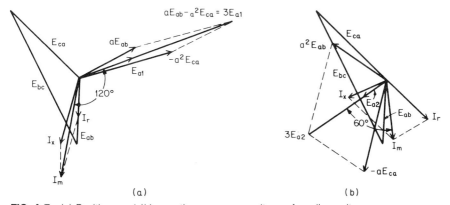

FIG. A-7 (a) Positive- and (b) negative-sequence voltages from line voltages.

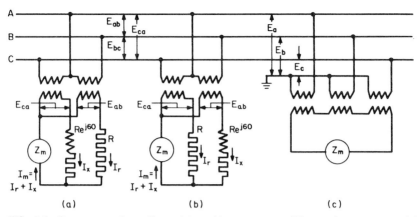

FIG. A-8 Sequence voltage filters: (a) positive sequence; (b) negative sequence; (c) zero sequence.

Multiplying Eq. (20) by a and Eq. (21) by $-a^2$ and adding, remembering that $e^{j60} = -a^2$,

$$aE_{ab} - a^2E_{ca} = 3E_{a1}$$

$$= aI_rR + aI_rZ_m + aI_xZ_m + a^4I_xR - a^2I_xZ_m - a^2I_rZ_m$$

$$= I_r(aR + aZ_m - a^2Z_m) + I_x(aZ_m + a^4R - a^2Z_m)$$

$$= I_r[aR + (a - a^2)Z_m] + I_x[a^4R + (a - a^2)Z_m]$$

Remembering that $a^4 = a$,

$$3E_{a1} = (I_r + I_x)[aR + (a - a^2)Z_m]$$

Therefore

$$I_m = I_r + I_x = \frac{3E_{a1}}{aR + (a - a^2)Z_m} \tag{22}$$

Writing this equation in another form by dividing both numerator and denominator by a, giving:

$$I_m = \frac{3a^2E_{a1}}{R + (1 - a)Z_m} \tag{23}$$

when Z_m is small enough to be neglected, this equation becomes:

$$I_m = \frac{3a^2E_{a1}}{R} = \frac{3E_{a1}}{R}e^{-j120} \tag{24}$$

which shows that for this case the current through the metering device lags the reference voltage by 120°.

When the connection (Fig. A-8a) is desired for voltage measuring purposes, it will often happen that Z_m is quite small. Hence the external impedance, which usually forms a large portion of the total impedance of the voltmeter, may be replaced by the impedances R and Re^{j60}.

The combination of impedances R and Re^{j60} was originally known as a *sequence network*. Owing to the possibility of confusing it with the term *network* as applied to the low-voltage ac network system, the combination of impedances R and Re^{j60} is now designated a *sequence filter*.

Other combinations of impedances may be used as well, as long as they fulfill the fundamental condition that the impedance in the phase which leads by 120° has a phase angle 60° greater than the other impedance. For example, in Fig. A-8a, the impedance Re^{j60} could be replaced by a resistance, and the resistance R by a capacitive impedance Re^{-j60}. This change, of course, changes the phase angle of the resultant current. Unless the phase angle of I_m meets certain requirements, as for example in the measurement of power, the impedances shown in Fig. A-8a are used.

Negative-Sequence Voltage Filter

Negative-sequence voltage can be measured in the same manner as positive-sequence voltage, except that the resistance branch and the impedance branch of the filter must be interchanged. In vector representation, negative-sequence voltages are represented with the same direction of rotation, but with two phases interchanged from their positive-sequence positions. From the physical standpoint this is equivalent to reversal of phase rotation.

Interchanging two phase leads in the positive-sequence filter will cause the network to indicate the voltage of opposite phase rotation, which is then the negative-sequence voltage. This may be accomplished on the line side of the potential transformers by interchanging any two leads. On the measuring side, the common connection is a return or neutral wire instead of a phase wire, so that the change on this side may be made only by interchanging the two impedances (refer to Fig. A-8b). To check these observations the analytical derivation may be made in a manner similar to that for the positive-sequence filter. It will be found that the negative-sequence voltage in terms of line voltage E_{ab} and E_{ca} is:

$$E_{a2} = \frac{a^2 E_{ab} - a E_{ca}}{3} \tag{25}$$

and

$$I_m = \frac{-3a^2 E_{a2}}{R} = \frac{3E_{a2}}{R} e^{j60} \tag{26}$$

If Z_m is negligible, the solution may be found in Fig. A-7b.

Zero-Sequence Voltage Filter

Since the zero-sequence component does not appear in the line-to-line voltages, the line-to-neutral voltages must be used. By definition, the zero-sequence voltage is given by the equation:

$$E_0 = \frac{E_a + E_b + E_c}{3}$$

The connection of Fig. A-8c obviously gives across a high-resistance meter $E_a + E_b + E_c$, which is $3E_0$.

Sequence Current Filters

As in the case of voltage filters, the determination of each of the three sequence current components requires the elimination of the other two components. In a three-phase three-wire system, without any neutral or ground connections, no zero-sequence current can flow, and the problem is somewhat simplified. Mathematically, this requires that:

$$I_{a0} = \frac{I_a + I_b + I_c}{3} = 0 \tag{27}$$

By definition,

$$I_{a1} = \frac{I_a + aI_b + a^2 I_c}{3} \tag{28}$$

Substituting for I_b from Eq. (27),

$$I_{a1} = \frac{1}{3}\left[(1 - a)I_a - (a - a^2)I_c\right] \tag{29}$$

Remembering that $1 - a = \sqrt{3}\, e^{-j30}$ and $a - a^2 = \sqrt{3}\, e^{j90}$, then

$$I_{a1} = \frac{1}{3}(\sqrt{3} I_a e^{-j30} - \sqrt{3} I_c e^{j90}) \tag{30}$$

Factoring out $j\sqrt{3}$,

$$I_{a1} = \frac{j\sqrt{3}}{3}(I_a e^{-j120} - I_c)$$

$$= \frac{j\sqrt{3}}{3}(a^2 I_a - I_c) \tag{31}$$

As would be expected, these equations are similar to the voltage equations and show that when the zero-sequence component is not present, the positive-sequence current can be expressed in terms of two line currents. Similarly, the negative-sequence current may be expressed in terms of two line currents. In the case of voltage networks, it was possible to obtain the required phase shift

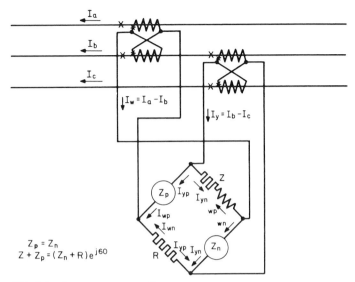

FIG. A-9 Sequence current filter.

by means of series impedance. In the current case this is not possible, since the current throughout the current-transformer series circuit is everywhere the same. Hence, parallel combinations of impedance must be used with the measuring coil in series with one branch.

When the system is grounded or provided with a neutral conductor, zero-sequence currents may flow in the phase conductors, which will affect the measuring currents. Hence, some means must be provided to remove the zero-sequence components. One of the most convenient means is to use pairs of current transformers connected in opposition instead of single-current transformers; see Fig. A-9.

From the figure,

$$I_w = I_a - I_b$$

By definition

$$I_a = I_{a1} + I_{a2} + I_{a0}$$

$$I_b = I_{b1} + I_{b2} + I_{b0}$$

and

$$I_w = I_{a1} + I_{a2} + I_{a0} - I_{b1} - I_{b2} - I_{b0}$$

From Eqs. (1), (2), and (3):

$$I_{b1} = a^2 I_{a1} \quad \text{and} \quad I_{b2} = a I_{a2} \quad \text{and} \quad I_{b0} = I_{a0}$$

Thus

$$I_w = I_{a1} + I_{a2} + I_{a0} - a^2 I_{a1} - a I_{a2} - I_{a0}$$
$$= (1 - a^2)I_{a1} + (1 - a)I_{a2} \tag{32}$$
$$= \sqrt{3}(I_{a1}e^{j30} + I_{a2}e^{-j30})$$

In a similar manner it may be shown that:

$$I_y = \sqrt{3}(a^2 I_{a1}e^{j30} + a I_{a2}e^{-j30}) \tag{33}$$

It will be observed from Eqs. (32) and (33) that the positive- and negative-sequence components of I_w and I_y are directly proportional to the corresponding components of I_a and I_b; i.e.,

$$I_{w1} = \sqrt{3} \, I_{a1}e^{j30} \tag{34}$$
$$I_{w2} = \sqrt{3} \, I_{a2}e^{-j30} \tag{35}$$

With the elimination of the zero-sequence component of current by the use of cross-connected current transformers,

$$I_{w0} = \frac{I_w + I_y + I_z}{3} = 0 \tag{36}$$

where

$$I_z = I_c - I_a = -I_w - I_y \tag{37}$$

By definition

$$I_{w1} = \frac{1}{3}(I_w + a I_y + a^2 I_z) \tag{38}$$

Substituting Eq. (37) in Eq. (38),

$$I_{w1} = \frac{1}{3}(I_w + a I_y - a^2 I_w - a^2 I_y)$$
$$= \frac{1}{3}(1 - a^2)I_w + (a - a^2)I_y$$
$$= \frac{1}{3}(\sqrt{3}I_w e^{j30} + \sqrt{3}I_y e^{j90})$$

Factoring out e^{j60},

$$I_{w1} = e^{j60} \left(\frac{\sqrt{3}}{3}I_w e^{-j30} + \frac{\sqrt{3}}{3}I_y e^{j30} \right) \tag{39}$$

From Eq. (34),

$$I_{a1} = \frac{I_{w1}}{\sqrt{3}e^{j30}} = \frac{1}{\sqrt{3}} \, I_{w1}e^{-j30} \tag{40}$$

Substituting Eq. (39) in Eq. (40),

$$I_{a1} = \frac{1}{\sqrt{3}} \, e^{j30} \left(\frac{\sqrt{3}}{3} \, I_w e^{-j30} \pm \frac{\sqrt{3}}{3} I_y e^{j30} \right)$$

$$= \frac{1}{\sqrt{3}} \, e^{j30} \left(\frac{I_w}{\sqrt{3}} e^{-j30} + \frac{I_y}{\sqrt{3}} \, e^{j30} \right) \tag{41}$$

By similar development,

$$I_a = \frac{1}{\sqrt{3}} \, e^{-j30} \left(\frac{I_w}{\sqrt{3}} \, e^{j30} + \frac{I_y}{\sqrt{3}} \, e^{-j30} \right) \tag{42}$$

Equation (41) indicates that a current proportional to the positive-sequence current may be obtained if a portion of I_w which lags I_w by 30° and a similar portion of I_y which leads I_y by 30° are taken through a measuring device. Likewise a current proportional to the negative-sequence current may be obtained by passing through a measuring device a portion of I_w which leads I_w by 30° and a similar portion of I_y which lags I_y by 30°. I_w and I_y may be divided into the necessary components by means of two impedances of equal magnitude but differing in phase by 60°, connected in parallel. Separate filters may be used to obtain the currents proportional to the positive- and negative-sequence currents or to allow simultaneous measurements of both currents. The most-used filter in the sequence network is shown in Fig. A-9; it must be designed with the following relation between the impedances to meet the requirements outlined above:

$$Z_p = Z_n \quad \text{and} \quad Z + Z_p = (Z_n + R)e^{j60}$$

The current I_w passing through the filter divides into two parts I_{wp} and I_{wn} equal in magnitude but differing in phase by 60°. I_{wp} lags I_w by 30° and I_{wn} leads

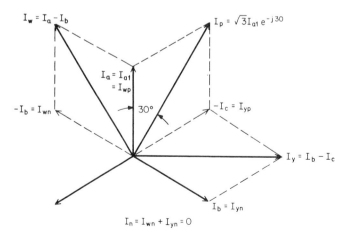

FIG. A-10 Sequence current filter—balanced currents.

I_w by 30°. Similarly the current I_y divides into two parts I_{yp} and I_{yn} equal in magnitude but differing in phase by 60°. I_{yp} leads I_y by 30° and I_{yn} lags I_y by 30°. The current through Z_p is $I_{wp} + I_{yp} = I_p$. From Eq. (41):

$$I_{wp} = \frac{I_w}{\sqrt{3}} e^{-j30} \quad \text{and} \quad I_{yp} = \frac{I_y}{\sqrt{3}} e^{j30}$$

Therefore

$$I_p = \frac{I_{a1}}{\frac{1}{\sqrt{3}}e^{j30}} = \sqrt{3}I_{a1}e^{-j30} \tag{43}$$

The current through Z_n is $I_{wn} + I_{yn} = I_n$. From Eq. (42)

$$I_{wn} = \frac{I_w}{\sqrt{3}} e^{j30} \quad \text{and} \quad I_{yn} = \frac{I_y}{\sqrt{3}} e^{-j30}$$

Therefore

$$I_n = \frac{I_{a2}}{\frac{1}{\sqrt{3}}e^{-j30}} = \sqrt{3}I_{a2}e^{j30} \tag{44}$$

Referring to Fig. A-10, which is a vector diagram of the currents through the filter when the line currents are balanced, the currents through the positive-sequence measuring coil (I_{wp} and I_{yp}) add to give a current I_p equal to $\sqrt{3}$ times the positive-sequence current and lagging it by 30°. The currents through the negative-sequence measuring coil (I_{wn} and I_{yn}) are equal and opposite and there-

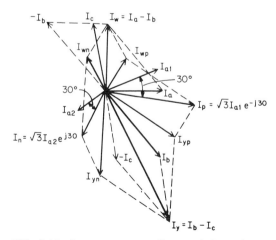

FIG. A-11 Sequence current filter—unbalanced currents.

fore cancel. It will be remembered that for balanced currents the negative-sequence current is zero.

Inspection of Fig. A-11, which is a vector diagram of the currents through the current filter when the line currents are unbalanced, will show that the current through the positive-sequence measuring coil again is equal to $\sqrt{3}$ times the positive-sequence current and lagging it by 30°. The currents through the negative-sequence measuring coil do not add to zero in this case but add up to equal $\sqrt{3}$ times the negative-sequence current and to lead it by 30°.

Thus the filter of Fig. A-9 produces, under balanced or unbalanced conditions alike, currents proportional to the positive- and negative-sequence currents existing in the system from which it is energized.

ECONOMIC STUDIES*

INTRODUCTION

Purpose

Economic studies are the means of evaluating the economic consequences of a particular proposal or of a number of alternate proposals for meeting a problem. Basic questions which continually face the management of any business are:

1. Will a venture be sufficiently profitable to justify the risk assumed in its undertaking?
2. Which of several ways of undertaking the venture will maximize the profits?

Scope

Economic studies may range from the extremely simple to the extremely complicated. In some cases, they may appear to be no more than the application of good common sense. The most important thing is the orientation which motivates a person to apply common sense or perform a more complicated evaluation of a situation.

Characteristics

No matter how complicated, economic studies all have certain definite characteristics:

---

1. Money to carry out every plan represents either:

   a. Annual Expense—Obtained from operating revenue; or

   b. Capital expenditure—Obtained from financing, reinvested depreciation reserve, reinvested earnings. In general, capital costs represent the initial purchase price of installed plant; or

   c. Both annual expense and capital expenditure.

2. Capital expenditures incur future annual expense.

3. The source usually available to a company to meet its annual expenses of operation, including taxes and obligations on its securities, is the revenue it receives from its consumers. Mathematically, therefore, the most economical of a number of plans (the one which will maximize profits) is the one which will require the minimum amount of additional revenue. A convenient way to conduct an economic study is to evaluate the effect of alternate proposals on the revenue requirements of the company.

4. Expenditures may take place (and thus affect revenue requirements) at different intervals over a period of time. The economic study must compare such expenditures on a consistent common basis.

5. The economics of alternate plans will generally be only one factor, although a major one, in the final selection of the most advantageous plan. Any differences in the non-tangible items of comparison, however, must be recognized and considered with economic differentials among the several plans. The assignment of a value for the effect of inflation may be arbitrary and best omitted from the calculations and considered a judgmental factor in the final recommendations. (The effects of inflation at several rates are contained in Table B-1; other rates may be interpolated.) The differences in the various plans should be pointed out so that the phases of each alternative may be fully evaluated.

ANNUAL CHARGES

The overall revenue requirement of a project, or the cost of doing business, is the sum of annual charges for:

1. Return on investment (stockholder, bondholder, etc.)

2. Depreciation (sinking fund, etc.)

3. Insurance expense

4. Property tax expense

5. Income tax expense

6. Operating and maintenance expense

7. Other taxes (e.g., on gross revenue)

The first five of these charges can usually, for convenience, be estimated as a percentage of original investment. The operating and maintenance charges

TABLE B-1   INFLATION FACTORS (COMPOUND INTEREST): $(1 + i)^n$

n	\multicolumn{9}{c}{Inflation rate—i in percent}
	2	3	4	5	6	7	8	9	10
1	1.020	1.030	1.040	1.050	1.060	1.070	1.080	1.090	1.100
2	1.040	1.061	1.082	1.103	1.124	1.145	1.166	1.188	1.210
3	1.061	1.093	1.125	1.158	1.191	1.225	1.260	1.295	1.331
4	1.082	1.126	1.170	1.216	1.262	1.311	1.360	1.412	1.461
5	1.104	1.159	1.217	1.276	1.338	1.403	1.469	1.539	1.611
6	1.126	1.194	1.265	1.340	1.419	1.501	1.587	1.677	1.772
7	1.149	1.230	1.316	1.407	1.504	1.606	1.714	1.828	1.949
8	1.172	1.267	1.369	1.477	1.594	1.718	1.851	1.993	2.144
9	1.195	1.305	1.423	1.551	1.689	1.838	1.999	2.172	2.358
10	1.219	1.344	1.480	1.629	1.791	1.967	2.159	2.367	2.594
11	1.243	1.384	1.539	1.710	1.898	2.105	2.332	2.580	2.853
12	1.268	1.426	1.601	1.796	2.012	2.252	2.518	2.813	3.138
13	1.294	1.469	1.665	1.886	2.133	2.410	2.720	3.066	3.452
14	1.319	1.513	1.732	1.980	2.261	2.579	2.937	3.342	3.797
15	1.346	1.558	1.801	2.079	2.397	2.759	3.172	3.642	4.177
16	1.373	1.605	1.873	2.183	2.540	2.952	3.451	3.970	4.595
17	1.400	1.653	1.948	2.292	2.693	3.159	3.727	4.328	5.054
18	1.428	1.702	2.026	2.407	2.854	3.380	4.026	4.717	5.560
19	1.457	1.754	2.107	2.527	3.026	3.617	4.348	5.142	6.116
20	1.486	1.806	2.191	2.653	3.207	3.870	4.635	5.604	6.727
21	1.516	1.860	2.279	2.786	3.400	4.141	5.071	6.109	7.400
22	1.546	1.916	2.370	2.925	3.604	4.430	5.477	6.659	8.140
23	1.577	1.974	2.465	3.072	3.820	4.741	5.915	7.258	8.954
24	1.608	2.033	2.563	3.225	4.049	5.072	6.388	7.911	9.850
25	1.641	2.094	2.666	3.386	4.292	5.427	6.899	8.623	10.83
26	1.673	2.157	2.772	3.556	4.549	5.807	7.451	9.399	11.92
27	1.707	2.221	2.883	3.733	4.822	6.214	8.047	10.25	13.11
28	1.741	2.288	2.999	3.920	5.112	6.649	8.691	11.17	14.42
29	1.776	2.357	3.119	4.116	5.418	7.114	9.386	12.17	15.86
30	1.811	2.427	3.243	4.322	5.743	7.612	10.14	13.27	17.45
31	1.848	2.500	3.373	4.538	6.088	8.145	10.95	14.46	19.19
32	1.885	2.575	3.508	4.765	6.453	8.715	11.82	15.76	21.11
33	1.922	2.652	3.648	5.003	6.841	9.325	12.77	17.18	23.23
34	1.961	2.732	3.794	5.253	7.251	9.978	13.79	18.73	25.55
35	2.000	2.814	3.946	5.516	7.686	10.68	14.90	20.41	28.10

should be separately estimated for each project. The tax (if any) on gross revenue must be calculated after all other charges are determined.

Return on Investment

A growing company must continually provide money for capital construction. In many cases, a large proportion of such funds are realized by sale of securities, bonds, debentures and stocks. These securities are purchased by people who believe that the future earnings of the company will provide a return on their investment commensurate with the hazards of the business and the nature of the security purchased. If the return provided is not sufficient to meet the expectations of investors when they analyze the risk involved, they will not invest in that firm. It is axiomatic then, that if a company is to be able to attract the necessary capital for continued expansion, it must maintain an adequate return on its invested capital.

Depreciation

The purpose of a depreciation allowance is to set aside a sufficient amount periodically (usually each year) to accumulate, over the life of the equipment, the original capital investment less net salvage.

There are a number of ways of taking account of depreciation; among the many types are two aptly named Straight Line Depreciation and Annuity Depreciation.

Straight Line Depreciation   The straight line method of calculating depreciation means that a fixed percentage is applied to surviving plant each year (usually monthly) to determine the accrual. The accrual rate is determined from the reciprocal of the average service life adjusted for salvage. This may be expressed by the equation:

$$D = \frac{1}{S}(1 - \text{SAL})$$

where
$$D = \text{straight line depreciation rate}$$
$$S = \text{average service life}$$
$$\text{SAL} = \text{salvage ratio}$$

Annuity Depreciation   It is also possible to express the annual charge for depreciation as an equivalent uniform annual charge. In cases where there is no salvage or dispersion (Iowa SQ dispersion), the annuity may be found in the future worth-to-annuity column in the compound interest table. This factor is determined from the equation:

$$\text{AA} = \frac{i}{(1 + i)^n - 1}$$

where
$$\text{AA} = \text{annuity depreciation rate}$$
$$i = \text{return as a percent of investment}$$
$$n = \text{number of years}$$

Dispersion is a factor to be considered in depreciable plant accounts. From actuarial studies, the nearest (Iowa) dispersion curve for each plant account has been previously determined. Thus to determine the annuity depreciation for a dispersed plant, the above equation is modified:

$$AA = \left(\frac{1}{\sum\limits_{n=1}^{m} \dfrac{R_n}{(1 + i)^n}} - 1 \right) \times (1 - SAL)$$

where, in addition to above:

$$m = \text{maximum or total life}$$
$$R_n = \text{mean annual survivor ratio}$$
$$SAL = \text{salvage ratio}$$

The annuity depreciation factors for a dispersed plant have been calculated for every (Iowa) curve; please refer to Fig. B-1. An elementary treatment of Depreciation, for illustrative purposes, is given in Table B-2.

There are many other ways of considering depreciation, and reference should be made to appropriate treatises on this subject.

Insurance

Unless specifically known, a value of 0.1% of original investment generally is sufficient to be used for insurance.

The four major forms of insurance carried to provide protection against damage to property are:

1. Fire insurance.
2. Boiler and machinery insurance covering accidental damage to such objects.
3. Coverage against damages due to motor vehicle collision, falling aircraft, storms, etc.
4. Insurance for general liability in excess of some value (e.g., $50,000).

The premium expense of such insurance, expressed as a percentage of total investment, is usually very small, and the value of 0.1% as an average annual charge adequately covers premiums on the insurance carried. In any special case where items of insurance make up a substantial portion of operating expense, they should be considered separately in the estimation of operating expense.

Property Taxes

Taxes on property fall into two classes: Special franchise or business taxes (applied to facilities on public property and to certain businesses); and, Real estate taxes (applied to facilities on private property). Plant classified as "Land Rights" (easements) or "Personal Property" such as tools, furniture and vehicles is not usually taxable. Depreciation is theoretically allowed on property, but in practice it is often not considered in computing taxes.

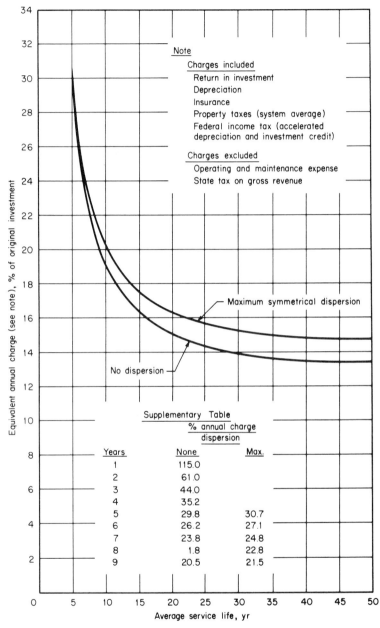

FIG. B-1 Equivalent annual charges as a percentage of original investment, assuming no salvage on project and 7 percent return on investment.

TABLE B-2 TREATMENT OF DEPRECIATION
$1000 capital investment—5 year life

Year	Investment at beginning	Depreciation	7% return	Total annual cost	Present worth Factor	Present worth Amount
A. Straight line depreciation						
1	$1,000	$200	$70	$270	0.935	$252
2	800	200	56	256	0.873	224
3	600	200	42	242	0.816	198
4	400	200	28	228	0.763	174
5	200	200	14	214	0.713	152
						1000
B. Very slow depreciation						
1	$1,000	0	70	70	0.935	65
2	1,000	0	70	70	0.873	61
3	1,000	0	70	70	0.816	57
4	1,000	0	70	70	0.763	54
5	1,000	1000	70	1070	0.713	763
						1000
C. Very fast depreciation						
1	$1,000	1000	70	1070	0.953	1000
2	0	0	0	0	0.873	0
3	0	0	0	0	0.816	0
4	0	0	0	0	0.763	0
5	0	0	0	0	0.713	0
						1000
D. Sinking fund depreciation (sinking fund factor from interest table)						
1	$1,000	174	70	244	0.935	228
2	—	174	70	244	0.873	213
3	—	174	70	244	0.816	199
4	—	174	70	244	0.763	186
5	—	174	70	244	0.713	174
						1000

Federal Income Tax

Federal income taxes are levied on taxable income as defined in applicable laws. The relationship of taxable income to revenue and to return on investment is illustrated by Fig. B-2a and b. Since rate of return on a project is the ratio of income available from the project to the net (depreciated) investment in the project, income tax must be calculated on the same basis of income or return.

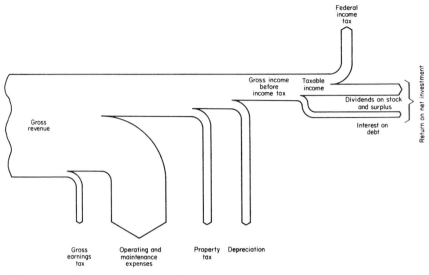

FIG. B-2 (a) Revenue and expense flow diagram.

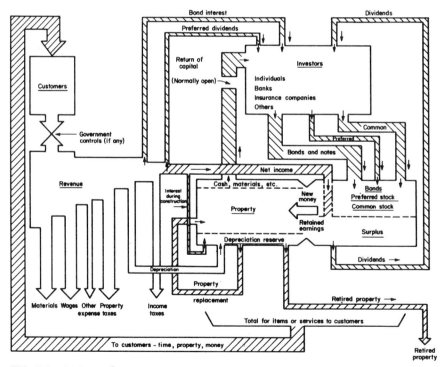

FIG. B-2 (b) Financial flowchart.

Operating and Maintenance Expenses

Operating and maintenance expenses are constituent parts of the total annual charge. As a general rule, operating and maintenance expenses cannot be expressed as a percent of the plant or unit of property investment since they do not vary directly with the investment cost. Expenses must be specifically estimated based on the individual project and must include applicable loadings as well as direct charges. In the comparison of alternate plans, costs common to the plans in the same year may be eliminated since their difference will be zero. Large non-recurring expenses must be evaluated on a present worth basis in the year of their occurrence.

Gross Earnings Tax

Some States (e.g., New York) levy earning taxes which are based on gross revenues. In evaluating alternate plans, the variation in this charge among plans will not affect the relative conclusions, and its consideration may be omitted unless total revenue requirements are desired.

BROAD ANNUAL CHARGE

For a complete study, it is necessary to evaluate the annual charges as discussed above applicable to a particular project. For many comparisons, enough accuracy can be obtained by using a more practical method employing broad annual charges. When the average service life exceeds twenty-five years, a broad annual charge of 15% of original investment may be used as a rough estimate of all charges exclusive of operating and maintenance expenses and gross earnings taxes. Figure B-1 shows the variation in total annual charge with service life and indicates the approximate basis for the 15% value.

This overall charge of 15% of original investment should not be applied to projects with a service life of less than 25 years. The annual charges on projects with a service life of less than 25 years increase rapidly as service life shortens, as shown in Fig. B-1. For such projects, a value determined from the upper curve of Fig. B-1 for the service life of the particular project will provide a reasonable first approximation of the annual charge.

TIME VALUE OF MONEY
Earning Power

Money has earning power. A dollar today is worth more than a dollar a year from now because of this earning power available through investment. The precise value of today's dollar in the future will depend upon the rate of interest earned on the invested dollar. Thus, one dollar today, invested at a 7% interest rate, will be worth $1.07 one year in the future. Conversely, $1.07 available a year from now has a present worth of $1.00. By using this concept, that money has an increasing value over a period of time, any expenditure in the future may be expressed in its equivalent "present worth" today. This principle is used to convert expenditures made at varying times to an equivalent value at any one given instant.

Conversions

Such conversions may be made by converting values to:

1. Present worth—the value today.
2. Future worth—value at any specified time in the future.
3. Annuity—a uniform series of payments over a period of time.

The result of spending capital money is a series of annual charges extending over the service life of the property in which the capital is invested. Some of these annual charges will be uniform every year and may be considered an annuity. Other annual charges will vary from year to year resulting in a non-uniform series; these can be converted to a uniform series.

Conversion factors at 7% interest for all these manipulations are provided in Table B-3.

There are a number of different ways of developing the conversion factors. The convention used in Table B-3 is that annuity payments and future worth values are evaluated at the end of periods (years) and present worth values are evaluated at the beginning of periods. Developed in this way, Table B-3 is in its most directly usable form since all payments are assumed to be made at the end of a year (December 31) throughout this study.

EXAMPLES

Eight conversions cover all cases and are summarized and illustrated in the following examples and worth-time diagrams.*

EXAMPLE B-1　Present worth to future worth (single amount at any date to single amount at any subsequent date)

You have just won $5,000, tax free. How much money will you have at the end of 10 years, if you invest it at 7% compounded annually?

Solution

See Fig. B-3a. The $5,000 is a present worth, the value 10 years hence is a future worth. The future worth is obtained by multiplying the present worth by the conversion factor "Present Worth to Future Worth" for 10 years from Table B-3:

Future worth in 10 years = $5,000 × 1.967 = $9,835

*Courtesy Long Island Lighting Co.

TABLE B-3 COMPOUND INTEREST TABLE

$i = 7\%$

	Lump sum		Uniform annual series			
	Present worth to future worth, $(1 + i)^n$	Future worth to present worth, $\dfrac{1}{(1 + i)^n}$	Annuity to future worth $\dfrac{(1 + i)^n - 1}{i}$	Future worth to annuity $\dfrac{i}{(1 + i)^n - 1}$	Annuity to present worth $\dfrac{(1 + i)^n - 1}{i(1 + i)^n}$	Present worth to annuity $\dfrac{i(1 + i)^n}{(1 + i)^n - 1}$
n						
1	1.070	0.9346	1.000	1.00000	0.935	1.07000
2	1.145	0.8734	2.070	0.48309	1.808	0.55309
3	1.225	0.8163	3.215	0.31105	2.624	0.38105
4	1.311	0.7629	4.440	0.22523	3.387	0.29523
5	1.403	0.7130	5.751	0.17389	4.100	0.24389
6	1.501	0.6663	7.153	0.13980	4.767	0.20980
7	1.606	0.6227	8.654	0.11555	5.389	0.18555
8	1.718	0.5820	10.260	0.09747	5.971	0.16747
9	1.838	0.5439	11.978	0.08349	6.515	0.15349
10	1.967	0.5083	13.816	0.07238	7.024	0.14238
11	2.105	0.4751	15.784	0.06336	7.499	0.13336
12	2.252	0.4440	17.888	0.05590	7.943	0.12590
13	2.410	0.4150	20.141	0.04965	8.358	0.11965
14	2.579	0.3878	22.550	0.04434	8.745	0.11434
15	2.759	0.3624	25.129	0.03979	9.108	0.10979
16	2.952	0.3387	27.888	0.03586	9.447	0.10586
17	3.159	0.3166	30.840	0.03243	9.763	0.10243
18	3.380	0.2959	33.999	0.02941	10.059	0.09941
19	3.617	0.2765	37.379	0.02675	10.336	0.09675
20	3.870	0.2584	40.995	0.02439	10.594	0.09439
21	4.141	0.2415	44.865	0.02229	10.836	0.09229
22	4.430	0.2257	49.006	0.02041	11.061	0.09041
23	4.741	0.2109	53.436	0.01871	11.272	0.08871
24	5.072	0.1971	58.177	0.01719	11.469	0.08719
25	5.427	0.1842	63.249	0.01581	11.654	0.08581
26	5.807	0.1722	68.676	0.01456	11.826	0.08456
27	6.214	0.1609	74.484	0.01343	11.987	0.08343
28	6.649	0.1504	80.698	0.01239	12.137	0.08239
29	7.114	0.1406	87.347	0.01145	12.278	0.08145
30	7.612	0.1314	94.461	0.01059	12.409	0.08059

EXAMPLE B-2 Future worth to present worth (single amount at any date to single amount at any previous date)

You have estimated that 10 years from now the unpaid mortgage on your house will be $9,835. How much money do you have to invest today at 7% interest to just accumulate $9,835 in 10 years?

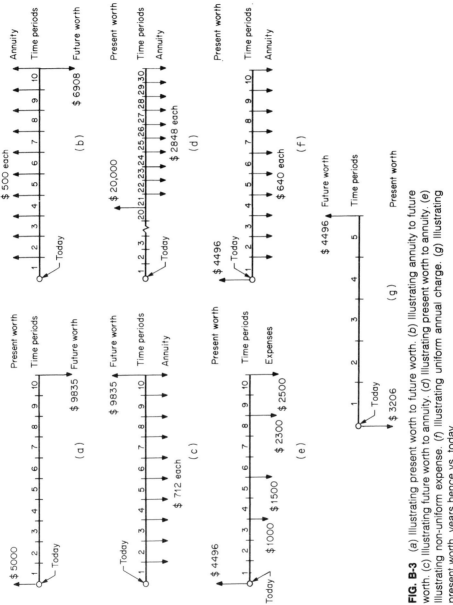

FIG. B-3 (a) Illustrating present worth to future worth. (b) Illustrating annuity to future worth. (c) Illustrating future worth to annuity. (d) Illustrating present worth to annuity. (e) Illustrating non-uniform expense. (f) Illustrating uniform annual charge. (g) Illustrating present worth, years hence vs. today.

Solution

The $9,835 is a future worth; the present worth of that amount is desired. From Table B-3, the conversion factor is 0.5083:

$$\text{Present worth} = \$9,835 \times 0.5083 = \$5,000$$

This is the reverse of Example B-1. The conversion factor for future worth to present worth is simply the reciprocal of the present worth to future worth factor. The worth-time diagram is the same as for Example B-1.

EXAMPLE B-3   Annuity to future worth (annuity over any period to single amount at end of period)

You plan to save $500 of your earnings each year for the next 10 years. How much money will you have at the end of the 10th year if you invest your savings at 7% per year?

Solution

The $500 each year is an annuity since it is a uniform amount each year. You wish to know the future worth. From Table B-3, the annuity to future worth factor, 10 years, is 13.816:

$$\text{Future worth of the annuity} = \$500 \times 13.816 = \$6,908$$

Note from Fig. B-3*b* that annuity payments are assumed to be made at the end of each time period. The conversion factor evaluates future worth at the same time that the last annuity payment is made.

EXAMPLE B-4   Future worth to annuity (single amount at any given date to annuity over any previous period ending at the given date)

If the unpaid mortgage on your house in 10 years will be $9,835, how much money do you have to invest annually at 7% interest to have just this amount on hand at the end of the 10th year?

Solution

See Fig. B-3*c*. The $9,835 is a future worth; the uniform amount (annuity) to set aside annually is desired. From Table B-3, the future worth to annuity factor, 10 years, is 0.07238:

$$\text{Annuity} = \$9,835 \times 0.07328 = \$712$$

EXAMPLE B-5   Present worth to annuity (single amount at any given date to annuity over any subsequent period starting at the given date)

You hold an endowment type insurance policy which will pay you a lump sum of $20,000 when you reach age 65. If you invest this money at 7% interest, how much money can you withdraw from your account each year so that at the end of 10 years, there will be nothing left?

Solution

See Fig. B-3*d*. The $20,000 can be considered the present worth at the end of the 10th year. From Table B-3, the present worth to annuity factor, 10 years, is 0.14238:

Annuity which may be withdrawn for 10 years = $20,000 × 0.14238

= $2,848

Note that direct use of the conversion factor assumes the first withdrawal to take place one period after the lump sum of $20,000 is received.

EXAMPLE B-6   Annuity to present worth (annuity over any period to single amount at start of the period)

You have estimated that for the first 10 years after you retire you will require an annual income of $2,848. How much money must you have invested at 7% at age 65 to realize just this annual income?

Solution

The present worth of an annuity for 10 years is desired. From Table B-3, the annuity to present worth factor, 10 years, is 7.024:

Present worth = $2,848 × 7.024 = $20,000

The worth-time diagram is the same as for Example B-5.

EXAMPLE B-7   Present worth non-uniform expenses to equivalent uniform annual charge

The maintenance expenses for the next 10 years on a piece of equipment are estimated as follows:

Year	Amount
3	$1,000
5	1,500
8	2,300
10	2,500

What is the present worth of these expenses? What is the uniform annual payment for 10 years equivalent to this non-uniform series? What does this mean?

Solution

See Fig. B-3*e*. The expense amounts are future worths in the year indicated. The present worth is desired. Future worth to present worth factors from Table B-3:

Year	Amount	Factor future worth to present worth	Present worth
3	$1,000	0.8163	$ 816
5	1,500	0.7130	1,070
8	2,300	0.5820	1,339
10	2,500	0.5083	1,271
Total			$4,496

The total present worth of these non-uniform series of expenses is $4,496.

The equivalent uniform annual series is obtained by applying the present worth to annuity factor for 10 years to the present worth. From Table B-3, present worth to annuity factor, 10 years, is 0.14238:

Equivalent uniform annual charge = $4,496 × 0.14238 = $640

(See Fig. B-3*f*.) This means that if you had $4,496 and invested it at 7%, you could withdraw the required amounts to meet exactly either the non-uniform series of expenses or pay out an equivalent amount of $640.

EXAMPLE B-8   Present worth some years hence to present worth today

Assume the expenses given in Example B-7 were to be associated with a piece of equipment to be installed 5 years from now. What is the present worth of the non-uniform expenses in that case?

Solution

See Fig. B-3*g*. The present worth previously obtained was the present worth for the expenses incurred in the 10 years following installation of the project. This is a present worth 5 years from now. In terms of today's present worth, it is a future worth 5 years away. The present worth today is obtained simply by converting the future worth in 5 years to a present worth. From Table B-3, the future worth to present worth factor, 5 years, is 0.7130:

Present worth today = $4,496 × 0.7130 = $3,206

PROCEDURE FOR ECONOMIC STUDIES

The procedures for commencing an Economic Study may be laid out in a sequence of steps:

1. The facts concerning the different plans that could be used to meet the requirements of the problem should be set down. The plans should be made as comparable as possible.

2. The capital expenditures which will be incurred under each of the plans and the timing of these expenditures should be determined. The amounts and

timing of operating and maintenance expenses must be estimated; allocations of cost to capital and expense must be adhered to.

3. A study period must be selected during which the revenue requirements incurred by the plans will be evaluated. In economic studies, it is seldom possible to find a study period which will precisely reflect the timing inherent in each of the plans under study. It will often be helpful to draw a diagram of the timing of capital and expense dollars for each of the plans in determining the study period. The study period chosen must be one determined on the basis of judgment. In every case, it must be sufficiently long to approximate the overall effects, over a long period of time, of the money reasonably to be spent for both capital and operating expenses.

4. The annual charges resulting from the capital expenditures in each phase must be calculated if broad annual charges cannot be applied. In considering alternate plans, items common to the several plans may be omitted from the calculations. The effect of temporary installations, salvage, and of the removal of equipment which can be used elsewhere on the system must be taken into account.

5. When annual revenue requirements are non-uniform, the present worth of the revenue requirements for each plan must be calculated. The most economical plan will have the lowest present worth of revenue requirements. In the case where annual revenue requirements are uniform throughout the study period, the plan with the lowest annual requirements will be the most economical.

6. The comparison of the economic differences among the plans may be made on the dollar differences among the present worths of the revenue requirements. If percentage difference is considered, the dollar differences may be misleading as, in conducting the study, charges which are the same in the several plans are generally omitted; this will distort the base upon which a percentage difference is derived.

7. A recommendation of the most advantageous plan must be made. The plan with the minimum revenue requirements would be recommended from an economic point of view. Other considerations may indicate the recommendation of one of the other plans despite higher revenue requirements.

CONCLUSION

Economic studies constitute perhaps the most important ingredient in the implementation of a project. In sum, the consideration of any undertaking must answer satisfactorily three basic requirements or questions:

1. Why do it at all?

2. Why do it now?

3. Why do it this way?

The answers to these can, in large part, be supplied by the results of economic studies.

THE GRID COORDINATE SYSTEM: TYING MAPS TO COMPUTERS*

INTRODUCTION

The grid coordinate system is the key that ties together two important tools, maps and computers. Maps are a necessity for the better operation of many enterprises, especially of utility systems. Their effectiveness can be increased manyfold by adding to their information data contained in other files. Much of the latter data are now organized and stored in computer-oriented files—on punched cards and on magnetic tapes, drums, disks, and cells. Generally, these data can be retrieved almost instantly by CRTs (cathode ray tubes) or printouts. The link that makes the correlation of data contained on the maps and in the files practical is the grid coordinate system.

Essentially, the grid coordinate system divides any particular area served into any number of small areas in a grid pattern. By superimposing on a map a system of grid lines, and assigning numbers to each of the vertical and horizontal spacings, it is possible to define any of the small areas by two simple numbers. These numbers are not selected at random, but have some meaning. Like any graph, these two coordinates represent measurements from a reference point; in this respect they are similar to navigation's latitude and longitude measurements.

Further subdivision of the basic grid areas into a series of smaller grids is possible, each having a decimal relation with the previous one (i.e., by dividing each horizontal and vertical space into tenths, each resultant area will be one-

---

*Reprinted (with modifications) from *Consulting Engineer,*® January 1975, vol. 44, no. 1, pp. 51–55. © Copyright by Technical Publishing, a company of the Dun & Bradstreet Corporation, 1975. All rights reserved.

hundredth of the area considered). By using more detailed maps of smaller scale, it is possible to define smaller and smaller areas simply by carrying out the coordinate numbers to further decimals. For practical purposes, each of these grid areas should measure perhaps not more than 25 ft by 25 ft (preferably less, say, 10 ft by 10 ft) and should be identified by a numeral of some 6 to 12 digits.

For example, by dividing by 10, an area of 1,000,000 ft by 1,000,000 ft (equivalent to some 190 miles square) can be divided into 10 smaller areas of 100,000 ft by 100,000 ft each, identified by two digits, one horizontal and one vertical. This smaller area can again be subdivided into 10 smaller areas of 10,000 ft by 10,000 ft each, identified by two more digits, or a total of four with reference to the basic 1,000,000-ft square area. Breaking down further into 1000- by 1000-ft squares and repeating the process allows these new grids to be identified by two more digits, or a total of six. Again dividing by 10 into units of 100 ft by 100 ft, and adding two more digits, produces a total of eight digits to identify this grid size. One more division produces grids of 10 ft by 10 ft and two more digits in the identifying number—for a total of 10 digits, not an excessive number to be handled for the grid size under consideration; see Fig. C-1.

This process may be carried further where applications requiring smaller areas are desirable; however, each further breakdown not only reduces the accuracy of the measurements, but also adds to the number of digits, which soon becomes unmanageable. Experience indicates that a "comfortable" system should contain 10 digits or fewer for normal usages.

While the decimal relation has been mentioned, other relations can be used, such as sixths, eighths, etc., or combinations, such as eighths and tenths, and others.

Standard References

To give these numerals some actual physical or geographical significance, they may be tied in with existing local maps, U.S. Geological Survey maps, coast and geodetic survey maps, state plane coordinate systems, standard metropolitan statistical areas, or latitude and longitude bearings. They may also be tied in with maps independent of all of these.

While reference to state and federal government systems lends some geographical significance, it produces identifying grid numbers with several additional digits. It is not necessary for any grid coordinate system to have this reference to a government system, but if it is desired, it is a relatively simple procedure to develop a computerized look-up program that can translate such coordinates.

Basic grid coordinate maps may be developed from the conversion of existing maps to a usable scale, if such maps are reasonably accurate and complete, both as to their geography and content. They may also be developed from exact land surveys, from aerial surveys, or from combinations of all of these.

Excellent maps are also available for most of the country. U.S. Geological Survey maps show latitude and longitude lines every few miles; they also show numerous triangulation stations with the latitude and longitude for each station

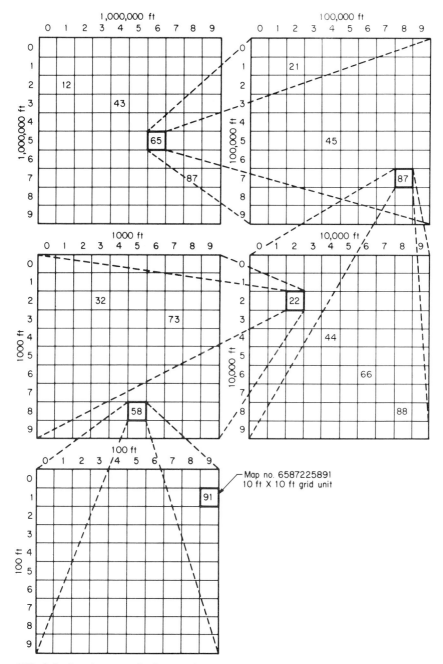

FIG. C-1 Development of grid coordinate system.

determined to an extreme degree of accuracy. Further, detail maps are available for practically every city and township, showing streets, houses, and lots. Despite the fine degree of accuracy of these maps, minor inaccuracies and discrepancies are bound to occur.

Earth's Curvature Errors occur in mapping the earth's curved surface on a flat map; see Fig. C-2. For example, in the case of the approximate 190-mi square mentioned previously, in the continental United States, the error introduced by this curvature, measuring from the center (95 mi in the longitudinal, or north-south direction) would probably not exceed 2 percent, a tolerable error. These errors need not be of great import, except in establishing match lines between maps. No gaps or overlaps should appear between adjacent maps, or between property or lot lines within a map. Tolerances of a few percent ordinarily are acceptable.

GRID COORDINATE MAPS

A grid coordinate map system should meet the following requirements:

1. It should include a simple and easily understood system of numerals for locating the data under consideration (numerals only; the x and y coordinates).

2. The grid areas should be small enough to be consistent with the purposes for which they are to be used (25 ft by 25 ft or less).

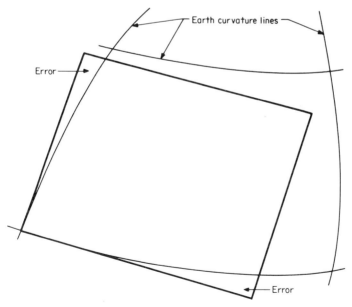

FIG. C-2 Error introduced in grid coordinate system by earth curvature.

3. The number of digits in the grid number should be held to a practical number so as not to become cumbersome and unwieldy (normally not more than about 10).

4. It should be designed to allow for expansion so that it will not have to be radically revised if unforeseeable expansion should occur.

5. It must provide reasonable accuracy (tolerances of a few percent), and it may or may not be tied in with some government or other established coordinate system.

6. Map sizes should be manageable (say, 24 or 36-in square).

7. Maps of different scales should be included in the system to accommodate different kinds of data (for circuit data, say 1000 or 500 ft; for details of facilities, say, 100 ft for overhead and 50 or 25 ft for underground).

8. A key map must show the entire grid area.

9. Optional is a grid atlas showing street locations with grid overlay.

In attempting to design a grid coordinate system for a very large area, it may be difficult to meet these requirements. In such instances it may be desirable and practical to divide the entire area into two or more convenient districts, establishing a separate grid coordinate system in each district and tying the separate systems together with match lines at the borders. A prefix letter or number may be used to identify districts, though this may not prove necessary in actual operation.

The size of individual maps should be large enough to encompass an area suitable for the purpose but small enough not to be unwieldy; sizes of 24 or 36 in square have proven practical. Maps of different scales are used for different purposes; for example, a 50- or 100-ft scale is used for dense or crowded areas; 300- or 500-ft for less dense or rural areas; and 500- or 1000-ft or even larger for district or overall area viewing. The series of scales used should be such that the larger-scale maps fit into those of smaller scale completely and evenly. Match lines of each sheet should fall on corresponding match lines of adjacent sheets.

The grid pattern applicable to each of the several scale maps may be printed directly on each map as light background lines, perhaps even in a different color, or printed on the back of the maps when they are reproduced. Alternately, a grid overlay can be applied to each map to be used when it is necessary to determine a grid coordinate for an item on the map. The actual grid number need not be printed on every item on every map unless desired. Such numbers assigned to key locations on each map normally suffice; the others may be determined from the grid background or overlay. To maintain their permanence and to minimize distortion from expansion and contraction because of changing humidity and temperature, the maps should be printed on a material such as Mylar, a translucent polyester-base plastic film; this is especially true of the base maps from which others of different scales and purposes are derived.

As much as practical, the data on the maps should be uncluttered and as legible as possible. It may be desirable in some instances to provide two or more maps (of the same scale) for several purposes; marks for coordinating these several maps, should it be necessary, may be included on each of the maps.

Maps may be further uncluttered by deliberately removing as much of the information on them as appears desirable and practical and consigning such information to files readily accessible by computer. In many instances, this information already is included or duplicated in such files, but may need to be labeled with the appropriate grid coordinate number. The use of CRTs and printouts makes this information available at will.

Application of Grid Coordinate Numbers

The grid coordinate number may be applied to each item of information contained in the computer-operated files by location. This may be done in several ways: manually, by machine, or a combination of the two.

The manual method is to superimpose or overlay the grid pattern on existing maps and manually assign numbers to each item to be processed. As mentioned previously, the grid pattern may be transferred to the master or original maps, and reproduced (or microfilmed) on the maps for the user; here numbers can be assigned directly from the map.

The machine method of grid number assignment employs an electronic scanning device called a *digitizer*. This machine includes a drafting table for map display and a cursor or pointer. The postal address and other fixed data are inserted on a punch card. For a particular map, the grid numbers of the map are set on the digitizer console. The digitizer assigns the x and y coordinates when the cursor is placed on a selected point and activated. These data are fed to a keypunch, which produces a punched card. In this method of producing the grid coordinate numbers, the digitizer enables additional refinement to be achieved, producing additional decimal numbers for the x and y coordinates. Hence, the ultimate grid area that can be measured can be one-tenth or one-hundredth, etc., of the basic unit area (1 ft by 1 ft, or 0.1 ft by 0.1 ft, for a 10-by 10-ft base area).

When numbers are assigned manually from maps, this degree of accuracy is not possible, nor is it necessary if the principal purpose of the grid coordinate system is to identify an item rather than a precise point. While the actual accuracy of such additional digits can be questioned, they provide a method of further subdividing a map for closer location of an item, but more important, they make possible a system of automatic mapping using the computer.

The grid coordinate number corresponding to the location of an item in question is added to its record and now becomes its computer *address*. In assigning these numbers to existing files, the digitizer generally can properly identify the location from a suitable map. As a practical matter, however, there will be some locations or descriptions that cannot be identified using the digitizer, and these may require manual processing and actual checking in the field; fortunately these usually constitute only a small percentage of the total records.

In the maintenance of such files, the grid coordinate numbers associated with changes in, or with the introduction of new items into, the records can be assigned manually by the originator of the record.

COORDINATE DATA HANDLING

As implied earlier, the grid coordinate system provides an easy and simple but, more important, a very rapid means of obtaining data from files through the use of the computer. In some respects, it assigns addresses to data in the same way as the ZIP code system in use by the postal service. The manner in which the grid number may be used is illustrated in the following examples; for convenience they refer to electric utility systems, although obviously they apply equally well to other endeavors employing maps and records.

Data contained on maps and records generally apply to the consumers served and the facilities installed to serve them. While maps depict (by area) the geographic and functional (electrical) interrelationship between these several components, the records supply a continuing history (by location) of each component item (consumers and facilities).

In the case of consumers, such data may include, in addition to the grid coordinate number, the name and post office address. Also a history of electric consumption (and demand where applicable), billing, and other pertinent data over a continuing period, usually 18 or 24 months. There may also be data on the consumer's major appliances; also the date and work order number of original connection and subsequent changes. The grid coordinate number of the transformer from which the consumer is supplied is included, as well as that for the pole or underground facility from which the service to the consumer is taken. Sometimes interruption data may be included. Other data may include telephone number, tax district, access details, hazards (including animals), dates of connection or reconnection, insurance claims, easements, meter data, meter reading route, test data, credit rating, and other pertinent information. Only a small portion of these data are shown on maps, usually in the form of symbols or code letters and numerals. In the case of facilities, such data may include, in addition to the grid coordinate number, location information, size and kind of facility (e.g., pole, wire, transformer, etc.), date installed or changed, repairs or replacements made (including reason therefor, usually coded), original cost, work order numbers, crew or personnel doing work, construction standard reference, accident reports, insurance claims, operating record, test data, tax district, and other pertinent information. Similarly, only a small portion of these data are shown on maps, usually in the form of symbols or code letters and numerals.

Data from other sources also may be filed by grid number for correlation with consumer and facility information for a variety of purposes. Such data may include government census data; police records of crime, accidents, and vandalism; fire and health records; pollution measurements; public planning; construction and rehabilitation plans; zoning restrictions; rights-of-way and easement locations; legal data; plat and survey data; tax district; and much other information that may affect or be useful in carrying out utility operations.

Obviously, all these data, whether pertaining to the consumer or to the utility's facilities, are not necessarily contained on one map or in one record only; indeed, there may be several maps and records involved, each containing certain amounts of specialized or functionally related data. All, however, may be correlated through the grid coordinate system.

Data Retrieval

Data contained in the files may be retrieved by means of the computer and may be presented visually by means of CRTs for one-time instant use, or by printouts and automatic plotting for repeated use over an indefinite time period. Data presented may be the exact original data as contained in one or more files, or extracted data obtained as a result of correlating data residing in one or more files, or a combination of both; such extracted data may or may not be retained in separate files for future use.

These data may be retrieved for an individual consumer or an individual item of plant facilities, or may be other data for a particular location. They may be retrieved for a group of individual locations, or for a particular area, small or large. The various specific purposes determine what data are to be retrieved and how they are to be presented. They also determine the programs and equipment required. Data thus retrieved then are used with data contained on the map to help in forming the decisions required. The decisions may include new data that can be reentered in the files as updating material, that can be plotted or printed for exhibit purposes, or that can be reentered on maps for updating or expanding the material thereon; all of these may be done by means of the computer.

The grid coordinate number is applied to utility facilities for ease of location and positive identification in the field. In the case of electric utilities, these may include services, meters, poles, towers, manholes, pull boxes, transformers, transformer enclosures, switches, disconnects, fuses, lightning arresters, capacitors, regulators, boosters, streetlights, air pollution analyzers, and other equipment and apparatus; also the location of laterals on transmission and distribution circuits.

OTHER APPLICATIONS

Similarly, for gas utilities, the applications of grid coordinate numbers may include mains, services, meters, regulators, valves, sumps, test pits, and other equipment; also the location of boosters, laterals, and nodes on the gas systems. For water systems, they may include mains, services, meters, valves, dams, weirs, pumps, irrigation channels, and other facilities. For telephone and telegraph communication systems, including CATV circuits, they may include mains, services, terminals, repeaters, microwave reflectors, and other items including poles, manholes, and special items.

Grid coordinate numbers also may find application in many other lines of endeavor: highway systems, railway systems, oil fields, social surveys (police, health, income, population distribution, etc.), market surveys (banks and industries), municipal planning and land use studies, nonclassical archeology, geophysical studies, and others where such means of location identification may prove practical.

The use of grid coordinates facilitates positive identification in the field; the numbers are posted systematically on facilities, such as streetlight or traffic standards, poles, and structures, and at corners or other prominent locations.

An atlas, consisting of a grid overlay on a geographical map, aids the field forces in locating consumers and plant facilities and provides a common basis for communication between office and field operating personnel.

The grid pattern permits the classical manipulation of data by individual grid sections or areas comprising several grid sections. In addition to the sample presentation of such data by means of CRT displays and typed printouts, data may be presented in the form of plotting in various graphical forms, in patterns indicating the distribution of data, the density of particular data, the accumulation of data within fixed boundaries, the determination of area boundaries for predetermined data content (the analysis of data within a given polygon), the calculation of lengths and distances between grid locations, and the mapping of facilities in acceptable detail—and all of these operations may be performed automatically by means of the computer.

Further, summaries and analyses employing the grid coordinate system may be more readily made and are susceptible to combination and consolidation, resulting in perhaps fewer and more comprehensive reports (eliminating the duplication of much needless data and the presentation of more complete and meaningful conclusions in one place).

In all of the foregoing discussion, the point must be made that all of the handling of data using the grid system may also be accomplished without the use of the grid system. It is apparent, however, that this latter method will in the vast majority of cases employ more effort in terms of work hours and will be more time-consuming, so as to render many applications impractical, even though their desirability may be great; in short, the grid coordinate system enhances the economics of data handling.

ECONOMICS

It is not to be denied that the introduction of the grid coordinate system will impose additional cost to the maps and records function. It is also evident that these costs will be offset by the decreased personnel requirements in the processing of data derived from the maps and records, especially when the computer may be made to take up a large part of this burden. Moreover, more refinement and a wider scope in processing of data are attainable.

The cost of implementing a grid coordinate system can be evaluated fairly accurately. Many factors will influence the final determination; these include the area of the system involved, the number of consumers and facilities, the condition of the basic and auxiliary maps and records, the number and scope of the applications desired, the extent of automation, and many other factors. A very approximate estimate may average perhaps about one day's revenue per consumer. Practical considerations associated with implementation may well dictate a period of several years, perhaps 5 years or even more, over which the expenditure will have to be made to accomplish the desired goals.

The offsetting savings from the introduction of a grid coordinate system, including those derived from the additional worth of the wider utilization, are difficult to pinpoint. It should be observed that while it is probable that a single application will not justify the adoption of the grid coordinate system, except in some unusual or special set of circumstances, it is also probable that the multiplicity of practical applications indicated will justify the relatively modest expenditure necessary for the conversion of present maps and records to the grid coordinate system.

The personnel requirements necessary to implement a grid coordinate system over a reasonable (short-term) period of time must be viewed together with the overall probable lessened longer-term in-house requirements. Since such a conversion is a one-time operation, it recommends itself admirably to the classical use of contractors having the special skills and experience. Further, such outside services are not apt to be diverted by crisis incidents prevalent in many enterprises.

One final observation. With the national consensus apparently pointing to an ultimate metric system for the United States to conform with world standards, the adoption of a grid coordinate system provides an excellent opportunity for its introduction with a minimum of conversion effort.

With the advent of the computer, it was inevitable that the grid coordinate system should be developed to provide a simple means of addressing the computer. The grid number provides the link between the map and the vast amount of data managed by the computer. This happy marriage of two powerful tools results not only in better operations but in improved economy as well. It is a must in the modernization of operations in many enterprises and especially in utility systems.

ONGOING RESEARCH AND DEVELOPMENT

From their inception, distribution systems have been improved to obtain greater safety, economy, and reliability. Research and development connected with the several elements of such systems have included not only new and better materials but their manufacture and construction, as well as their installation and operation. Many of these activities were, and are, carried on by individual utility companies and the manufacturers, often in cooperation with each other. More recently, the Electric Power Research Institute (EPRI) was created not only to coordinate and undertake some of the projects, some as joint ventures on the part of utilities and manufacturers, but to act as a clearinghouse for the assembly and dissemination of information.

The list of projects under EPRI supervision numbers in the hundreds and includes every phase of electric utility operations. Only a small, representative group covering the distribution system is outlined below; a single item, however, may include a number of interrelated separate projects tending toward the same goal. The data for these have been furnished by EPRI and are reproduced here with its permission; EPRI's generosity is greatly appreciated and acknowledged.

POLES

Development projects include power poles made from fly-ash-derived foam glass, and from crystallized fly ash (both from power plant residue), as possible materials meeting the structural and insulating requirements for overhead distribution lines, and cross arms made from the latter material. Also under study is the use of wood particles bonded together with synthetic resins, and reinforcing materials utilized to obtain desired strengths.

Investigation into the use of plastics, reinforced concrete, and laminated wood for poles and structures continues.

Research continues for better preservative materials, and for a better understanding of the fundamental chemical and biological processes of wood decay; also for methods to obtain a deeper penetration and diffusion of the preservative material. This includes the study of the effectiveness of fumigants, and the determination of types of decay and the location and developmental patterns of fungi. Also under study are certain types of volatile fungicides—the diffusion of their vapors, their effects on the vegetation around poles, and their effects on the strength properties of the wood.

TREE GROWTH CONTROL

In the area of tree growth, there are studies of the basal application of growth-inhibiting chemicals (e.g., the chlorfluorenol methyl esters, CEME for short) to trees; studies to determine which chemicals are most effective, what dosage and frequency of application are required, what time of year is best for application, and what additions may aid absorption through the tree bark; and studies of the effects of rainfall and its long-term influence on tree survival. Other efforts are aimed at developing practical methods of application, such as bark banding, pressure injection, and spraying. Studies also exist to obtain a better understanding of the causes of tree-to-tree variability and the effects of repeated applications, and to screen growth retardants for new species.

INSULATING MATERIALS

Outdoor polysil insulation, a new material which appears superior to porcelain both electrically and mechanically at reduced costs, is being developed; field tests are included. The possibility exists that this material may also be used for pole and substation construction.

Technical and economic feasibility studies are under way for the use of conducting polymers, metal-filled polymers, expanding polymers, new ceramics, moisture-impermeable plastics, and encapsulants for electrical insulation.

Dielectric gases or gas mixtures with characteristics superior to sulfur hexafluoride (SF_6) for insulation and arc interruption are being investigated and developed.

Studies are being made to improve cellulosic electrical insulation by upgrading thermal, dimensional, and mechanical stabilization and mechanical strength, and by reducing hygroscopicity. These will also determine changes necessary in the paper-making process: thermal treatments, chemical modifications, and additives.

Laser application is under development for the detection of voids and contaminants in polyethylene-insulated cables for real-time manufacturing quality control, for the inspection for voids, flaws, and contaminants, and for the identification and characterization of defects in cable insulation.

Development is under way of a dielectrically graded insulation system by modifying a highly filled resin with various resistive- and capacitive-grading particle materials, in order to redistribute dielectric stresses more uniformly on insulators.

TRANSFORMERS

Amorphous alloys which have soft magnetic properties for use in transformers are being developed. The major advantages would be a significant reduction in transformer core loss and lower manufacturing costs. Alloy composition and postquenching treatments to improve magnetic properties, saturating flux density, and stability are being determined. The development of manufacturing methods is also under study.

Studies are under way for the development of passive hot spot detectors for measuring hot spot temperatures, for the use of passive temperature-sensing devices, and for the use of different connection media between sensor and readout devices. Included is the development of an optical temperature sensor based on a ruby glass sensor that changes optical density with temperature and is connected by fiber optics to the exterior of the tank. The object of all of these is to achieve better utilization of transformer overload capability.

Techniques are being developed for the chemical destruction and physical separation of PCBs (polychlorinated biphenyls, which are reputed to be carcinogenic) from transformer oils and askarels.

Designs of distribution transformers are being considered for eliminating oil from the insulating materials by using high-temperature solid inorganic materials. The designs also include a core assembly cooled by radiation, primary and secondary windings of aluminum foil, and an insulation system of low-temperature-melting glass materials, fiberglass cloth, ceramic tubes, and high-temperature paint. Another system under consideration is the encapsulation of the transformer core and coil assembly in epoxy. These designs will require the development of new manufacturing techniques.

The design of a transformer suitable for direct burial in the ground (rather than installed in enclosures), thus eliminating the cost of a vault, is also under consideration.

FUSES

Development of a film-type fuse to provide improved alternatives to present types of fuses is being studied. The new type of fuse uses metallic film deposited on an insulating substance as the fuse element. Studies include extending the film-type fuse technology now available for voltages up to 600 V to progressively higher voltage and current ratings, ultimately to 38 kV and 250 A. Fabrication techniques are also being studied.

Studies are under way for the development of an improved current-limiting fuse using a novel "explosive charge" technique for initiating, timing, and controlling the fault, current-limiting, and clearing characteristics of the fuse. The object is to overcome the incompatibility of high continuous currents (above 100 A) with low "let-through" short-circuit currents in existing current-limiting fuses.

FAULT LOCATION

Means are being studied for the detection of fallen overhead 12-kV (and above) distribution conductors; these include several proposed relaying schemes capable of detecting restricted faults.

Studies are also under way to improve the methods for rapid location of cable faults, including the use of attenuation of acoustic signals, of tracing current, of capacitor discharge, of resistive bridges, and of combinations of these. Techniques for the automatic disconnection of the faulted section are included in these studies.

UNDERGROUND CONSTRUCTION

The development of a water-jet concrete-cutting system is under consideration to provide for a faster, quieter, and lower-cost method than present conventional cutting methods. The new method will be capable of cutting through 8-in-thick concrete, asphalt, aggregate, or a composition of these.

Studies are under way for the development of an improved pavement breaker (for use on frozen ground, concrete, rock formations, and other pavements), using pulse-type water cannon, microwave precracking or preweakening, and silent pulsed-water jet fragmentation at pressures up to 100,000 lb/in^2.

An improved boring system is under study, using an advanced hydraulic water-jet system for boring horizontal holes.

Studies are under way for improved plows, including vibrating plows, for installing underground cable. The improved plows are to be compact and highly maneuverable in all types of soil, with separate control, cable-tension, cable-bending control, and blade-offset features.

CABLE

The use of sodium as a conductor has been tested in the field and has not proved entirely practical; further research is indicated.

The use of polyvinyl chloride (PVC) and polyethylene (PE) for both insulation and sheath purposes has proved practical, and their extension to cables operating at 69 kV is under development.

SWITCHING DEVICES

Studies are under way for the development of switch-gear mechanisms to meet the requirements of advanced circuit breakers, grounding switches, and current limiters. The studies will encompass the analysis of operator ratings for various switching devices; the comparison of solid and liquid propellants, and of electric and percussion primers; and the design, manufacture, and testing of advanced mechanisms.

LOAD CONTROL

A digital instruction control and protection system, including different communications media for substations, is being developed.

Consideration is being given to load control by means of automatic peak load reduction, demand limitation, and dropping of loads on mutually agreed-upon priorities and schedules, and by means of radio control of air conditioning, house heating, water heaters, and other large loads.

ELECTRONIC METERING AND CONTROL

Development and testing of a multiple-function electronic watthour meter are under way. Load-management systems now require consumer meters that can store digital data in several registers for time-of-use meter reading, can perform consumer load control, and can provide digital logic and data storage registers. Work progresses on a single-phase electronic meter containing programmable logic for automatically controlling loads.

Methods using solid-state electronic and microprocessor technology are being developed for automatic remote meter reading, for load management and distribution control, and for using distribution lines as primary communication channels. Application will include capacitor switching, TCUL control, and distribution line fault sectionalizing. Solid-state watthour and VAR-hour metering, including demand data, are also being developed.

Also under development is a single system for time-of-day metering, automatic meter reading, revenue security (antitampering) monitoring, consumer cut-on and cutoff, fault tracing, switch status and control, capacitor and regulator control, feeder load dispatch, control of various residential consumer loads, load surveys, and collection of operating data.

U.S. AND METRIC RELATIONSHIPS

Metric to U.S.			U.S. to metric		

Length

1 mm	=	0.03937 in	1 in	=	25.4 mm
1 cm	=	0.3937 in	1 in	=	2.54 cm
1 m	=	39.37 in	1 in	=	0.0254 m
1 m	=	3.2808 ft	1 ft	=	0.3048 m
1 m	=	1.094 yd	1 yd	=	0.9144 m
1 km	=	0.6214 mi	1 mi	=	1.609 km

Surface

1 mm^2	=	0.00155 in^2	1 in^2	=	645.2 mm^2
1 cm^2	=	0.155 in^2	1 in^2	=	6.452 cm^2
1 m^2	=	10.764 ft^2	1 ft^2	=	0.0929 m^2
1 m^2	=	1.196 yds^2	1 yd^2	=	0.8361 m^2
1 hectare	=	2.471 acres	1 acre	=	0.4047 hectare
1 hectare	=	0.00386 mi^2	1 mi^2	=	258.99 hectare
1 km^2	=	0.3861 mi^2	1 mi^2	=	2.59 km^2

Volume

1 cm^3	=	0.061 in^3	1 in^3	=	16.39 cm^3
1 m^3	=	35.314 ft^3	1 ft^3	=	0.0283 m^3
1 m^3	=	1.308 yd^3	1 yd^3	=	0.7645 m^3
1 liter	=	0.0353 ft^3	1 ft^3	=	28.32 liters
1 liter	=	61.023 in^3	1 in^3	=	0.0164 liter
1 liter	=	1.0567 qt	1 qt	=	0.9463 liter
1 liter	=	0.2642 gal	1 gal	=	3.7854 liters
1 m^3	=	264.17 gal	1 gal	=	0.0038 m^3

Weight

1 g	=	0.0353 oz	1 oz	=	28.35 g
1 kg	=	2.2046 lb*	1 lb	=	0.4536 kg
1 T metric	=	1.1023 net tons**	1 net ton	=	0.9072 T (metric)

Compound Units

1 kg/m	=	0.6720 lb/ft	1 lb/ft	=	1.4882 kg/m
1 kg/cm	=	14.223 lb/in	1 lb/in^2	=	0.0703 kg/cm^2
1 kg/m^2	=	0.2048 lb/ft^2	1 lb/ft^2	=	4.8825 kg/m^2
1 kg/m^3	=	0.0624/ft^3	1 lb/ft^3	=	16.0192 kg/m^3
1 kg·m	=	7.233 ft·lbs	1 ft·lb	=	0.1383 kg·m
1 kW	=	1.340 hp	1 hp	=	0.746 kW
1 kg·m/cm^2	=	46.58 ft·lb/in^2	1 ft·lb/in^2	=	0.0215 kg·m/cm^2

*Avoirdupois.

**1 ton = 2000 lb.

INDEX

ABOUT THE AUTHOR

Anthony J. Pansini, with over half a century of working experience, is one of the most highly regarded authorities in the field of electrical transmission and distribution.

After receiving his professional degrees (B.S. and E.E.) from Cooper Union, Mr. Pansini began with what is now known as Consolidated Edison in New York. Later, at the Long Island Lighting Company (LILCO) for over 30 years, his responsibilities included engineering planning, design, and operations. He has been consultant to many other utilities, REAs, and co-ops throughout the U.S.A. and Mexico. And he has taught professional-level courses at both LILCO and colleges.

The author of more than 30 articles for such influential journals as *Transmission & Distribution* magazine, *Electrical World*, and *Electrical Engineering*, Anthony J. Pansini currently makes his home in Waco, Texas. This is his ninth book.